The theory of classical dynamics

The theory of classical dynamics

J. B. GRIFFITHS

CAMBRIDGE UNIVERSITY PRESS

Cambridge

London New York New Rochelle

Melbourne Sydney

CAMBRIDGE UNIVERSITY PRESS
Cambridge, New York, Melbourne, Madrid, Cape Town, Singapore, São Paulo, Delhi

Cambridge University Press
The Edinburgh Building, Cambridge CB2 8RU, UK

Published in the United States of America by Cambridge University Press, New York

www.cambridge.org
Information on this title: www.cambridge.org/9780521237604

First published 1985
This digitally printed version 2008

A catalogue record for this publication is available from the British Library

Library of Congress Catalogue Card Number: 84-7000

ISBN 978-0-521-23760-4 hardback
ISBN 978-0-521-09069-8 paperback

CONTENTS

PREFACE

The purpose of this book is to describe in detail the theory known as classical dynamics. This is a theory which is very well known and has a large number of important practical applications. It is regarded as one of the most basic scientific theories, with many other theories being direct developments or extensions of it. However in spite of its acknowledged importance, it is not always as well understood as it ought to be.

In the English speaking world at least, the emphasis in the teaching of classical dynamics is largely on its relevance to idealised applications. Thus most traditional mechanics textbooks are example orientated, and students are required to work through large numbers of artificial exercises. Other textbooks are available which emphasise the mathematical techniques that are appropriate in the applications of the subject. These approaches are of course of great importance. An understanding of this subject can only be achieved by working through numerous examples in which the theory is applied to practical problems. However, I feel that a discussion of the fundamental concepts of the theory and its basic structure has largely been neglected.

Classical dynamics is far more than an efficient tool for the solution of physical and engineering problems. It is a fascinating scientific theory in its own right. Its basic concepts of space, time and motion have fascinated some of the greatest intellects over many centuries. My aim in writing this book is to encourage students to think and to gain pleasure from their intellectual studies, as well as provide them with an essential tool for the solution of many practical problems.

The theory of classical dynamics is one of the most basic of all scientific theories. Almost all modern physical theories include concepts carried over from it. So my aim is to describe the theory in detail, paying particular attention to the basic concepts that are involved, as well as to the way in which the theory is applied.

Writing in the latter part of the twentieth century, it is possible to use not only recent studies in the philosophy of science, but also the additional insights into the subject of dynamics that have come from the development of the modern theories of relativity and quantum mechanics. It is in fact one of my concerns in this book to indicate the limitations of the classical theory, and hence also to specify those areas in which it may confidently be applied.

I have omitted a discussion of the historical development of the subject. There already exist many excellent volumes covering this material. I include here only an analysis of the mature theory in the present form in which it is understood and applied.

Although the main purpose of this book is to expound the theory of classical dynamics, I do not believe that this can adequately be done without considering a number of examples. Accordingly, I have included a number of separate sections of worked examples, which are followed by exercises that are left to the reader to complete. To obtain a deep understanding of the theory it is necessary both to consider real situations and to work through more idealised exercises of the type included here.

With the combination here of the theory with examples and exercises, I hope that this book will be found appropriate as a recommended or supplementary text for university courses in both mathematics and physics departments. A certain amount of mathematical notation has necessarily been used throughout the book, and this has accordingly been set at the level of a first-year undergraduate course in mathematics or physics.

INTRODUCTION

The classical theory of dynamics is that scientific theory which was developed through the seventeenth, eighteenth and nineteenth centuries to describe the motion of physical bodies. The theory was originally developed in two distinct parts, one dealing with terrestrial bodies such as a projectile or a pendulum, and the other dealing with celestial bodies, in particular the planets. These two apparently distinct subjects were first brought together by Isaac Newton whose book, known as the *Principia*, is rightly acknowledged as the first complete formulation of the theory that is now referred to as *classical dynamics*.

Since those early days the theory has been thoroughly developed and extended so that it can now be applied to a very wide range of physical situations. Some of the early concepts have been clarified, and others have been added. But, although the theory has now been formulated in many different ways, it is still essentially the same as that originally proposed by Newton. The most significant advances have in fact been those associated with the development of new mathematical techniques, which have subsequently enabled the theory to be applied to situations which previously had proved too difficult to analyse.

Now, as is well known, a revolution occurred in scientific thinking in the first part of the present century. Classical theories were disproved, and exciting new theories were put forward. The theory of relativity was suggested in order to explain the results obtained when the speed of light was measured. Such experimental results could not be accounted for using classical theories, but the new theory involved replacing the concepts of space and time that had been accepted for centuries. Then a new theory of gravitation was put forward that involved the concept of a curved space–time. This was a great shock to a generation that had been schooled on Euclid. Yet the new theory was shown to be in agreement with observations, while the classical theory was not. Also, a whole new theory of atomic structure with subatomic particles was developed. This was the theory of quantum mechanics, which has led to a much deeper understanding of the nature and structure of matter. Old concepts were modified or abandoned and a whole new science emerged.

These new theories, however, are extremely complicated mathematically and in order to apply them to the motion of reasonably sized bodies, it is necessary to use many approximations. In most situations of this type, the mathematical approximation of the new theories is identical to that which is also obtained from the much simpler theory of classical dynamics. In such situations it is therefore appropriate simply to continue to apply the classical theory. In practice it can thus be seen that the new theories of relativity and quantum mechanics, far from replacing classical theory, actually support it. In addition, since they also clarify the limiting situations in which the various theories become significantly different, they clarify the range of physical situations in which the classical limit can be applied. They therefore enable the classical theory to be applied with greater confidence within this range.

It is the purpose of this book to carefully explain the theory of classical dynamics. This is to be done by describing in detail the concepts that are required in the theory and the way in which it can logically be applied. The two main reasons for this are the obvious ones. Firstly, since the classical theory continues to be applied in a vast number of different situations both in science and engineering, it is necessary that the theory should be understood and applied appropriately and correctly. Secondly, most of the concepts that were introduced first in classical dynamics such as mass, energy and momentum, are taken over and used in many other theories. For this and other reasons, the theory of classical dynamics is regarded as one of the most fundamental of all scientific theories. This being so, it is very important that its methodology and basic concepts should be properly understood.

In order to understand any theory, it is imperative that it should be stated as clearly and accurately as possible, and the most appropriate way to state a scientific theory is in axiomatic form. Once the basic axioms of a theory are stated, its implications can be deduced using the standard methods of deductive logic and mathematical reasoning.

Such an approach displays both the essential character of the theory and the way in which it is to be applied.

It is important though to realise that the axioms of a scientific theory are not 'self-evident truths'. They are, rather, simply the assumptions or conjectures upon which the theory is based. They are the foundational statements which are taken on trust and used without question in the development and application of the theory. However, since they are essentially conjectural, they may be questioned, and it is sometimes instructive to consider replacing them by alternatives which would give rise to different theories making different sets of predictions.

In order to emphasise their character, the basic axioms of classical dynamics are stated in this book as a series of *assumptions* and *definitions*. From these the basic results of classical dynamics can be deduced. The most important of these are stated formally here as a series of *propositions*. Since the theory can be summarised in terms of these statements, they are separately listed for convenience in the appendix.

Now the purpose of any scientific theory is to describe what actually happens in the real world. So the more accurately a theory describes real events, the closer it may be assumed to be to the truth, and the greater the confidence that can be placed in its foundational assumptions. Because of its great success, considerable confidence has been placed in the foundational axioms of classical mechanics. These relate particularly to the assumed nature of space and time, to the concept of force, and to the particle and many-particle representations of real bodies. However, with the greater success of the modern theories of relativity and quantum mechanics, which each have some basic axiom which differs significantly from those of classical theory, some of the foundational axioms of classical dynamics have been shown to be false, and certain alternatives are now considered to be closer to the truth.

From the above discussion it can be seen that it is most convenient to adopt an axiomatic approach in order both to describe the theory of classical dynamics, and also to indicate the points of departure which lead to the theories of relativity and quantum mechanics. However, the main aim here is to expound the theory as clearly as possible rather than to present an axiomatic formulation of it. Thus no attempt has been made to demonstrate the usual requirements of axiomatic systems that they be consistent, independent and complete. In fact in at least one case the basic assumptions are not independent. However, it is hoped that the essential character of the theory is clearly demonstrated.

The theory of classical dynamics has been developed over the centuries so that it is now applicable to a very wide range of physical phenomena. It has in addition been directly extended to cover such other subjects as classical electrodynamics and magneto fluid mechanics. These and many other theories can also be formulated in terms of the generalised theories of analytical dynamics. The subject of this book, however, is only the classical theory of dynamics. The extensions of the theory are not considered. The theories of Lagrangian and Hamiltonian dynamics are therefore described here only in terms of their origins in classical mechanics, and no attempt is made to develop them to their full generality. It is, however, necessary to treat these theories in some detail both because of their use in applications, and because of the additional insight that they give into certain aspects of the classical theory.

1

The Newtonian method

As with other scientific theories, theories of dynamics aim to contribute to an understanding and explanation of phenomena that occur in the real world. In the theories of dynamics, it is the general subject of motion that comes under investigation. The aim is to describe how objects move, and to suggest physical reasons as to why they move. In particular they should provide methods for analysing or predicting the motion of specific bodies, and also possibly suggest techniques for controlling the motion of some objects.

The theory known as classical or Newtonian dynamics, the subject of this book, is one such theory. Before describing it in detail, however, it is convenient to describe in general terms the way in which the theory is used. This is the purpose of this first chapter.

1.1. The technique of mathematical modelling

The basic method by which any theory of dynamics is applied can be described in terms of three distinct phases. The first phase consists essentially of constructing a simplified model. This is an idealised imaginary representation of some physical situation in the real world. In phase two this theoretical representation is analysed mathematically and its consequences are deduced on the basis of some assumed theory. Finally, in phase three, the theoretical results of phase two are interpreted and compared with observations of the real physical situation. This whole process can of course be repeated many times using different initial representations or different basic theories. However, it is usually only repeated until the results obtained at phase three are considered to be sufficiently accurate for the purpose that was originally envisaged. This general method of approach is that known as mathematical modelling, and is the basic method of all so-called applied mathematics. However, since the basic theory being applied also regulates the whole modelling process, it is appropriate to describe this process here in a little more detail in terms of the particular way in which it is used in conjunction with the theory of classical dynamics.

As an initial illustration of this method, we may consider how the motion of a planet can be described and its position at any time predicted. In the construction of a mathematical model it may be appropriate to begin with to regard the planet as a point mass, to assume that it is only acted on by a gravitational force directed towards the sun, and to regard the sun as a fixed point. Such a model can then be analysed mathematically and, using the Newtonian theory of gravitation and Newton's laws of motion, it is found to predict that the planet will follow an elliptic orbit about the sun. There are, however, a whole family of possible ellipses, and so a series of observations are required, first to determine the parameters which specify the particular ellipse, and then to check that the planet does in fact continue to follow such an ellipse, at least approximately. Once this is confirmed, it is possible to predict its position at any later time within a certain estimated error bound. This whole process can then be repeated using more realistic models. For example, the planet and sun can be represented as bodies of finite size, the sun may be considered to move, and the effects of other planets or other perturbing forces can also be introduced. It is also possible to change the basic theory and to attempt to follow the same procedure using the general theory of relativity. Using this approach it is ultimately found to be possible to determine the position of any planet at any time with a remarkable degree of accuracy.

With this illustration in mind, it is now appropriate to consider the separate phases of this approach in greater detail, paying particular attention to the way in which it is applied to the analysis of the motion of real bodies.

In phase one the physical situation being considered is represented by an idealised mathematical model. Such a representation is purely theoretical. It transfers attention from the real physical world to some imaginary construct in the mathematical domain where it may be analysed in terms of mathematical equations. It is therefore necessary that the representations of the various parts of the physical system should be stated in terms that can be expressed mathematically. It is usually found, however, that if a fully detailed representation is considered the mathematics becomes unwieldy and the model cannot be analysed. It is therefore essential that the model adopted is a much simplified and idealised representation of its real counterpart. The art of mathematical modelling lies in choosing a model that is sufficiently simple to be analysed easily, but yet is sufficiently detailed to accurately reflect the behaviour of the real physical system. In practice, however, a number of models are usually considered. Very simple models may first be used to describe the general qualitative behaviour of a system, or to provide initial approximations. Then progressively more detailed models may be considered until the degree of mathematical sophistication required becomes out of proportion to the improvement of results achieved.

Even at this initial stage it is necessary to decide which theory is going to be applied, since the choice of theory usually also affects the choice of possible representations. Here it is assumed that the theory to be applied is that of classical dynamics, and so the representation is stated in terms of mass distributions, forces, frames of reference, initial velocities and perhaps internal or external constraints. These are all regulated by the theory and will be discussed in later chapters.

The choice of classical mechanics as the appropriate basic theory immediately specifies the type of mathematical models to be considered, and the range of features of the real situation that are ignored. For example, a physical body is usually represented in this theory solely in terms of its assumed mass distribution. Other properties, such as its colour, are assumed to be totally irrelevant. Even properties such as its temperature, or chemical composition, are ignored unless they are thought to affect its mechanical properties. It can thus be seen that this approach leads to a theoretical model that is necessarily only a restricted or partial representation of its real counterpart.

To further simplify the model it is always necessary to theoretically isolate the systems being considered from all other aspects of the real-world situation that are not considered to be relevant to its motion. This aspect is similarly regulated by the theory. For example, as far as classical mechanics is concerned, the theory indicates immediately the significant forces which may be assumed to act on the system. Other less significant forces need not be included in the model. Thus, in discussing the motion of a planet, the position of its moons, the other planets and the stars may be ignored, at least initially.

Finally, having isolated a particular system, it is still necessary to simplify or idealise its representation in order to analyse it mathematically. For this reason it is often appropriate to consider such fictions as the point masses, light inelastic strings, smooth planes or rigid bodies. The planet, for example, may initially be represented as a single particle. Together these various stages of idealisation essentially consist of sets of simplifying assumptions which in applications enable particular physical systems to be represented simply in precise and manageable mathematical terms.

Once such an idealised model has been constructed, it is then analysed mathematically. This is the second phase of the modelling process. In it the basic theory is applied explicitly. In this case it is the theory of classical dynamics that is to be applied, and this is usually stated in terms of equations of motion. Although there are alternative formulations, the equations of motion of classical dynamics can usually be expressed as, or at least result in, sets of ordinary differential equations. The various terms of these equations and the parameters they contain are all determined by the theory when applied to the particular model. In fact it is often convenient to initially describe the model in terms of certain characteristics or expressions that can be substituted directly into the equations of motion. This again illustrates the point that the basic theory affects the way in which a physical system is represented in the model.

At this point it may be assumed that the mathematical model has been described in terms of sets of ordinary differential equations. Apart from sometimes having to evaluate expressions or characteristics to substitute into these equations, the main effort in this phase is to attempt to integrate the equations and to investigate the mathematical properties of their solutions.

Ever since the theory was first put forward, much of the research associated with it has been directed towards the development of mathematical techniques that are particularly appropriate to the type of equations that occur in applications of the theory. In spite of this, however, it still usually happens in practice that complete analytic solutions of the equations of motion cannot be obtained. Attention is therefore directed towards the possible occurrence of first integrals, or constants of the motion, such as the energy integral or integrals of momentum. Particular techniques have been developed in order to derive such integrals.

In addition, various approximation techniques may also be applied. These may be of two types. They may simply involve the numerical integration of the equations of motion as they stand, for particular initial conditions. The errors that arise in such calculations are purely of a mathematical nature and can easily be estimated. The alternative approach is to relate the approximation technique more directly to the physical situation. This essentially involves a further simplification of the model in which a parameter which is known to be small is temporarily assumed to be zero. If an analytic solution of this simplified model can be obtained, the parameter may then be reintroduced and a better approximate solution obtained using perturbation techniques.

In the final phase of the modelling process, the mathematical results obtained in the model are reinterpreted in terms of the real physical system under consideration. In some ways this can be seen as the opposite process to the initial construction of the model. Mathematical constructs are again related to physical objects in the real world, and the behaviour of the model can be compared with observed physical processes.

The results obtained from a mathematical model are referred to as predictions. Essentially they predict what the physical system would do if it behaved exactly like the model. If it is possible to observe the motion of the physical system, then these observations may be compared with the predictions, and if these are roughly in agreement the model is said to be satisfactory.

In practice, however, the comparison of predictions and observations is a little more complicated, since the mathematical model, and the predictions derived from it, usually contain a number of unspecified parameters. These have to be determined before specific predictions can be made. This can frequently be achieved simply by observing the physical system for a period of time and choosing the parameters to match the model to the system. Often, however, this process is more complicated, and, since this forms an essential step in the testing of scientific theories, it is appropriate to consider it here in greater detail.

1.2. The testing of models and theories

As has been stated above, the results or predictions obtained from mathematical models are usually described in terms of a number of arbitrary parameters. These arise in a variety of ways. To start with, certain quantities or parameters are frequently introduced in the theory itself. Examples of such are the mass or moment of inertia of each body, or Newton's constant of gravitation. These are not always specified explicitly in the initial construction of the model, although some may be calculated directly from it. In addition, the integration of the equations of motion may also result in the introduction of further arbitrary constants.

It is in fact usually regarded as advantageous for models to contain a number of unspecified parameters, since this enables them to represent a range of physical situations. Models are often deliberately constructed in this way, although they do not aim at total generality. That is the aim of the basic theory. Scientific theories are developed to cover as wide a range of physical processes as possible. It is the application of such theories to particular types of situations within that range that yields theoretical models. Thus the purpose of the model is to represent a certain class of situation within the broader range that is covered by the theory.

The behaviour of general models like this which contain arbitrary parameters can be compared with the corresponding range of physical situations. However, to obtain more than a general qualitative agreement special cases have to be considered. In order to do this it is necessary to specify the particular values of the parameters which correspond to each particular case being considered. The various parameters, however, are determined in a number of different ways.

Some of the parameters can frequently be determined directly from observation of the physical situation. For example, the initial positions and velocities of certain parts of the system can often be observed and their values used to specify some of the parameters. In other cases it is necessary to observe the motion of the system over a period of time before some parameters are specified. In such cases the comparison of the model with observations is used to clarify or specify the model rather than test it. Finally, there are also cases in which certain parameters have to be estimated from separate experiments based on other, or related, theories and their associated models. Sometimes it is necessary to perform such experiments. For example, the mass of a small object can sometimes be estimated separately by taking it aside and weighing it. Alternatively, the generally accepted results of others may be used. For example, the mass of the planets, or the value of Newton's constant of gravitation, are now considered to be well known.

It should, however, be emphasised that the particular values of the parameters that are inserted in the model are only estimates of the corresponding physical quantities. It is only possible to determine them up to a certain specified degree of accuracy. Thus the mathematical models based on the theory of classical dynamics are only capable of providing an approximate representation of a particular system. However, since the accuracy of each parameter can be determined, it is possible to predict the behaviour of the system to within certain specified error bounds.

Such predictions can now be compared with the actual observed behaviour of the physical system. However, the observations cannot be made with perfect accuracy either. Observed quantities can only be estimated to within certain specified error bounds.

Observations of a particular system can now be compared with the predictions obtained from a particular model. If these are found to be in agreement within the estimated errors, then the model is clearly satisfactory for this particular case. Once the model has been corroborated for one such case, it may then be tested for other particular cases. Finally, if the model has been found to be satisfactory for a number of situations of a particular type having parameters within a certain specified range, it may then be used with a certain amount of confidence for other situations of the same type within that range. It is possible to use the model to predict what would happen for various values of the parameters in the acceptable range. Such predictions may then be assumed to be correct, even though the physical situation may not have actually been observed for every particular case.

It is of course also possible to use an established model to describe situations in which the values of some parameters lie outside the range for which the model has been tested. The results obtained in such cases should, however, only be accepted with a certain reservation until the extrapolation has been tested.

At this point it is convenient to consider what should be done if the predictions of a model do not agree with observations of the physical system it is supposed to represent.

In such cases it is first necessary to check the accuracy of the observations, the accuracy of the estimates for any parameters including initial conditions that are required in the model, and also the accuracy to which any calculations are performed. The possible errors arising in each of these cases can be calculated, and we must now consider the possible situation in which predictions and observations still do not agree, even after allowing for these possible errors.

At this point it may also be assumed that the disagreement is not due to any trivial mistake, but that repeated observations of a particular case consistently indicate a significant difference between predictions and observations. It is therefore only possible to conclude that the model is unsatisfactory. The problem now is to find the unsatisfactory aspects of the model, and to attempt to correct them.

Now, when a particular model is unsatisfactory in the sense described above, it is almost always found that this is due to some oversimplification in its original construction. Sometimes the initial assumptions are wrong, or at least are not sufficiently accurate. In such cases more accurate initial assumptions should be tried. In other cases it is often found that some significant factor has been ignored. In all such cases it is necessary to return to phase one of the modelling process and to consider a more accurate model.

Sometimes it is convenient to retain the initial model as an approximate representation. It frequently happens that, in order to obtain a model that is perfectly satisfactory, terms have to be included which make the mathematical analysis extremely complex. In such situations simpler approximate models are usually used. Even though they are known to be technically unsatisfactory, they may still give answers that are sufficiently accurate for some purposes. Approximate models of this type are particularly useful once it is known, at least in principle, how to correct them. In such cases it is possible to estimate the effect of the terms that have been omitted. Thus the errors which arise in the model as a result of the simplification can be explicitly determined.

Although it only rarely occurs in practice, it is still necessary to consider the remaining possibility that the observations of a system may not agree with the predictions obtained from a model, even after the model has been corrected and made as accurate as possible. In such a case it can only be concluded that it is the basic theory that is at fault, and an alternative theory should be considered.

It is this final possibility that enables scientific theories to be put to the test. However, before considering any particular theory to be disproved or falsified, it is necessary always to check thoroughly that the disagreement does not arise from the oversimplification of the model. The particular way in which this approach can be used to test the theory of classical dynamics is described in more detail in section 3.5. A further example in terms of the Newtonian theory of gravitation is also described in section 5.5.

Now, part of the scientific method is that every theory, and the models based upon it, should be tested as thoroughly as possible. This is done in principle as described above. However, this does not mean that endless experiments have to be performed to check each model for more and more values of its associated parameters. Instead, the theory is tested in two ways. The first method is to attempt to reduce the error bounds in both the model and the observations. This quest for greater and greater accuracy is made possible by technological advances in both observational instruments and experimental apparatus. The second method, which is also associated with it, is the quest to extend the range over which the models are tested.

For a theory to be generally accepted it should be capable of providing models which, in principle at least, are entirely satisfactory. In addition, it should be applicable to a wide range of physical situations. Both of these features are demonstrated by the theory of classical dynamics. To start with it is applicable to an exceptionally wide range of physical problems, both terrestrial and celestial, and this range can easily be extended to many situations that are outside the usual subject area of mechanics. In addition, within this range, it is capable of describing what happens in the real world to at least the accuracy that can be achieved by the most advanced instruments.

However in some extreme situations the theory of classical dynamics is found to fail. This occurs when considering systems in which the velocity of some components becomes a significant fraction of the speed of light, or when the size of the bodies considered are so small that they must be considered as atomic or subatomic particles. In these situations new theories have been established.

These new theories of relativity and quantum mechanics provide a different approach to the whole modelling process. However, when they are applied to situations in the range in which the classical theory has been so successful, they become so complicated that the mathematical analysis of the models usually becomes impossible. In this situation further simplifications have to be made, and it is usually found that the resulting models constructed in this way are identical to those which would have been obtained using the simpler classical theory. The reason for this is that the new theories claim to be deeper theories which are closer approximations to reality. It is therefore necessary that all the tests which previously have been considered to corroborate the classical theory, should also corroborate the new theories. So the new theories are required to approximate to the classical theory in the appropriate range, and should only deviate from it where it is found to be unsatisfactory. Thus it can be seen that the classical theory may continue to be used, and that the new theories also clarify the range of situations in which it can be applied with confidence.

1.3. The primitive base of classical dynamics

Having briefly described the approach of mathematical modelling, it is appropriate to consider in a little more detail how the model is related to the actual physical situation in the real world. On the one hand we have physical objects that can be seen and handled, and on the other we have mathematical symbols and equations.

In fact there is no unique way in which these two distinct types of object can be related. They exist in totally different worlds. Yet in order to apply the scientific method some correspondence must be assumed. This can in fact be achieved using a certain number of primitive concepts. These form the foundational base upon which the theory is built.

The foundational concepts of classical dynamics are those of space, time, a physical body and force. These concepts are not defined exhaustively in the theory, as are, for example, angular momentum or kinetic energy. The question: 'What is kinetic energy?' can be answered satsifactorily in purely mechanical terms, once some basic aspects of the theory are understood. On the other hand the question: 'What is time?' or 'What is a physical body?' immediately involve deep philosophical problems, and cannot even be considered without stepping outside the narrow subject area of classical dynamics.

It can be seen that these foundational concepts involve some assumed representation of certain aspects of the real world, that can be initially taken on trust and used to develop theories. They are undefined but meaningful concepts which enable aspects of the real world to be analysed scientifically.

The foundational concepts can themselves be the subject of a deeper level of scientific or metaphysical theories. However, they are not concepts that are taken from elsewhere and simply inserted as the basis of a theory. Rather, they are taken as vague and ill-defined concepts and given a precise interpretation. The philosophical problems are then temporarily forgotten and they are used in the theory in terms of exact mathematical models. Finally, the successes and failures of the theory in its many applications provides a commentary on the appropriateness of the assumptions concerning these foundational concepts. In this way it can be seen that the theory of classical dynamics, for example, contributes very significantly to an understanding of space and time, of the nature of matter and of force.

The foundational concepts of a theory are primitive assumptions on which the theory is based. Thus if a theory is found to be unsatisfactory, critical attention may be directed towards the foundational concepts, and alternatives may be suggested. The modern theories have in fact been obtained in this way. The theories of relativity are based on a different concept of space and time, and the theory of quantum mechanics uses a different concept of a material body.

Now, one of the most remarkable features of the Newtonian theory of dynamics is that it is based on a particularly simple set of foundational concepts. In this it appears to follow common sense and the intuition based upon it. Indeed, the simplicity of its basic concepts has significantly influenced its general acceptance, and also enables it to be very widely taught and applied. However, it has also induced a certain unconcious reluctance to consider alternative foundations, as many discover when they first approach the subject of relativity.

The primitive concepts of classical dynamics are particular representations of space, time and physical bodies, and an intuitive concept of force. These will all be discussed in the following chapters, but a few brief comments are appropriate here.

The concept of a physical body to be adopted is the intuitive one that is based on common sense. If an apparently solid object can be distinguished from its immediate surroundings, and if it moves or can be moved without changing its character, then it can be regarded as a particular physical body. This description is of course very vague, and it would need to be greatly clarified before it would satisfy a philosopher. Nevertheless it is sufficiently meaningful to enable the motion of real physical objects to be described by a scientific theory.

Now the purpose of any theory of dynamics is to describe the way in which objects move. It must explain both how a body can occupy different positions at different times, and why it occupies the positions it does rather than any others. It must describe the motion of physical bodies in terms of the change of their locations in space as time progresses.

It can immediately be seen that in this whole description are intuitive concepts both of space and of time. In his original formulation of the theory, Newton introduced concepts of absolute space and absolute time. These, however, are found to be unessential to the theory, and will not be introduced or discussed in this book. Instead, it is sufficient just to assume that space is Euclidean in character. This is in fact the common assumption that is used in everyday life. The concept of time is similarly adopted following everyday usage as a unique linear progression that is independent of space.

Having adopted a Euclidean representation of space, it is possible to identify particular 'frames of reference' relative to which the position and motion of physical bodies can be observed and recorded. Using this approach it is possible to describe the motion of real objects. However, far more than this is required of a theory of dynamics. It is also required that it should explain why an object has followed a particular motion, and that it should be capable of predicting what the future motion of the object may be. In order to do this another concept has to be introduced, namely that of force.

The concept of force is introduced to describe anything which causes a body to move. However, in order to develop a theory of dynamics, this concept must be made considerably more precise. In addition, the various types of forces that are introduced may themselves be the subjects of deeper scientific theories. All this will be discussed later in chapter 3.

With the aid of these primitive concepts, the motion of physical objects in the real world can be modelled in terms of theoretical objects moving in a mathematical space, and the results and predictions of the model can be interpreted and checked by observation. In the Newtonian theory of dynamics, the way in which bodies move in space is related to the forces that are assumed to act on them. Thus, models can be set up using a description of the bodies and their motion. The theory can be used to determine the forces that are acting, and these can then be investigated. However, the theory is more frequently used the other way round. Models are usually set up using a description of the bodies and an assumed set of forces which are considered to act on them. The theory can then be used to predict the possible motions of the system on the basis of the assumed forces. The results and predictions obtained may subsequently be checked if required.

Once appropriate representations of the above-mentioned primitive concepts have been adopted, the concepts of mass and particle can also be introduced. A particle is simply an idealisation of a small body, and mass is a concept that is introduced in the theory and which is found to provide a particular unique characteristic of each physical body. With these additions the basic equation of particle dynamics can be introduced: namely, that relative to a suitable frame of reference the resultant force which acts on a particle is equal to its mass times its acceleration. This is the basic postulate of the theory of classical dynamics. From this point onwards the entire subject appears in some ways to be simply an exposition or development of this single equation.

The theory developed in this way can immediately be extended to include the associated theories of fluid dynamics and continuum mechanics, without any need to modify or extend its primitive base. For these subjects further derived concepts have to be introduced such as streamlines, viscosity, turbulence, stress and strain. In addition, other subjects such as classical electrodynamics and magneto fluid mechanics can also be developed from classical dynamics by extending the primitive base to include the additional concepts of charge and electric and magnetic fields.

Finally, it must again be emphasised that the modern theories of relativity, quantum mechanics and statistical mechanics use different representations of the primitive concepts described above. In this sense they are built on different foundations. But in each case they include many concepts that are introduced in classical dynamics.

1.4. The character of the theory

It is often asserted that the theories of classical dynamics and relativity are deterministic, but that the theory of quantum mechanics is probabilistic. From this it is sometimes concluded that nature itself must ultimately be probabilistic. A few comments on this debate are appropriate here.

There are two basic points to be noted as far as the theory of classical dynamics is concerned. The first is that the variables that are considered in the theory are quantities such as those which define the position or velocity of certain specified points of a system. It is implicitly assumed that, in principle, these variables can be directly determined physically to any required degree of accuracy. The second point to be noted is that the equations of motion contained in the theory are ordinary differential equations relating these variables. Exact solutions of these equations exist, at least theoretically. Whether or not analytic expressions for these solutions can be obtained, unique solutions containing a number of arbitrary parameters exist in principle. These arbitrary parameters may also be determined, at least in principle, by a sufficient set of initial conditions. These can usually be expressed in terms of the initial positions and velocities of parts of the system. So, if the initial conditions were known exactly, and if the forces acting on the system were also known exactly, then the theory would lead to an exact and unique prediction of the motion. If objects in the real world actually behaved in this way, it would mean that the position of any object at any time could be determined exactly provided its position and velocity at some earlier time and the forces acting over the intervening period were all known exactly. Such motion is said to be deterministic.

It must be pointed out, however, that in practice it is not possible to determine either initial conditions or forces exactly. Measurements and observations can at best only estimate quantities within certain error bounds. Uncertainties necessarily occur in the values of all the parameters that are required in the application of the theory. Thus, even if systems in the real world behaved in a deterministic way, exactly as described in the theory of classical dynamics, it would not be possible in principle to corroborate this. The configuration of a system at any time can only be predicted to be within certain error bounds.

The character of the theory of quantum mechanics is, in this respect, in direct contrast to that of classical dynamics. In modern quantum theory the basic functions considered are interpreted as measures of the probability that a system exists in a certain configuration. Thus, the whole character of the theory is probabilistic. In fact, one of the basic propositions states that it is not possible for the exact position and velocity of a system to be known simultaneously at any time. Some uncertainty must be included, even in principle.

The contrasting character of these two theories is well known. Discussion of these points has sometimes led on to questioning whether the motion of real physical systems is ultimately deterministic or probabilistic. Does motion itself necessarily include uncertainties, or do the uncertainties only arise through observations and measurements? Certainly, uncertainties must arise through the processes of observation or experiment, since all measurements necessarily involve disturbing the system under consideration, even if this only involves bouncing a photon off it. There is, however, no way of answering the question as to whether objects in the real world ultimately move in a deterministic or probabilistic way. A deterministic theory can be used to approximate an essentially statistical motion. Alternatively, a statistical theory can always be used to approximately represent a deterministic motion. Thus no ultimate answers can be given. All that needs to be said is that the quantum theory contains a deeper and more accurate theory of matter than the classical theory. No theory should ever be regarded as ultimate truth.

2

Space, time and vector notation

The subject of classical dynamics deals with the motion of bodies in space. Its aim is to provide models which are capable of accurately describing the way in which bodies change their position in space as time progresses. Thus the starting point for a study of dynamics must be a set of initial assumptions about the nature of space and time. These are primitive or foundational concepts which are necessary for the development of any theory of dynamics. Newton himself found it necessary to discuss these concepts in the first chapter of his *Principia*, although it is not necessary here to introduce his concepts of an absolute space and absolute time.

Writing at the end of the twentieth century, it is possible to assess the theory of classical dynamics in the light of the modern theories of relativity and quantum mechanics. The classical theory has ultimately been refuted, and the new theories indicate those aspects that require modification. According to the theories of relativity some of the weakest points of the classical theory are in fact its assumptions about space and time. It is therefore appropriate here to clarify the way in which these concepts are used in classical dynamics, and to contrast this with their use in the theories of relativity, before going on to describe the theory itself. We shall also take the opportunity in this chapter of introducing the vector notation which is so useful in classical dynamics as a consequence of its initial assumptions about space and time.

2.1. Space

The basic concept of space is a very difficult one to comprehend or describe. However, it is not necessary to go into a deep philosophical discussion of it here. In the subject of classical dynamics we are concerned with the motion of bodies, and so it is only necessary for a concept of space to provide a way in which the location of such bodies can be described mathematically. The concept of space can in fact be regarded simply as the set of possible positions that the component points of a body may occupy. A body is said to move in space if it has different locations or configurations at different times.

The first problem that arises concerns the basic need to specify the position of any point in space. The problem occurs because the position of a point can only be defined relative to a set of other points. Position is therefore a relative concept. This leads to the philosophical question as to whether space itself is a relative concept, or whether it is in some sense absolute. Fortunately, however, it is not necessary here to discuss and answer this question. All that is required is to indicate how this problem is dealt with in practice. In fact, in practice the problem is posed in a different form.

All observations and measurements are necessarily relative. They are measurements taken relative to the observer or experimenter. The observer chooses his own origin and coordinate system, which is referred to as a *frame of reference.* This is an imaginary construction of a physical coordinate net which pervades the appropriate region of space. The position of any point relative to the observer and his frame of reference can then be determined uniquely in terms of the coordinates. The problem comes when the observer then tries to relate his measurements to those of others using different origins and coordinate systems. This leads to the subject of relativity, which is the theory which relates the measurements of different observers. This subject is covered in chapter 4. At this point it is only necessary to emphasise that in practice we choose an arbitrary origin and frame of reference and work relative to that. The fundamental question as to whether or not space itself is absolute is outside the scope of the theory of classical dynamics.

Another problem arises with the theory of measurement itself. For example, it is known theoretically that the diagonal of a unit square is $\sqrt{2}$ units long, but it is not possible to measure it as being exactly that. In fact, no distance can ever be measured exactly. All that can be done is to measure approximate distances, although, with the aid of more sophisticated equipment, the degree of approximation can sometimes be reduced.

Now, since we are necessarily only dealing with approximations, it is not possible to know whether or not all lengths are equally possible. This leaves open the question of whether space is continuous or quantised. If space itself were continuous, all positions would be possible. Alternatively, if space is a set of disconnected points, then only certain positions would be permitted. On a large scale, space appears to be continuous, but it is not clear that its continuous nature would persist under extreme magnification. All that can be said is that space appears to be continuous down to a scale of at least 10^{-13} cm, and it does not seem to be possible to determine any position with greater accuracy than this.

Although it is not possible to determine whether space itself is ultimately continuous or quantised, the classical approach is to make the assumption that it is continuous. This is one of the foundational assumptions in constructing a model of the real world. This assumption is a perfectly reasonable one for the theory of classical dynamics since this is applied to the motion of bodies of a reasonable size. For bodies of dimensions less than 10^{-8} cm, the classical theory of dynamics breaks down anyway and quantum mechanics should be applied. The assumption of the continuity of space therefore does not lead to any significant error in the application of this theory. However, the assumption does have specific advantages. To start with, it permits the use of the mathematics of real numbers for the calculation of distances. It also permits the application of the mathematical techniques of calculus. The assumption of continuity can thus be seen to be arbitrary but justified in the development of a theory because it does not lead to significant errors and because it permits simpler and more precise calculations.

Another important assumption to be made about space is that it has three dimensions. The number of dimensions of a space may be defined here as the minimum number of parameters required to define the position of a point in the space relative to a fixed origin and a specified frame of reference. The dimensionality of a space is one of its invariant properties. That physical space has three dimensions is well known, and can be seen in that an unconstrained body can be moved in three independent directions neglecting rotations.

The above points may be summarised by saying that space may be represented as a three-dimensional manifold. The term manifold is a technical term that is used to imply continuity. It describes a precise mathematical structure which permits differentiation to be defined, but does not distinguish intrinsically between different coordinate systems. However, a formal statement of this assumption will be delayed until a further aspect of it has been clarified.

2.2. Time

If the concept of space is a difficult one to describe, the concept of time is even more so. Time is a concept of which we all have an intuitive notion, but which is almost impossible to explain. It is a concept which contains an element of progression. By it we classify events as relatively 'before' or 'after'. Time is the way in which we order events as they happen. But it is more than this. It is also related to our conciousness as individuals, and so our intuitive concept of time, though difficult to explain, is sometimes an accurate guide to a precise formulation.

Time is most frequently described by an analogy with one spatial dimension. For example, we readily speak of a 'long time' or of going 'forward' in time. This analogy between space and time is often very useful, but it has to be used with care as the concepts are essentially different. A body may be in the same place at different times, but cannot be in different places at the same time. After measuring a distance it is always possible to go back and check that measurement at a later time if that is required, but it is never possible to return to check a measured time interval. Once a time interval has passed it can never be returned to, but a particular region of space can frequently be revisited. We have a measure of control over our location in space, we can move ourselves and we can move other objects, but we have absolutely no control over our movement in time. Time passes, and we can do nothing about it.

Also in contrast to space, time has a unique direction associated with it. The future is qualitatively different from the past. If a film is accidentally shown backwards, this is usually immediately obvious from the unfamiliar character of the motions that may be shown. Our memories recall the past, not the future. These examples and many others that could be quoted demonstrate the fact that there is a unique direction associated with the passage of time.

The differences between space and time must be kept in mind, but it is convenient to proceed with a discussion of certain aspects of time in a similar way to the previous discussion of space.

Again, the first problem is to consider whether time is a relative or an absolute concept. Clearly, it is a concept that is applicable to all individuals, and in fact also to all real objects in the universe. Time intervals can be measured relative to all individuals or objects. Thus the measurement of time is a relative concept. However, the way in which times recorded by different observers are related is very complicated. To start with, the actual sequence of events in time may differ between two observers. For example, the time interval between a flash of lightning and the associated clap of thunder is variable. One observer may record a second flash of lightning before the first clap of thunder, but another observer situated somewhere else may record the opposite order of events. Now this particular change of order can easily be explained, and the two observers may agree on a unique time sequence for events occurring in their vicinity. The classical method is now to proceed to the assumption that there is a unique time sequence associated with the world as a whole. This assumption is made purely on the evidence of closely associated observers and therefore involves an immense extrapolation. However, this is one of the basic assumptions of classical mechanics. It is assumed that there is a unique time sequence covering the whole of space. Time therefore is regarded as being totally independent of space and independent of any particular observer. In this sense, time is regarded as an absolute concept, although the origin and scale of time are dependent on each particular observer or clock and are therefore relative.

This assumption of the uniqueness of time sequences, which is basic to classical dynamics, is in fact the weakest point of the theory. Time sequences have now been shown to be nonunique for observers that are moving relative to each other. The theory of special relativity has been developed to describe such situations and, in this theory, time sequences are relative concepts and consequently no two events can be uniquely simultaneous. However, although the concept of time considered in the theories of relativity is a closer approximation to reality, the mathematics associated with it is notoriously complex, and the confusion that has surrounded it has generated numerous supposed paradoxes. Thus the assumption that time sequences are unique, because of its convenience and simplicity, continues to be made in classical dynamics as in most other scientific theories.

By analogy with space, time also is regarded in classical mechanics as being continuous. In observations of the real world, time does appear to 'flow' continuously. However, there are no means available at present of determining ultimately whether time is continuous or quantised. If time were quantised, the world would appear like a sequence of images on a cinema screen. All that can be said is that time appears to be continuous at least down to the scale of 10^{-24} seconds, which is more than sufficient a limit for the applications of classical dynamics.

It can be seen from the above that time as it is regarded in classical dynamics can be represented by a single continuous parameter, which is all that is required to represent time over all space and for all possible observers. Time intervals in this theory are described uniquely in terms of such a parameter.

It must now be considered how this parameter should be defined. It turns out that the parameterisation is not totally arbitrary. Rather, it seems that the real physical world singles out a preferred parameter which is unique up to linear transformations. It is appropriate here to be guided by intuition, so that time intervals which are intuitively regarded as equal are measured as being approximately equal in the adopted parameterisation.

Traditionally, time has been parameterised by the motion of the earth. The time the earth takes to orbit about the sun is defined as a year, and a day is defined as the period of rotation of the earth about its axis with respect to the sun. These definitions are consistent with intuition which regards each such successive period as being of approximately the same duration. By way of contrast, had linear increments in radioactive decay been regarded as a suitable measure of time, then the period of each day would continually decrease at an ever faster rate contrary to intuition. The more useful parameterisations are those which regard free oscillations or free rotations as having approximately constant period. This is not only in accord with intuition, but is also amazingly self-consistent irrespective of the mechanism of oscillation or type of body which is rotating. This appears to be the natural parameterisation that is singled out by the physical world. Thus the most appropriate choice of time parameter is that defined by the most free natural oscillations that can be found. Nowadays this appears to be the natural oscillations within an atom, and so by international agreement a standard atomic clock is adopted and time is parameterised in seconds according to it. To be specific, since 1967, a second has been defined as 9 192 631 770 periods of the unperturbed microwave transition between the two hyperfine levels of the ground state of Cs^{133}. This convention enables time intervals to be measured quantitatively with such accuracy that it is even possible to measure the rate at which the earth's rotation is slowing down. However, although nature appears to single out this parameter, the adoption of it is a convention and the choice of scale in seconds is arbitrary.

We are now in a position to formally summarise the basic assumptions about time that are made in the theory of classical dynamics.

Assumption 2.1. *Time is represented as a one-dimensional manifold defined over all space and parameterised in seconds as internationally defined.*

The assumed independence of time and space and the assumption of continuity, as implied by the technical term 'manifold', permit representations of the position of a point in space and other quantities to be differentiated with respect to the agreed time parameter. This is an important facility which is frequently used. The time parameter is usually denoted by the symbol t.

2.3. Scalar and vector notation

In the real-world different types of physical quantity are observed. Some physical quantities have a property of magnitude and this appears to be the only property they have. Such quantities are called *scalar quantities.* Examples of such are mass, temperature and energy. However, many quantities found in the physical world are not scalar. Some, for example, in addition to magnitude have also a direction associated with them. Examples of such quantities are velocity, acceleration, force and magnetic field. These are known as *vector quantities.* Other types of quantity also need to be considered in a scientific description of the real world, such as tensor quantities. Further types, particularly, appear in the study of subatomic particles. However, in this study of classical dynamics it is only necessary to consider scalar and vector quantities.

Scalar quantities have only the property of magnitude. They can therefore be represented by a single real number. Such a representation is referred to as a *scalar.* Not all scalar quantities are constants. Some may vary with time, such as the temperature of a body. The scalar representation must therefore vary with the time parameter. It can be regarded mathematically as a *scalar function.* This defines a real number representing the scalar quantity for all values of the given parameter. The scalar function may be discontinuous, continuous or differentiable according to the particular application. Some scalar quantities may also vary with position. For example, temperature and density are quantities which can be different at different places and different times. Such quantities are represented by *scalar fields.* These define a real number representing the scalar quantity at all points in a region of space and for all times in an approximate period. Scalar fields are therefore generally functions of position and of time. They also may be continuous or discontinuous according to the application. In many applications, scalar fields are found to be independent of time and can therefore be regarded as functions of position only.

In a similar way, mathematical objects called *vectors* are used to represent the vector quantities that are observed in the real world. These too may vary with space and time, and may be continuous or discontinuous according to applications. However, they have additional properties to scalars and therefore require a more complicated mathematical representation. A physical vector quantity has associated with it both a magnitude and a direction. Vectors must therefore represent both of these features.

The simplest definition of a vector is as an arrow, whose length represents the magnitude and which is pointing in the direction to be represented. If the vector quantity varies with time, then both the magnitude and direction of the arrow must vary accordingly. If the vector quantity varies with position, then the vector must be a function of position in space. This case is described by a *vector field* which defines appropriate arrows at all points in space. When visualising such a field it must be understood that each arrow exists only at one point in space, which is usually taken to be the tail of the arrow.

In such a representation vectors can be added by the so-called 'parallelogram' or 'triangle' law in which the sum is defined as the diagonal of the parallelogram constructed from the two arrows. This also permits any vector to be decomposed into a number of components if this is required. It is important to emphasise that this rule is consistent with the way in which vector quantities appear to sum in the real world. However a certain feature of it needs to be clarified.

In adopting a geometrical rule for the addition of vectors, the properties of an underlying geometry have been implicitly assumed. In this case classical Euclidean geometry has been used. Physical vector quantities have effectively been represented by what are called Euclidean vectors.

A three-dimensional Euclidean space can be defined as one which can be spanned by three cartesian coordinates x, y, z such that the distance between any two points (x_1, y_1, z_1) and (x_2, y_2, z_2) is given by $((x_1-x_2)^2+(y_1-y_2)^2+(z_1-z_2)^2)^{1/2}$. In such a space the distance and direction between any two points are uniquely determined. Thus any arrow representing a vector quantity can be defined by the two points at its tip and tail. If a set of cartesian coordinates are now introduced with an origin at the tail of the arrow, then the single point representing the tip of the arrow gives a unique representation of the vector quantity. A vector can therefore alternatively be defined in terms of an ordered set of three numbers. These can be interpreted as the cartesian coordinates of a point having a unique distance and direction from the origin of a Euclidean space. It is a generalisation of this definition of a vector that is usually given in textbooks on linear algebra, together with the mathematical rules for the manipulation of such quantities. However, it should be emphasised that vectors are mathematical objects that have an objective meaning which is independent of any particular coordinate system.

It must also be emphasised that, whichever definition is used, a vector exists only at a single point. The Euclidean space that is defined at that point is only a mathematical artifice that permits the algebraic manipulation of vectors such as addition, decomposition into components, and differentiation.

In the analysis and applications of classical dynamics, the mathematics of Euclidean vectors is particularly useful. It is not necessary to explain the mathematical techniques here, but it is appropriate to clarify one or two points of vector notation that are used in this book. The convention is followed that vectors are denoted by letters in bold print such as $\boldsymbol{A}$ or $\boldsymbol{v}$.

It is convenient first to separate the concepts of magnitude and direction that are contained in vectors. This can be achieved by denoting the magnitude of a vector $\boldsymbol{a}$ by the scalar a or $|\boldsymbol{a}|$, and a unit vector in the direction of $\boldsymbol{a}$ by $\hat{\boldsymbol{a}}$. The vector can then be written as

$$\boldsymbol{a} = a\hat{\boldsymbol{a}}$$

which distinguishes the concepts of magnitude and direction.

In application it is frequently required to extract a scalar from two vectors according to the rule

$$\boldsymbol{a} \cdot \boldsymbol{b} = ab \cos \theta$$

where θ is the angle between the directions of the two vectors. This is known as the *scalar* or *inner product* of two vectors, and is denoted by placing a dot between them. It can be seen that with a unit vector $\boldsymbol{a} \cdot \hat{\boldsymbol{b}}$ is the component of $\boldsymbol{a}$ in the direction $\hat{\boldsymbol{b}}$. This rule of multiplication is useful in a considerable number of physical situations.

A different rule for multiplying vectors that arises naturally in a number of physical situations can be defined by

$$\boldsymbol{a} \times \boldsymbol{b} = ab \sin \theta \hat{\boldsymbol{c}}$$

where θ is the angle between the vectors $\boldsymbol{a}$ and $\boldsymbol{b}$, and $\hat{\boldsymbol{c}}$ is a unit vector perpendicular to both $\boldsymbol{a}$ and $\boldsymbol{b}$ and directed such that $\boldsymbol{a}$, $\boldsymbol{b}$, $\hat{\boldsymbol{c}}$ are ordered in a right-handed sense. This rule is known as the *vector product* rule, and is denoted by placing the symbol × between the two vectors. The vector product is itself a vector. It should be emphasised that this product is defined solely because of its usefulness in applications. It does not arise as naturally from the mathematics as do the scalar or inner product and the outer product which there is no need to define here.

It may also be pointed out that the evaluation of both scalar and vector products is particularly simple using the cartesian component notation. It is, however, inappropriate to describe further vector methods here, although some knowledge of vector field theory will be assumed in one or two sections later in the book.

2.4. Position and Euclidean space

It is now important to consider whether or not position is a vector quantity. Intuitively, position does not appear to have the qualities of magnitude and direction and therefore is not an obvious vector quantity. However, a coastguard can state the position of a ship in terms of the direction of a sighting and an estimated distance. Thus it appears to be possible to regard a relative position as a vector quantity. First, it is necessary to choose a frame of reference with an origin from which position is to be defined. Then the relative position of any object can be defined by a direction and a distance. In this way the position '10 km south-west of the coastguard' can be regarded as a vector quantity.

There are, however, a number of problems associated with the above classification of relative position as a vector quantity. The magnitude and direction are defined only at the agreed origin. The meaning is this – that a lifeboat travelling south-west from the coastguard would reach the ship after it had travelled 10 km. This appears satisfactory, but difficulties soon appear. How, for example, would one define the position of New York from London? An aircraft could be given a direction and a distance. At each point on his journey the pilot could check that he was on course using his compass and could calculate the distance he has travelled. But over such a distance the curvature of the earth's surface is significant. The pilot would in fact have had difficulty choosing his course to start with. If he had a model of the globe he could have stretched a string between the representations of London and New York and chosen that as his course. This would have given the shortest route to fly, but to achieve it the compass bearing would be continuously changing. In this the vector position of New York from London is irrelevant. This illustrates that the above vector representation of position only has meaning locally to the origin, and can only be formulated when there is a unique agreed method of specifying both direction and distance.

The two examples used in the above paragraphs were both essentially two-dimensional. The ship was restricted to the surface of the sea, and the aircraft was presumed to fly at a constant altitude. It was seen that a vector representation of position was easily defined when the two-dimensional surface was approximately flat, but that difficulties occurred over distances containing significant curvature.

The question that must now be considered is whether or not space is curved. The surface of the globe is curved when considered in the three dimensions of space. In a similar way it must be considered whether or not space itself has some curvature when it is considered to be embedded in a hypothetical four-dimensional space.

In order to answer this question it is first necessary to consider how curvature could be detected. This can easily be seen in two dimensions. Take any three points on the flat surface of a desk. Determine the straight lines between the points by stretching string between them. Measure the length of the lines and the angles between them. Within the accuracy of the measurements, these lengths and angles will satisfy an appropriate generalisation of Pythagoras's theorem which is based on Euclidean geometry. To see the effects of curvature also take three points on the spherical surface of a football. Determine lines between the points by stretching string between them. In this case the measured lengths of the lines will not satisfy Pythagoras's theorem for the measured angles between them. The three angles will in fact add up to more than two right-angles.

A creature constrained to live in a two-dimensional world could thus determine, at least in principle, whether or not his world is curved in a higher-dimensional embedding space. He could even conduct experiments, such as measuring triangles, to estimate the curvature of his space. Similarly creatures such as ourselves living in a three-dimensional space can estimate, at least in principle, the curvature of our space in some hypothetical four-dimensional embedding space. Such experiments have been carried out and the conclusion has been reached that the curvature of the space we live in is negligibly small over reasonable distances, there being no direct means of measuring curvature over astronomical distances.

The basic assumption concerning the nature of space may now be stated formally.

***Assumption* 2.2.** *Space may be regarded as a three-dimensional Euclidean manifold.*

The assumption that space is three-dimensional and continuous was introduced in section 2.1. The further point added here is that space is also assumed to be flat or Euclidean. This of course is an assumption, but is justified since it is both consistent with observations and also permits calculations to be very much simpler than they would have to be if a small curvature had to be allowed for.

One of the main advantages of this assumption is that it permits a very simple representation of relative positions in terms of Euclidean vectors. The cartesian coordinates of any point in space can themselves be taken to represent its position vector with respect to the origin and coordinate axes.

Assumption 2.2 is in fact basic to classical dynamics. Until this century it was never questioned, but in Einstein's general theory of relativity space is permitted to be curved and a more general initial assumption is introduced. Einstein's theory regards space and time on an equal footing and introduces the four-dimensional entity space–time. Strictly speaking, it is the four-dimensional manifold space–time which is regarded as being curved. This permits not only the curvature of space, but particularly the interdependence of space and time. Although this theory appears to provide a more accurate model of the real world, it is extremely difficult to use. Its four-dimensional pseudo-Riemannian geometry requires advanced mathematical techniques. By contrast, classical dynamics, which uses three-dimensional Euclidean geometry and an independent time parameter, is very simple. However, it must be stressed that these are working assumptions which are not justified when considering very large volocities and very strong gravitational fields. In either of these cases the theories of relativity must be applied.

2.5. Velocity and acceleration

Now that the basic assumptions about space and time have been clarified, the motion of bodies in space may be considered. It is necessary to describe the way in which the position of a body in space may vary with time. To do this it is convenient to introduce the familiar concepts of velocity and acceleration.

The first point to emphasise here is that velocities and accelerations are real objective quantities. Strictly speaking, we only deal with relative velocities and relative accelerations, but this should be understood since position itself is only a relative concept, and velocities and accelerations are measurements of the rate of change of position. In our mathematical model of the real world it is easy to define velocities and accelerations objectively. Velocity is the time rate of change of position, and acceleration is the time rate of change of velocity. These definitions are dependent upon the assumptions of the continuity and independence of space and time. They have an obvious real interpretation. Similarly, in the real world velocities and accelerations are real measurable quantities. Even if space and time were ultimately quantised, real meaning could still be given to them. It is possible to measure the approximate velocity of an object shown on a cinema film which shows only ten frames per second. If time were quantised into intervals of 10^{-24} seconds, this would have negligible effect on the measurement of velocities and accelerations, and certainly would not alter their ontological status.

Let us now define the terms formally.

***Definition* 2.1.** *The velocity of a point relative to a given frame of reference is the rate of change of its position vector relative to that frame, with respect to time.*

***Definition* 2.2.** *The acceleration of a point relative to a given frame of reference is the rate of change of its velocity relative to that frame, with respect to time.*

As a consequence of assumption 2.2 the position of any point relative to an arbitrary frame of reference can be represented as a Euclidean vector. The velocity and acceleration of that point are therefore also represented by vectors. However, velocity and acceleration are both locally defined vector quantities and could still be represented by vectors even if the assumption of Euclidean space were relaxed.

If the position of a point relative to a given frame of reference is denoted by $\boldsymbol{r}$, then the velocity $\boldsymbol{v}$ and acceleration of that point relative to the frame are

$$\text{Velocity } \boldsymbol{v} = \lim_{\delta t \to 0} \frac{\boldsymbol{r}(t+\delta t) - \boldsymbol{r}(t)}{\delta t} = \dot{\boldsymbol{r}}$$

$$\text{Acceleration} = \dot{\boldsymbol{v}} = \ddot{\boldsymbol{r}}$$

where a dot above a vector denotes its derivative with respect to time.

It is of great importance to distinguish between the concept of velocity as defined above and the more familiar concept of speed. Velocity is a vector quantity. It must always have associated with it both a magnitude and a direction. Speed on the other hand may be defined simply as the magnitude of velocity. It is a measure of how fast an object is moving without any reference to the direction in which it is going. Speed is therefore a scalar quantity, and contains only part of the information that is required in order to express a velocity.

The term acceleration unfortunately causes a great deal of confusion. As defined above, acceleration is a vector quantity. It must therefore have both a magnitude and a direction. The confusion arises because in ordinary language the rate of change of speed is also called acceleration, but this is not the definition adopted here. Acceleration is the rate of change of velocity and is a vector. Its component in the direction of the instantaneous velocity is the rate of change of speed, but if the direction of the velocity changes, then there must also be a component of acceleration perpendicular to the velocity vector. For example, if an object moves in a circle with constant speed, then the direction of its velocity is continuously changing, and the acceleration vector is constant in magnitude but its direction is continuously changing so that it is always pointing towards the centre of the circle. It is most important to emphasise that if a body is not moving with constant speed in a straight line, then it must be accelerating, and the acceleration may have components both in the direction of the velocity vector and in some other direction perpendicular to it.

The terms angular velocity and angular acceleration are also used in the following chapters and it is convenient to include a definition of these at this point.

The concept of angular velocity is used to measure the rate at which something rotates. Consider for example a body that is free to rotate about a defined axis. At any time the orientation of the body can be defined in terms of the angle through which it has rotated. The rate at which that angle changes with respect to time is called the angular speed. The angular velocity is a vector which has this magnitude and is directed along the axis of rotation in the direction such that a positive change of angle corresponds to a rotation in the right-handed sense. Thus the angular velocity of an object defines not only its rate of rotation, but also the direction about which it rotates. Angular acceleration is simply the rate of change of angular velocity and, consequently, is also a vector quantity. These definitions may now be stated formally.

***Definition* 2.3.** *Angular velocity is the vector whose magnitude is the rate of rotation with respect to time, and whose direction is that of the axis about which rotation takes place in a positive sense.*

***Definition* 2.4.** *Angular acceleration is the rate of change of angular velocity with respect to time.*

These quantities are of course relative in the sense that rotation can only be measured with respect to a fixed or agreed set of directions. Relative to such an agreed frame, any rigid body or other frame of reference may be rotating. If it is rotating about an axis defined by the unit vector $\hat{\boldsymbol{a}}$ with angular speed $\dot{\theta}$, then its angular velocity is defined by

$$\boldsymbol{\omega} = \dot{\theta}\hat{\boldsymbol{a}}$$

and if $\hat{\boldsymbol{a}}$ is a constant vector, then θ measures the rotation at any time. However, if the axis of rotation is itself changing, then $\boldsymbol{\omega}$ is simply the instantaneous angular velocity.

It is important to emphasise that the angular acceleration may not be in the same direction as the angular velocity. Its component in this direction measures the rate at which rotation is speeding up, but if the direction of the rotation is also changing, then the angular acceleration has a component perpendicular to this direction as well.

$$\dot{\boldsymbol{\omega}} = \ddot{\theta}\hat{\boldsymbol{a}} + \dot{\theta}\frac{\mathrm{d}}{\mathrm{d}t}\hat{\boldsymbol{a}}$$

2.6. Kinematics: examples

It is necessary here to obtain expressions for acceleration in a number of commonly used coordinate systems. These will then be applied in subsequent examples, both in this and in later chapters.

Example 2.1: *Cartesian coordinates*

It has become conventional to denote unit vectors aligned with the cartesian coordinate axes Ox, Oy and Oz by $\boldsymbol{i}$, $\boldsymbol{j}$ and $\boldsymbol{k}$ respectively. Using this notation the position vector of a point relative to these axes can be written in the form

$$\boldsymbol{r} = x\boldsymbol{i} + y\boldsymbol{j} + z\boldsymbol{k}$$

Since the basis vectors $\boldsymbol{i}$, $\boldsymbol{j}$ and $\boldsymbol{k}$ are constant in both magnitude and direction at all points, expressions for the velocity and acceleration of the point are

$$\dot{\boldsymbol{r}} = \dot{x}\boldsymbol{i} + \dot{y}\boldsymbol{j} + \dot{z}\boldsymbol{k}$$

$$\ddot{\boldsymbol{r}} = \ddot{x}\boldsymbol{i} + \ddot{y}\boldsymbol{j} + \ddot{z}\boldsymbol{k}$$

Clearly $\ddot{x}$, $\ddot{y}$ and $\ddot{z}$ are the components of the acceleration of the point in the corresponding coordinate directions.

Example 2.2: *Plane polar coordinates*

When the motion of a point P in a plane is described in terms of polar coordinates r and θ, it is convenient to introduce two mutually orthogonal unit vectors $\boldsymbol{e}_r$ and $\boldsymbol{e}_\theta$ which at any point are aligned with the coordinate directions (see figure 2.1). In terms of the alternative cartesian coordinates in the plane, these basis vectors can be expressed as

$$\boldsymbol{e}_r = \cos\theta\boldsymbol{i} + \sin\theta\boldsymbol{j}, \qquad \boldsymbol{e}_\theta = -\sin\theta\boldsymbol{i} + \cos\theta\boldsymbol{j}$$

Clearly, these vectors do not point in a fixed direction at all points in the plane. In fact, it can easily be shown that as the point $P(r, \theta)$ moves in the plane

$$\dot{\boldsymbol{e}}_r = \dot{\theta}\boldsymbol{e}_\theta \qquad \dot{\boldsymbol{e}}_\theta = -\dot{\theta}\boldsymbol{e}_r$$

Fig. 2.1 $\boldsymbol{e}_r$ and $\boldsymbol{e}_\theta$ are unit vectors in the directions of increasing r and θ at the point P.

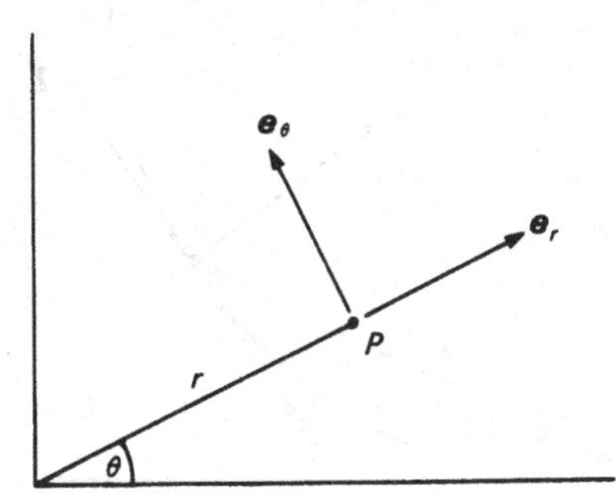

Now, the position of P relative to the origin can be written as

$$\boldsymbol{r} = r\boldsymbol{e}_r$$

and, differentiating this using the above identities, gives the expressions for the velocity and acceleration of P in the form

$$\dot{\boldsymbol{r}} = \dot{r}\boldsymbol{e}_r + r\dot{\theta}\boldsymbol{e}_\theta$$
$$\ddot{\boldsymbol{r}} = (\ddot{r} - r\dot{\theta}^2)\boldsymbol{e}_r + (2\dot{r}\dot{\theta} + r\ddot{\theta})\boldsymbol{e}_\theta$$

The latter expression can alternatively be written in the form

$$\ddot{\boldsymbol{r}} = (\ddot{r} - r\dot{\theta}^2)\boldsymbol{e}_r + \frac{1}{r}\frac{\mathrm{d}}{\mathrm{d}t}(r^2\dot{\theta})\boldsymbol{e}_\theta$$

Example 2.3: *Intrinsic coordinates*

When considering the motion of a point P in a plane, it is sometimes convenient to use the intrinsic coordinates s and ψ, where s is the distance along the path, and ψ determines the direction of motion by specifying the angle between the tangent vector to the curve and some fixed direction. In such situations it is convenient to use the unit tangent to the curve $\boldsymbol{T}$, and the unit normal $\boldsymbol{N}$ (see figure 2.2).

Since s is the distance along the curve, the speed of the point P is $\dot{s}$. Also, since the point P instantaneously moves in the tangent direction, the velocity of P is

$$\dot{\boldsymbol{r}} = \dot{s}\boldsymbol{T}$$

From this identity it can clearly be seen that the tangent vector can be expressed as

$$\boldsymbol{T} = \frac{\mathrm{d}\boldsymbol{r}}{\mathrm{d}s}$$

It can also be seen that

$$\dot{\boldsymbol{T}} = |\dot{\psi}|\boldsymbol{N}$$

Fig. 2.2. $\boldsymbol{T}$ and $\boldsymbol{N}$ are unit vectors tangent and normal to the curve at P.

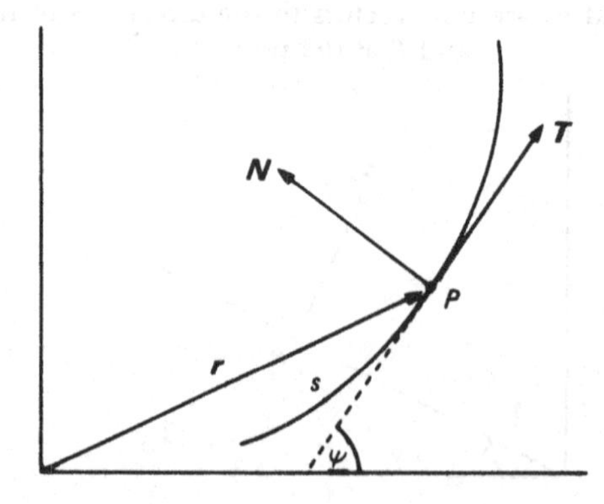

where the modulus sign has been introduced because it is conventional for the unit normal vector $\boldsymbol{N}$ to be directed always towards the concave side of the path. Clearly, this quantity measures the rate at which the velocity of P changes direction. This is related to the curvature of the path which is defined by

$$\kappa = \left|\frac{\mathrm{d}\psi}{\mathrm{d}s}\right|$$

Now, denoting the speed of P by v, and using the result that $\dot{\boldsymbol{T}} = \kappa v \boldsymbol{N}$, the velocity and acceleration of P can be expressed in the form

$$\dot{\boldsymbol{r}} = v\boldsymbol{T}$$
$$\ddot{\boldsymbol{r}} = \dot{v}\boldsymbol{T} + \kappa v^2 \boldsymbol{N}$$

Example 2.4

A point P moves with constant speed V on the equiangular spiral $r = a\,\mathrm{e}^{\theta \cot \alpha}$, where a and α are constants. Find the velocity and acceleration of P using plane polar coordinates.

Differentiating the equation of the path gives

$$\dot{r} = r\dot{\theta} \cot \alpha$$

and, substituting this into the standard expression for velocity gives

$$\dot{\boldsymbol{r}} = r\dot{\theta}(\cot \alpha \boldsymbol{e}_r + \boldsymbol{e}_\theta)$$

Now, it is the magnitude of this that is the constant V, and therefore

$$V = r\dot{\theta} \operatorname{cosec} \alpha$$

Thus

$$\dot{\theta} = \frac{V}{r} \sin \alpha$$

$$\dot{r} = V \cos \alpha$$

$$\ddot{\theta} = -\frac{V^2}{r^2} \cos \alpha \sin \alpha$$

$$\ddot{r} = 0$$

Substituting these into the standard expressions for velocity and acceleration (see example 2.2) gives

$$\dot{\boldsymbol{r}} = V(\cos \alpha \boldsymbol{e}_r + \sin \alpha \boldsymbol{e}_\theta)$$

$$\ddot{\boldsymbol{r}} = \frac{V^2}{r} \sin \alpha(-\sin \alpha \boldsymbol{e}_r + \cos \alpha \boldsymbol{e}_\theta)$$

Clearly, the velocity of P is always inclined at a constant angle α to its position vector, thus demonstrating the equiangular character of this spiral. It can also be seen that the acceleration of P is orthogonal to its velocity, a condition that is required in view of its constant speed.

Exercises

2.1 A wheel of radius a rolls with constant speed V along a straight track. Show that the speed of a point on its rim is given by $V\{2(1-\cos(\theta+\alpha))\}^{1/2}$, where θ is the angle turned through by the wheel and α is a constant. Show also that the acceleration of the point has magnitude V^2/a and is always directed towards the centre of the wheel.

2.2 A point P is constrained to move along the equiangular spiral $r = a\,\mathrm{e}^{\theta\cot\alpha}$, where a and α are constants, in such a way that its radius vector rotates with constant angular velocity. Denoting the speed of P by v, show that its acceleration has magnitude v^2/r and is directed at an angle 2α to the position vector.

2.3 If a two-dimensional curve is expressed in cartesian coordinates $y = y(x)$, use the identities $\tan\psi = \mathrm{d}y/\mathrm{d}x$, $\cos\psi = \mathrm{d}x/\mathrm{d}s$, etc. to show that the curvature at any point is given by

$$\kappa = \frac{\mathrm{d}^2y}{\mathrm{d}x^2}\left\{1+\left(\frac{\mathrm{d}y}{\mathrm{d}x}\right)^2\right\}^{-3/2}$$

2.4 If the position vector of a moving point is expressed either as a function of the distance it has moved s, or as a function of time t, show that the curvature of its path is either

$$\kappa = \left|\frac{\mathrm{d}^2\boldsymbol{r}}{\mathrm{d}s^2}\right| \quad \text{or} \quad \kappa = |\dot{\boldsymbol{r}}\times\ddot{\boldsymbol{r}}|\,|\dot{\boldsymbol{r}}|^{-3}$$

2.5 If the position vector of a point at any time is given by

$$\boldsymbol{r} = \tfrac{1}{2}g\sin\alpha(\cos\alpha\cos t^2\boldsymbol{i}+\cos\alpha\sin t^2\boldsymbol{j}-t^2\sin\alpha\boldsymbol{k})$$

where g and α are constants, interpret its motion in terms of the speed, velocity, acceleration and the curvature of the path.

3

Force, mass and the law of motion

We come now to consider the more familiar subject matter of the theory of classical dynamics as we turn to the concepts of force and mass, and the way in which such concepts are used to analyse and predict the motion of a physical body. However, it is necessary first to introduce the concept of a particle.

3.1. The concept of a particle

In classical dynamics we consider the motion of real physical objects. These are sometimes referred to as bodies. The aim is to develop a method whereby the motion of such bodies can not only be analysed, but also predicted. Unfortunately, the physical bodies whose motions are to be considered are usually fairly complex objects. It is therefore necessary to start by making some simplifying assumptions about such bodies in order that they may easily be represented in a theoretical model.

In some circumstances it turns out to be comparatively simple to build a theoretical representation of physical bodies. For example, when a stone is thrown into the air, or when considering the motion of a planet in the solar system, the main interest is usually in the general position and velocity of the object, or its general linear motion, rather than its orientation and angular motion. In such situations the information given or required about its motion can usually be stated in terms of the position at different times of some representative point of the body, such as some kind of centre. In practice this turns out to be surprisingly easy. In fact the detailed size and shape of the object is often irrelevant to a general consideration of its linear motion, and so the physical body can be replaced in a theoretical model by some idealised representation that is much easier to deal with mathematically.

This convenience of being able to consider an idealised type of body whose angular motion can be ignored, leads to the concept of a particle and to the development of the subject of particle dynamics. This is the subject of this and the next few chapters. After that, a more realistic representation of a body can be developed in terms of the concepts introduced in particle dynamics. It is then possible to develop a theory for the motion of finite bodies whose orientation and angular motion is also considered to be important, or even of prime importance.

Consider first a body whose angular motion can be ignored in a certain situation. Its linear motion or velocity as a whole is its only important dynamical property. Its size and shape are largely irrelevant to its motion. In a mathematical model, therefore, the size of the body can be artificially reduced, leaving its dynamical properties unaltered. This leads to the following definition of a particle, which is now given, together with its related assumption.

***Definition* 3.1.** *A particle is a mathematical idealisation of a small body. It is considered to have the same dynamical properties as the body but occupies zero volume so that it exists only at a point in space.*

***Assumption* 3.1.** *A body, or part of a body, whose dimensions are small enough to be neglected in any situation can be represented by a particle.*

These statements require a certain amount of clarification. A particle is usually defined as a 'point mass', or an object with mass but zero volume. Such definitions are not wrong, but the concept of mass has not yet been defined here. The term 'dynamical properties' has therefore been used in the definition instead. This term has been preferred to the ambiguous 'quantity of matter' since it also emphasises that the particle has all the properties of the small body it is designed to represent except rotation and volume.

The importance of the concept of a particle derives from the fact that it is regarded as existing only at a single point in space at any time. Its position can therefore be represented by a unique vector and its velocity and acceleration can consequently be unambiguously defined.

The theory developed in this book is entirely classical. The theories of relativity or quantum mechanics are not introduced.

However, a knowledge of some quantum theory is relevant at this point since it has a bearing on the applicability of the concept of a classical particle. It is now known that matter is not continuous, but is composed of atoms and molecules of specific construction. These atoms and molecules are themselves composed of elements known as protons, neutrons and electrons, which are referred to as 'elementary particles'. This confusion of notation is unfortunate. In some situations atoms and elementary particles can be represented by particles in a classical theory, but this is not always possible. It is most important to emphasise that the concept of a particle defined above is purely an idealisation in a mathematical model, and is not in any way a real physical object.

According to assumption 3.1 any body can be represented by a particle if its dimensions are small compared to the scale of the system being considered. Thus a stone thrown into the air, or the earth orbiting about the sun can be represented by particles. Even a body the size of the sun can be represented by a particle when we are considering its motion in the galaxy. However, there is a lower bound to the size of object that can be represented by a particle. Since we are only considering classical bodies we cannot go down to dimensions of the order of the atomic. At such dimensions assumption 3.1 is inappropriate.

3.2. The law of inertia

We now come on to develop the laws of classical particle dynamics. These are traditionally given as the three laws stated by Newton in his *Principia.* However the logic of these statements may easily be criticised. Ernst Mach[†], for example, has given a thorough criticism of the Newtonian formulation. He also suggested an alternative logical formulation of the laws of motion, but unfortunately this too may be criticised. The formulation that is given here in the following assumptions and definitions closely follows that of Newton, but it is hoped that they have been given in a form that satisfies most of Mach's criticisms. The main aim at this point, however, is to give a clear exposition of the foundations of the subject, rather than a formal axiomatic formulation. The various axiomatic formulations of the subject are discussed at the end of the chapter.

[†] E. Mach, *The Science of Mechanics* (English translation, Open Court 1902).

It is appropriate to start with Newton's first law of motion, which can be stated in the following way.

***Assumption* 3.2.** *There exists a frame of reference relative to which the natural motion of a particle is one of constant or zero velocity.*

This statement requires some explanation. It clearly includes an assumption about the existence of a particular reference frame. Historically, such a frame has been associated with the concept of an absolute space. However, in the development of the theory of dynamics it is found that such a frame is not singled out uniquely. Rather, there exists a whole class of such frames which move relative to each other with constant velocity. These will be referred to as *inertial* frames of reference. Thus the concept of a unique absolute space does not need to be introduced. However, it is convenient initially to assume the existence of one such frame. The whole class of equivalent inertial frames will be clarified later.

The remaining part of assumption 3.2 means that, if a particle were left alone in space without being influenced by any other object, then, relative to the assumed reference frame, it would continue in a state of rest or of uniform motion in a straight line. Putting it another way, considered relative to an inertial frame of reference, a particle does not accelerate without an external cause.

The importance of this part of the assumption is simply that it is not motion that requires a cause but change of motion. Historically, the change of outlook resulting from the introduction of this assumption was a critical turning point in the development of the theory of classical dynamics. It is found that the natural state of a body is not one of rest, as had previously been supposed, but one of constant velocity. Thus it is not velocity as such that needs to be explained but acceleration.

Although this assumption is found to be consistent with observations in the real world through the development of the theory of classical dynamics, it is by no means obvious. Since the motion of most terrestrial objects usually ends in a state of rest relative to the earth, it is understandable that past generations of scientists regarded that state as the natural one. It could alternatively be argued that the natural motion of terrestrial objects is one of constant acceleration in a vertically downwards direction. However, the concept that it is the rate of change of acceleration that requires a cause does not yield a consistent theory of dynamics. In fact, it is the assumption that it is acceleration that needs to be explained that yields a consistent theory. However, to reconcile this with the observed motion of terrestrial objects, it has to be used in conjunction with a theory of gravitation.

Assumption 3.2, which is known as the law of inertia, is one of the foundational statements upon which the whole subject of classical dynamics is based. it states that, relative to a certain privileged class of reference frames, all particles would naturally move with constant velocity unless they are compelled to change their motion by some external influence. This is the basic idea from which both the concept of force and the law of motion are developed. However, once the law of motion is obtained, it is found to include assumption 3.2 as a special case. This assumption is not therefore one of the independent axioms of particle dynamics. Rather, it is an initial or preliminrary assumption which is further clarified as the subject is developed, and is then used only to define the concept of an inertial frame of reference.

3.3. The concept of force

The assumption has already been introduced that, relative to a certain privileged class of reference frames, a particle does not accelerate without an external cause. When a particle does accelerate, it is appropriate to follow Newton and to state that this is due to the action of a force. Relative to an inertial frame, every body continues in a state of rest or uniform motion in a straight line unless acted on by a force. Clearly, therefore, forces are considered to be those things that cause particles to accelerate.

At this point, no method has been suggested for identifying an inertial frame of reference. Thus it must initially be assumed that any reference frame that may have been chosen for convenience is not necessarily an inertial frame. Now, if a particle were observed to be at rest relative to an inertial frame, it would be observed to be accelerating relative to a noninertial one. It is therefore necessary, at least initially, to extend the concept of a force to include all causes of observed accelerations including those that appear to arise from a particular choice of reference frame. At a later stage it is found to be possible to identify the particular components of acceleration, and hence the extra forces, that arise in this way. This will be discussed again in more detail later in this chapter and also in chapter 4. It is thus necessary to define the general concept of force in the following way.

***Definition* 3.2.** *Relative to an arbitrary frame of reference, any cause of an acceleration of a particle is called a force.*

It should be emphasised that this statement only defines force in the terms of the mathematical models that are constructed using the theory of classical dynamics. The concept of force remains a primitive concept in this theory. It is not possible to exhaustively define the real forces that are considered to act in the physical world. Mathematical models are therefore constructed or interpreted in terms of a concept of force that is based both upon intuition and experience. However, since the theory has been extensively used with great success over three centuries, most of the forces that are found to occur in the real world are now very well understood.

It is now necessary to consider in greater detail what is meant by a force. It should be understood that it is something that causes an acceleration, but that thing is also regarded as having a physical meaning of its own. Here it is both convenient and legitimate to be guided by intuition and to say that a force is something like a push or a pull. When an unresisted object is pushed it is induced to accelerate. Thus when objects push against each other it is understood that there are forces acting between them. In this way the concepts of pressure and tension may be developed. However, an intuitive concept of force only includes such forces as are induced or transmitted by the contact of one body with another. Such forces are called *impressed forces*.

Clearly, if an unconstrained object is acted on by an impressed force, then it must accelerate. However, it has been assumed that a particle does not accelerate unless it is acted on by a force. Forces are therefore regarded as the causes of all accelerations. This leaves open the possibility that forces may occur which are not impressed forces. A body may sometimes be observed to be accelerating while nothing is pushing or pulling it. In the theory of classical mechanics such forces have to be admitted, but, since they are not related to pushes or pulls, it is convenient to distinguish them from impressed forces. All such forces which cannot be classified as impressed forces are referred to as *fictitious forces*. They are the hypothetical forces that are required to explain accelerations for which no direct physical cause can be observed. Such forces, however, also require an explanation. This will be given later for the most common types of fictitious force that are found to occur in practice.

The distinction between these two types of force is fairly clear. A force is described as impressed if it is related to an immediate physical action arising from the contact of one body with another. Otherwise it is described as fictitious. Impressed forces are usually related to pressures and tensions and can be understood fairly intuitively. They are involved in the collision of two particles, for example. Friction also is an impressed force since it involves the contact of two bodies.

The concept of a fictitious force, on the other hand, is a difficult one to grasp, but may be clarified by considering a number of examples. For instance, a child on a roundabout feels the effect of a force tending to throw him off. When the roundabout is rotating faster he has to hold on more tightly. He does not see anyone or anything trying to push him off, yet he has to exert a real impressed force in order to prevent being thrown off radially. The unseen force trying to push him off is a fictitious force known sometimes as the centrifugal force. There is no physical or material cause associated with it, it is merely the effect of a rotating frame of reference.

Similarly, an astronaut in a space vehicle being accelerated by a large rocket feels the effect of a fictitious force. What he actually feels is the real impressed force that his seat exerts on his body. But his seat does not push him across the space ship. Rather, he appears to be pushed into his seat by an unseen hand and the impressed force from his seat merely balances this unseen one. The unseen force is a fictitious force which is the effect of his relative acceleration.

Gravitational forces are identical in nature. When a stone is thrown into the air there is no string pulling it back to the earth. Similarly, there is no string holding the moon in its orbit. We say that the acceleration of these bodies is caused by a fictitious force which is called gravitation. Sometimes people argue that gravitation must be a real impressed force because it is something that we feel. But this is not so. All we ever feel are the impressed forces, such as that exerted by the floor on our feet, which prevent us being accelerated towards the centre of the earth. There is no way of distinguishing locally between gravitation and a relative acceleration. This is one of the basic postulates of the general theory of relativity, but it is also assumed in classical dynamics. Gravitation is regarded as a hypothetical or fictitious force.

The distinction between impressed and fictitious forces is quite clear in the above examples. Impressed forces are those which arise through contact between bodies. Fictitious forces include those due to gravitation or the adoption of a noninertial frame of reference. However, in practice this distinction is not always so clear, and indeed is not particularly important.

In the practical application of the theory of classical dynamics a large number of different types of force need to be considered. Initially it is convenient to develop the concept of a force from the more easily understood types of impressed force. In this way a force can be considered initially to be something like a push. Then, as the theory is developed, more and more particular kinds of force may be included. These may have been poorly understood at first, but further investigation has generally revealed either their source or the mechanism through which they act. For example, the development of the theories of statistical mechanics and quantum mechanics has led to a much deeper understanding of the way such forces as air resistance, pressures, tensions, friction and radiation pressure act.

The need to introduce the concept of a fictitious force arises because forces are frequently observed through their effects in situations in which no acting mechanism can be observed. Such forces still need to be explained. Those that arise simply because of the relative motion of the frame of reference can be easily understood, or explained away. Others, such as gravitation, are explained in terms of their sources. In this latter case, as also in the case of electromagnetic forces which are not considered in this book, it is convenient to introduce the concept of a hypothetical force field. This will be described in chapters 5 and 6. Such a concept can be used to describe the fictitious forces that are observed in the real world, even though the way in which they act remains a mystery. In this way it can be seen to be possible to start with a simple intuitive concept of a force and to develop it to include the more advanced concept of a force field. In the deeper theories of quantum mechanics it is the field concept of force that is regarded as the more fundamental.

Force, however, remains a primitive concept. In a mathematical model it is simply that which causes particles to accelerate. However, in order to relate the model to situations in the real world, in either stages of construction or interpretation, certain assumptions need to be introduced about the particular forces that are considered to occur. Thus it is not possible to state a general formal assumption about the nature of force at this stage, but it should be understood that in every application of the theory, an assumption about the forces that are acting is always made, at least implicitly. Such an assumption is a basic axiom in any application and must be treated logically as such.

One further point needs to be clarified here. Consider, for example, a particle moving with constant velocity in outer space, and a space ship accelerating past it. To an observer in the space ship the particle would appear to be accelerating. The cause of the observed acceleration is the fictitious force which is the effect of the relative acceleration of the observer. In a sense, therefore, the force is caused here by an acceleration rather than the other way round. However, it is the relative acceleration of the observer which gives rise to the fictitious force, and the fictitious force can still be regarded as causing the observed acceleration of the particle, at least formally. The difficult question of relative motion is considered later, but it is essential at this point to regard the effects of relative acceleration in terms of fictitious forces in order that inertial frames of reference can be determined at a later stage.

3.4. Mass and the magnitude of a force

We now move on to consider the magnitude of a force and the concept of mass. These are defined as follows.

***Definition* 3.3.** *When a force acts on a particle the force is regarded as acting in the direction of the acceleration induced in the particle and its magnitude is taken to be proportional to that acceleration.*

***Definition* 3.4.** *The constant of proportionality implicit in definition* 3.3 *is called the mass of the particle.*

Definition 3.3 is purely a convention. It associates a unique direction to the concept of force, and also associates a magnitude. Force is therefore regarded as a vector quantity. Mass is introduced as a scalar coefficient. However, the magnitude of the force is still ambiguous. The idea behind the definition is that if the force acting on the particle was suddenly doubled, then the acceleration would be suddenly doubled. The force acting on a given particle is measured by the acceleration it induces.

The constant of proportionality, at this stage merely labelled 'mass', is not yet given an objective meaning. Neither has a method for determining the mass of a given particle yet been described. If a certain mass were to be ascribed to a particular particle, it would be possible to determine the magnitude of the force acting on it. It would simply be its observed acceleration multiplied by its mass value.

At this stage it is still not clear whether or not the introduction of the mass parameter will be a useful one in practice. For example, for the sake of simplicity it is appropriate to consider always putting the mass value equal to unity, and so to omit the concept of mass altogether and to define a force as being always equal to the acceleration it induces. However, although such an approach may seem sensible to describe fictitious forces, it is found not to yield a fully consistent theory. Neither is it possible *a priori* to assert that a particular body has a unique mass value. This is clearly the second possibility to consider, and it is in fact this approach that enables a consistent theory to be developed. In order to test this approach it is necessary to consider different forces acting at different times on the same body, but this of course is of no help unless the magnitudes of those forces are known beforehand.

In order to avoid a circular argument at this stage, it is necessary to consider a number of particles and the forces that act on them. In some circumstances it is possible to follow intuition and to attempt to act on each of the particles with the same force. If this could be achieved, the measured accelerations of each particle could be used to determine their comparative masses. Then after subjecting all the particles to a number of different forces it would be possible to test whether or not their measured comparative masses remained constant.

Intuitively, this is the best approach to use, but in practice it requires the assumption that certain forces which act on different particles are equal in magnitude. Now that the theory is fully developed and well understood, it is found that the following statement provides the best formal way of introducing this assumption about the magnitudes of forces acting on different bodies, and it hence provides a way of comparing the masses of distinct bodies.

***Assumption* 3.3.** *When two particles interact, the mutual forces acting upon each other are always equal in magnitude and act in opposite directions.*

It can be seen that this statement is identical to Newton's third law. Action and reaction are equal and opposite. This is one of the basic assumptions of classical mechanics.

Now consider a situation in which two particles interact with each other. Denote the forces acting on each particle by $\boldsymbol{F}_1$ and $\boldsymbol{F}_2$, their induced accelerations by $\boldsymbol{a}_1$ and $\boldsymbol{a}_2$ and their mass coefficients by m_1 and m_2. Assume that an inertial frame of reference is being used and that no other forces are acting on the particles. Then from assumption 3.3 we have

$$\boldsymbol{F}_1 = -\boldsymbol{F}_2$$

and each force is defined in terms of the acceleration it induces according to definitions 3.3 and 3.4. Therefore

$$m_1\boldsymbol{a}_1 = -m_2\boldsymbol{a}_2$$

or

$$\frac{m_1}{m_2} = -\frac{a_2}{a_1}$$

Thus the mass ratio of the two particles is determined by their mutually induced accelerations. Because of its importance, let us state this formally.

***Proposition* 3.1.** *The mass ratio of any two particles is the negative inverse ratio of their mutually induced accelerations in situations where they interact.*

As has been shown, this follows directly from assumption 3.3 and the definition of force. This now provides a method, at least in principle, of determining the mass ratio of any two particles. It need not worry us that it is impossible in practice to observe the free interaction of two particles and so measure their mass ratio. It is not necessary to isolate two given particles, place them in outer space and then measure their mass ratio on the assumption that an inertial frame of reference is being used. The whole theory of classical dynamics is based on a number of assumptions such as assumption 3.3. From these we build up a detailed mathematical model. It is only by comparing the complete model with the real world that the set of assumptions can be corroborated. Each assumption does not have to be justified separately.

It has been well emphasised by Mach that it is only the concept of mass ratio that has an objective meaning. Mass on its own has no objective meaning except as a comparison with other masses. This need cause no trouble. We are familiar with such concepts as the 'cash value' of an object. This would be meaningless if only one object in the world has a cash value. The term only has a useful meaning because we compare one object with another and so give many objects a cash value. It is exactly the same with mass. We can choose one particle as having a standard mass, and then by comparing it with other particles the mass value of all particles can be determined.

Although it is only mass ratio that has an objective meaning, an intuitive idea of the concept of mass can be developed. If two particles of different mass are acted on by forces of equal magnitude, their mass ratio is the inverse ratio of their accelerations. If one particle accelerates faster than the other its mass value is proportionately less. This leads to the idea that the mass value of a particle is a measure of that particle's resistance to being accelerated. It appears to be a fundamental property of material objects that they do not accelerate unless they are forced to. Some objects, however, are more difficult to accelerate with impressed forces than others. What appears to be the inherent power of a body to resist being accelerated by impressed forces is the property we call mass.

It is the lack of a proper definition of mass which is the main weakness of Newton's axiomatic formulation of dynamics, and although the above argument was given by Mach there is still considerable confusion over the meaning of the term. Some people still try to define mass as a 'quantity of matter' or the 'number of elementary particles'. They say that mass is a fundamental property of elementary particles and so it is only necessary to count these component particles to obtain the mass of larger bodies. This may be true, but the question as to what is the mass of an elementary particle still remains. The answer to this question is that given above; namely that the mass, even of an elementary particle, is a measure of its resistance to being accelerated, and that this can be evaluated objectively only as a mass ratio.

The uniqueness of the mass value of any particle can not be deduced from the theory so far, but has to be obtained from an additional assumption.

***Assumption* 3.4.** *The mass ratios of particles are unique constants, independent of the way they may be obtained, and can be calculated directly or indirectly.*

This remarkable statement appears to be in complete agreement with observations to the greatest degree of accuracy that can presently be achieved, although the concept of mass is modified slightly in the theories of relativity.

There are many different ways of determining the mass of a body, and each method is found to be completely consistent with all of the others to within the usual experimental errors. It is also found to be unnecessary to compare all masses with a single standard. Secondary standards can be used equally well. For example, it is possible to evaluate the mass of household objects either by comparison with secondary standards on a balance, or by observing the extension of a calibrated spring from which it is suspended. Each method gives consistent results, provided the measurements are made under the correct conditions. The mass value of any body appears to be an important and unique characteristic of it.

The magnitude of a force acting on a particle is now uniquely defined as the product of the mass value of the particle and its induced acceleration. In SI units forces are measured in 'newtons' and the agreed standard mass is called a 'kilogram'. A unit of force is thus defined as follows. *A force of one newton is that which would accelerate a mass of one kilogram at one metre per second per second.* Alternative standards of mass are frequently used. For example, the mass of the sun or the mass of an electron are commonly taken as standards when they are appropriate.

Forces have been regarded as vector quantities, but it is still necessary to make an assumption concerning the addition of forces.

***Assumption* 3.5.** *When a particle interacts with a number of particles simultaneously, the mutual forces induced are independent of each other.*

This assumption states that forces satisfy the vector law of addition. This is the well-known parallelogram law for the addition of forces. When two forces act on the same particle no interactive force needs to be considered. It follows that any force can be resolved into appropriate components.

3.5. The law of motion and its application

We are now in a position to state the basic law of classical dynamics. This is given in the following form

***Proposition* 3.2.** *The motion of any particle satisfies the equation*

$$m\ddot{\mathbf{r}} = \mathbf{F}$$

where $\mathbf{F}$ *is the sum of all forces acting on it, both impressed and fictitious,* m *is the mass of the particle and* $\ddot{\mathbf{r}}$ *its acceleration relative to some frame of reference.*

This proposition is referred to as Newton's law of motion. It follows immediately from the above assumptions and definitions.

Consider any particle whose mass value is known. Forces acting on it induce accelerations. Both forces and accelerations obey the same vector law of addition, and summing over all forces gives the above law.

It should be emphasised that this law is considerably more than a definition of a force in spite of its similarity to definitions 3.2 and 3.3. It is a law that is designed to be used in practice. It is the basic law of classical dynamics, and its importance can hardly be over emphasised in view of the success with which it has been used over the past three centuries.

It is necessary, however, to consider the way in which it is used in some detail. Clearly, it can be used to determine the net force acting on a particle. If its mass is known and the acceleration can be measured at any instant, then the net force acting at that instant can be determined from the law of motion. However, one main purpose of the law is to predict the future motion of bodies. To achieve this it is not the force which is to be determined but the acceleration. Then, from a knowledge of the future acceleration, the future positions and velocities can be calculated. The way in which we proceed therefore is to guess what the forces acting in the future are likely to be. These forces are then substituted into the law of motion which then becomes an ordinary differential equation whose solution, subject to the given initial conditions, becomes our prediction of what the motion is going to be. When used in this way the law of motion is often referred to as the 'equation of motion'.

It is important to emphasise that the law of motion is usually applied in conjunction with an additional assumption concerning the nature of the forces acting. The predicted motion is therefore a consequence of this additional assumption, as well as the assumptions of the classical theory. If the predicted motion is not followed in practice, then it usually turns out to be the assumption about the forces which was in error. However, in the majority of the applications of classical dynamics it turns out to be easy to guess the relevant forces. We can look at the past motion and use the law of motion to determine the forces that have been acting. We can then assume that the same forces will continue to act and use the law of motion as an equation to predict the future motion. With experience we become familiar with the forces to be considered, such as gravitation and friction. We thus learn to use the law of motion with confidence. It must be emphasised here though, that with every application of the law of motion to make predictions, an assumption is always made concerning the forces that are acting.

Consider now some particular physical situation involving the motion of a body that can be represented as a particle. We may have some intuitive knowledge of the situation, so that we can make a reasonable set of assumptions about the forces that are acting. The law of motion can now be used to make a prediction that can be compared with physical observations. If our initial assumptions were satisfactory, then the predictions and observations should agree to within the accuracy of our calculations and observations. If, however, the predictions and observations disagree, even after allowing for the computational and observational inaccuracies, we do not immediately regard the whole theory as wrong. Rather, we go through the following routine to try to obtain agreement.

Rule **(1).** *Look for any additional force that might have been neglected.*

Rule **(2).** *Accelerate or rotate the frame of reference to achieve agreement.*

Rule **(3).** *Invent a new fictitious force to satisfy the equation.*

Now consider how these rules work in practice.

Consider a stone thrown into the air. Its acceleration at any instant can be measured by plotting its path. This is found to be approximately a parabola and the law of motion is satisfied for a constant vertical gravitational force as expected. Now repeat the experiment but replace the stone by a ping-pong ball, giving it the same initial velocity. This time the path will deviate from the parabola. The theory is not wrong. Rather, we use rule (1) and observe that air resistance has been neglected. With this additional force we find that the law of motion is satisfied to within experimental accuracy achieved.

A spectacular application of rule (1) was made by Adams and Leverrier in their investigation of the unexplained irregularities in the motion of the planet Uranus. They suggested that there might be another planet beyond Uranus which was causing the observed irregularities and were even able to predict the position of the unknown planet. Such a planet, now called Neptune, was subsequently discovered in 1846 almost exactly in the position predicted. However, we must also report the unsuccessful application of rule

(1). No extra planet orbiting close to the sun has yet been observed which could explain the anomalies in the orbit of the planet Mercury.

Consider now Foucault's pendulum experiment. A large pendulum is set swinging in a plane and it is observed that the plane of the pendulum slowly rotates contradicting the prediction of the law of motion for a simple pendulum. If, however, the motion of the pendulum bob is measured relative to a frame of reference based on the sun rather than on the earth, then the equation of motion is satisfied. Rule (2) has been used to allow for the rotation of the earth.

Consider again the stone thrown into the air. It is observed to move in a parabolic path, yet there are no impressed forces acting on it. Rule (3) has to be used to invent a constant fictitious force acting vertically downwards in order to satisfy the equation of motion. This force of gravitation is now so familiar that we would include it anyway, but at some stage it had to be invented purely to satisfy the equation. Similarly, a fictitious force has to be invented to account for the elliptic path of the planets. It will be shown in chapter 5 that this must be an inverse square law force directed towards the sun.

These three rules are essential to the successful application of classical dynamics, yet as they stand they ensure its uselessness. On the face of things they all appear to be attempts to look for additional forces that may be used to explain an observed discrepancy. Usually this will work. However, it is essential to leave open the possibility that the whole theory may be wrong, and in this case the law of motion would not be satisfied. This possibility is effectively ruled out by an unrestrained application of the above rules. With them the theory of classical dynamics can be used to explain every theoretical motion, not only those that are physically possible. For any motion, realistic or unrealistic, the equation of motion could always be satisfied. In this case it would not be possible to think of any experimental result that would falsify or disprove the theory. It would always be possible either to artificially accelerate the frame of reference or to invent a new fictitious force so that Newton's equation of motion could be satisfied. In this way classical dynamics would cease to be a scientific theory. The value of a scientific theory is measured by the possibilities that it excludes, and with the unrestrained use of these rules classical dynamics excludes nothing.

Ideally, a theory of dynamics should permit motions that are physically possible and exclude motions that physically cannot occur. Of course this cannot be achieved completely for the theory of classical dynamics since the whole theory has ultimately been falsified. However, it is still necessary to formulate the theory in as falsifiable a way as possible, and to achieve this the above rules must be restricted.

In practice, the approach used is well understood. The method closely follows intuition, and since it concerns the creative stage of theory building, it is difficult to state in a formal logical way. The point is that in the Newtonian approach to classical dynamics, motion of a body is considered to be caused by a force. The next step is then to look for the cause of that force. It is considered that all forces should be explained in one way or another. In other words there must always be some physical cause for any accelerated motion.

It is therefore to be understood that rules (1) and (3) above are not to be used to introduce arbitrary unexplained forces. When either impressed or fictitious forces are introduced to explain observed accelerations they must be presented in the form of a proper scientific theory. For example, it may be stated that when a certain type of particle is in a certain type of situation it will experience a force given by a particular expression. The mechanism by which such a force acts may not be understood, but the statement should be testable by further observations and experiments. It is not acceptable to introduce a force which is considered to act only on one particular particle on a single occasion.

Consider again how rule (3) is used in practice. When a stone is thrown into the air, its motion satisfies Newton's law only if rule (3) is used to include a fictitious vertical force. Now, it is neither necessary nor acceptable to invent a new fictitious force for every individual stone. Rather, it may be noticed that all stones are subject to a similar vertical force. A mathematical model and theoretical structure can in fact be developed which predicts that all particles situated near the earth's surface are acted on by such a force whose magnitude can be explicitly stated. Such a theory of gravitation can be further generalised to include planetary motion, and can then be developed to include the motion of any particle anywhere in the universe. This will be explained in chapter 5. However, at this point it is only necessary to observe that a general theory can be developed to apply to all particles. In all subsequent applications it is this theory that is used and not rule (3).

It is tempting to think that the nature of all the forces that actually occur in the real world are now understood, at least at the classical macroscopic level. However, such a judgement may still be a little premature. Rule (3) should therefore be retained, but in practice it is only used to introduce and explain forces such as gravitation which are nowadays well understood.

Rule (2) is used in a similar way. If all the forces acting on a particle were known, then, provided the theory were correct, its motion relative to an inertial frame of reference would satisfy Newton's equation. However, relative to a noninertial frame there would be a discrepancy in the equation of motion which could be resolved in one of two ways. The first possibility simply involves inserting the fictitious forces that are associated with motion in that reference frame. However, to do this effectively requires a knowledge of the acceleration and angular velocity of the frame of reference relative to an inertial frame. The alternative method involves a transformation to a new frame of reference which may be accelerating or rotating relative to the first in such a way that the motion in this frame satisfies Newton's equation. It should be noticed in this approach that the new frame is of course an inertial frame. The method is to transform from the noninertial frame to an inertial one. Clearly, if the acceleration and rotation of the noninertial frame relative to an inertial frame are known, then either method may be used. If this is not known, then again either approach may be attempted using the technique of trial and error, as suggested by rule (2). If this is successful, then what has been achieved is not only that the equation of motion is satisfied but also that an inertial frame of reference has been found.

Rule (2) is thus used in practice to determine initially which frame is to be regarded as inertial. For this purpose it is essential. The methods used will be discussed in more detail in the next chapter. However, once an inertial frame of reference has been established rule (2) should not be used again.

The above arguments may be summarised by saying that rules (2) and (3) will be used to introduce and explain certain well-known aspects of the theory but that, in practice, rule (2) cannot be used to resolve an anomaly. If a situation arose in which a mathematical model for the motion of a particle based on classical dynamics consistently disagreed with observations, then all the parameters of the model would have to be checked and rule (1) would also be used to see whether any possibly occurring forces had been neglected. However, in practice only well-known and well-understood forces would be considered. In the past this approach has almost always been found to work satisfactorily.

The remaining possibility, however, should still be considered. If the discrepancy persisted, even after considerable effort to resolve it using known forces, then an uncomfortable choice would remain. Either the theory should be regarded as having been faslified, or rule (3) should again be used to introduce a new type of force that would be capable of explaining the observed motion. The option that would be chosen in practice would depend largely on the situation. If, for example, a complete alternative theory were available which satisfactorily explained the observed motion, then it would be appropriate to regard the motion as falsifying the classical theory. The new theory would then be adopted. In the absence of such a new theory, however, a new type of force introduced using rule (3) would be seriously considered. However, it would still be regarded with suspicion until it could be expressed in a satisfactory theoretical framework that was corroborated by a number of independent observations.

3.6. Rectilinear motion and projectiles: examples

Before considering a number of examples of particle dynamics, it is first necessary to consider some of the particular mathematical techniques that will be required in obtaining integrals of the equation of motion. Some of these can be demonstrated in situations in which the particle is constrained to move in a straight line.

It is therefore appropriate initially to consider the motion of a particle which is free to move in one dimension. Generally, it may be assumed that the forces acting on it may depend on its position, its velocity, and also time. Its equation of motion can therefore be expressed in the general form

$$m\ddot{x} = F(x, \dot{x}, t)$$

The problem is now to solve, or at least to find a first integral of, this equation. There are, however, no general analytical techniques that can be used at this stage. The methods of approach depend heavily on the character of the force function F, particularly since this is usually nonlinear. In large classes of particular cases, however, standard techniques can be applied. These are as follows.

Case A: when $F = F(\dot{x}, t)$ or $F = F(t)$; i.e. when the force function is independent of position.

In this case we put

$$v = \dot{x}$$

and the equation of motion reduces to the first-order equation

$$m\frac{\mathrm{d}v}{\mathrm{d}t} = F(v, t)$$

which it may be possible to solve analytically, for example by the technique of separation of variables. If an analytic solution of this equation can be obtained, then v is known, at least implicitly, as a function of t. It may then be possible to integrate this expression to obtain x as a function of t. If both of these stages are possible, then a complete integral of the original equation can be obtained.

Case B. When $F = F(x, \dot{x})$ or $F = F(x)$; i.e. when the force function is independent of time.

In this case we also put

$$v = \dot{x}$$

but in this case we write the acceleration as

$$\ddot{x} = v\frac{\mathrm{d}v}{\mathrm{d}x}$$

The equation of motion thus reduces to first order as

$$mv\frac{\mathrm{d}v}{\mathrm{d}x} = F(x, v)$$

which again may be integrable analytically to give v as a function of x. A second integration to give x as a function of t may then be possible at this stage.

Case C. When $F = F(x, t)$; i.e. when the force function is independent of velocity.

In this case no general simplification is possible, as the equation of motion cannot be reduced to a single first-order differential equation.

Case D. When $F = F(\dot{x})$; i.e. when the force function depends only on velocity.

In this case the techniques described in cases A and B can both be attempted. If the integrations can be performed, expressions for v as functions of t or x, respectively, are obtained in each case. It is therefore necessary at an early stage to decide which of these is the most appropriate for the problem under consideration.

Example 3.1

Consider the case of a particle projected vertically upwards. It may be assumed that it is acted on by the approximately constant gravitational force mg, and also a resistive force which is proportional to the square of its speed. Writing the resistance term as $-mkv^2$, where k is a constant, and ignoring any other forces, the equation of motion is

$$m\ddot{x} = -mg - mkv^2$$

where x is the height of the particle above the point of projection.

This equation is included in case D above, and so the methods of A and B may each be used. In method A the equation of motion becomes

$$\frac{dv}{dt} = -(g + kv^2)$$

Separating the variables and integrating gives

$$\int \frac{dv}{\left(\frac{g}{k} + v^2\right)} = -k \int dt$$

The indefinite integral of this is

$$\frac{1}{\sqrt{(kg)}} \tan^{-1}\left(\sqrt{\left(\frac{k}{g}\right)}v\right) = -t + \text{const}$$

where the constant can be evaluated from some boundary condition such as a knowledge of the initial velocity. From this or, alternatively, using the definite integral, it can be seen that the time taken to reach the maximum height is

$$\frac{1}{\sqrt{(kg)}} \tan^{-1}\left(\sqrt{\left(\frac{k}{g}\right)}u\right)$$

where u is the initial speed of projection.

Alternatively, to obtain v as a function of x rather than t, method B may be used. In this case the equation of motion becomes

$$v\frac{dv}{dx} = -(g + kv^2)$$

which can be separated in the form

$$\int \frac{v\,dv}{\left(\frac{g}{k} + v^2\right)} = -k \int dx$$

and hence it follows that

$$\frac{1}{2}\log\left(\frac{g}{k} + v^2\right) = -kx + \text{const}$$

From this it can be seen that the maximum height attained after projection with initial speed u is

$$\frac{1}{2k}\log\left(1+\frac{ku^2}{g}\right)$$

It may be noticed at this point that, using either approach, the indefinite integrals obtained can be further integrated to give the height x of the particle at any time t. The general expression for this, however, is rather lengthy.

Clearly, the general character of the motion is that the speed of the particle decreases continuously until, after reaching the maximum height specified above, it starts to fall. The equations above describe only the upward motion of the particle. In the downward motion, the resistance acts in the upwards direction, and to describe this case the sign of the resistance term in the above equation of motion has to be changed. Alternatively, it is sometimes convenient to introduce a new coordinate which increases in the downwards direction. In this case it is the sign of the gravitational force which changes and the effect is the same.

Example 3.2

Consider the case of a particle projected vertically upwards from the surface of the moon. In this case there is no atmosphere and therefore no resistance to the motion. The dominant force is the gravitational force, and if this is considered to be the only force that acts on the particle it is possible to include the variations of this force with the height above the surface. Now, it is known that the gravitational force on any planet or moon varies inversely as the square of the distance from the centre. The equation of motion of the particle is therefore

$$m\ddot{r}=-kr^{-2}$$

where r is the distance of the particle from the moon's centre and k is a constant. This constant, however, is more conveniently expressed in terms of the acceleration due to gravity g_0 on the surface and the mean radius a of the moon ($mg_0=ka^{-2}$). The equation of motion thus takes the form

$$\ddot{r}=-g_0\frac{a^2}{r^2}$$

which is of the type described in case B above. Writing this as

$$v\frac{\mathrm{d}v}{\mathrm{d}r}=-g_0\frac{a^2}{r^2}$$

a first integral can immediately be obtained in the form

$$v^2=2g_0a\left(\frac{a}{r}+c_1\right)$$

where c_1 is a constant which is given in terms of the initial projection speed

u by

$$c_1 = \frac{u^2}{2g_0 a} - 1$$

A number of properties of this motion can be deduced immediately from this integral, since the left-hand side is a perfect square. For example, if c_1 is positive (i.e. if $u^2 > 2g_0 a$) it is clear that the velocity must remain positive for all values of r. In this case the particle clearly escapes from the moon's gravitational field. On the other hand, if c_1 is negative (i.e. if $u^2 < 2g_0 a$), then the right-hand side is positive only for a certain range of values for r. In this case there exists a certain specified height at which the velocity of the particle is zero. Motion to greater heights is not possible and the particle returns to the moon.

The above integral can now be written as

$$\frac{\mathrm{d}r}{\mathrm{d}t} = \pm\sqrt{\left(2g_0 a\left(\frac{a}{r} + c_1\right)\right)}$$

where the positive sign must be chosen in this case since the velocity is initially positive. The negative sign should be chosen in the return motion to the moon's surface. Separating the variables in this case gives

$$\int \frac{\mathrm{d}r}{\sqrt{\left(\frac{a}{r} + c_1\right)}} = \sqrt{(2g_0 a)} \int \mathrm{d}t$$

which integrates to give the transcendental equation

$$\sqrt{r}\sqrt{\left(1 + \frac{c_1 r}{a}\right)} - \sqrt{\left(\frac{a}{c_1}\right)} \sinh^{-1} \sqrt{\left(\frac{c_1 r}{a}\right)} = \sqrt{(2g_0)} c_1 t + c_2$$

where c_2 is a constant given by

$$c_2 = \sqrt{a}\sqrt{(1 + c_1)} - \sqrt{\left(\frac{a}{c_1}\right)} \sinh^{-1} \sqrt{c_1}$$

This is the complete integral of the equation of motion. The value of r at any given time can be obtained from it using simple numerical techniques.

Example 3.3: *simple projectile*

Consider the motion of a particle in a constant vertical gravitational field. The standard projectile problem is first to assume that the particle is given some initial velocity, and then to make a prediction of the subsequent motion. As an initial approach it is appropriate to neglect resistance and any other forces, and to assume that the only force acting is the constant gravitational force.

It is convenient to choose a frame of reference which is fixed relative to the earth's surface, and in which the z axis is chosen to be vertically upwards. The constant gravitational force acting on the particle which has mass m is then given by $-mg\mathbf{k}$. It is also convenient to choose the x axis of the reference frame to be in the direction of the horizontal component of the initial velocity. Thus, if the particle is projected with speed V at an angle α to the horizontal, the initial velocity is given by $V(\cos\alpha\boldsymbol{i}+\sin\alpha\boldsymbol{k})$ (see figure 3.1). At this point it may be assumed that the frame of reference chosen in this way is at least approximately an inertial frame. (The small perturbations caused by the actual rotation of such a frame are considered later in example 4.1.)

With this notation and set of assumptions the equation of motion is

$$m\ddot{\boldsymbol{r}}=-mg\boldsymbol{k}$$

and this can simply be integrated twice, using the initial conditions to give the solution

$$\boldsymbol{r}=Vt(\cos\alpha\boldsymbol{i}+\sin\alpha\boldsymbol{k})-\tfrac{1}{2}gt^2\boldsymbol{k}$$

The equation of motion and its solution can alternatively be expressed in terms of their cartesian components as follows

$$\ddot{x}=0 \quad\Rightarrow\quad x=Vt\cos\alpha$$
$$\ddot{y}=0 \quad\Rightarrow\quad y=0$$
$$\ddot{z}=-g \quad\Rightarrow\quad z=Vt\sin\alpha-\tfrac{1}{2}gt^2$$

This solution clearly predicts that in the subsequent motion the particle remains in the same vertical plane $y=0$. In addition the x and z coordinates are given explicitly as functions of time. Eliminating t between these expressions gives the equation of the trajectory in the form

$$z=\tan\alpha x-\frac{g}{2V^2}\sec^2\alpha x^2$$

which clearly indicates that the particle is expected to follow a parabolic path.

It can immediately be seen from this solution that, apart from the initial point when $t=0$ and $x=0$, the particle is again at the same horizontal level $z=0$ when $t=2(V/g)\sin\alpha$ and $x=2(V^2/g)\sin\alpha\cos\alpha$. The latter equality clearly expresses the horizontal range of the projectile as

$$R=\frac{2V^2}{g}\sin\alpha\cos\alpha$$

It may be noticed that the same horizontal range is obtained for the same initial speed of projection, if the angle of projection is either α or $\pi/2-\alpha$.

Fig. 3.1. Initial conditions for the simple projectile problem.

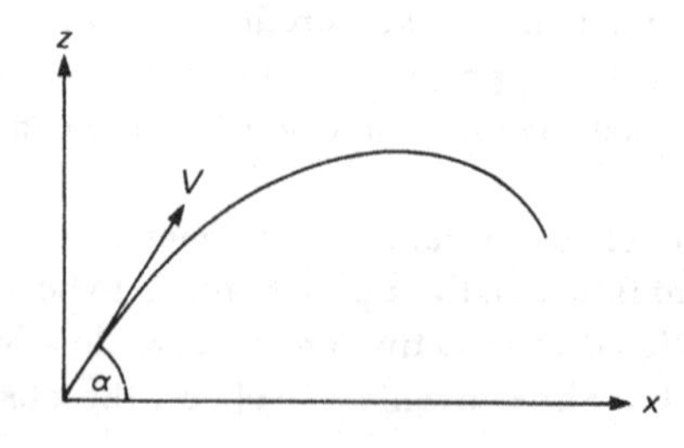

In addition, the maximum possible range is V^2/g, and this is achieved with a projection angle $\alpha = \pi/4$.

Example 3.4: projectile with resistance

It is appropriate now to attempt to improve the model described in the previous example by including a resistance to the motion. The magnitude of the resistance may be assumed to be of the form mkv^2, where k is a constant parameter. Then, denoting the unit tangent to the trajectory at any point by $\boldsymbol{T}$, the equation of motion for the particle can be expressed as

$$m\ddot{\boldsymbol{r}} = -mg\boldsymbol{k} - mkv^2\boldsymbol{T}$$

It can immediately be seen that there is no component of force in a direction perpendicular to the instantaneous plane of motion. It may therefore be concluded that the projectile remains in the same vertical plane $y = 0$, but with the additional term the remaining cartesian components of this equation do not yield expressions that can be integrated. In this case it is convenient to use the intrinsic coordinates s and ψ described in example 2.3 (see also figure 3.2).

The tangential and normal components of the equation of motion on their own, however, are still not integrable, but the horizontal component can now be written as

$$\ddot{x} = -kv^2 \cos \psi$$

This equation can now be simplified by introducing the horizontal component of the velocity. Denoting this by u, we have $\dot{x} = u$ and $v = u \sec \psi$, and the

Fig. 3.2. The path of a projectile with resistance. $\boldsymbol{T}$ and $\boldsymbol{N}$ are unit vectors tangent and normal to the path at any point. s is the distance travelled and ψ is the inclination of the tangent to the horizontal.

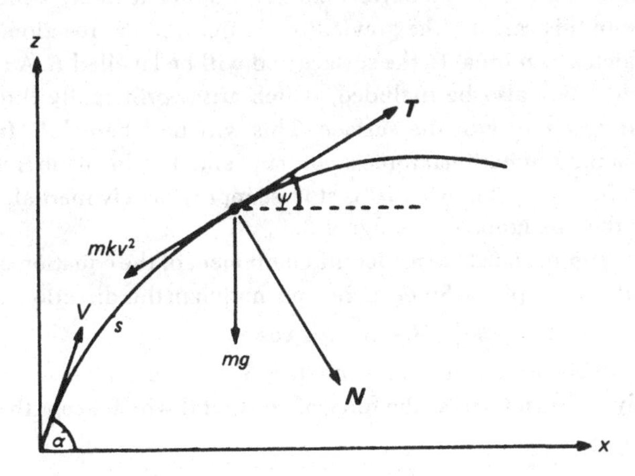

equation takes the form

$$u\frac{\mathrm{d}u}{\mathrm{d}x}=-ku^2\sec\psi$$

Now, since $\sec\psi=\mathrm{d}s/\mathrm{d}x$, this equation can be separated using variables u and s to give

$$\int\frac{\mathrm{d}u}{u}=-k\int\mathrm{d}s$$

Using the boundary conditions, and reintroducing the speed, this integrates to give the horizontal component of the velocity as

$$v\cos\psi=V\cos\alpha\,\mathrm{e}^{-ks}$$

This is clearly a decreasing function as should be expected from the general character of resisted motion.

This first integral enables the speed to be expressed in terms of the intrinsic coordinates s and ψ. Using this result the second integral can now be obtained using the normal component of the equation of motion

$$m\kappa v^2=mg\cos\psi$$

where $\kappa=-\mathrm{d}\psi/\mathrm{d}s$, the negative sign appearing since ψ is clearly a decreasing function and $\boldsymbol{N}$ is directed towards the concave side of the path. After substituting the above result, this equation can now be separated to give

$$-V^2\cos^2\alpha\int\sec^3\psi\,\mathrm{d}\psi=g\int\mathrm{e}^{2ks}\,\mathrm{d}s$$

Hence, using the initial conditions, the intrinsic equation of the trajectory is

$$\frac{g}{kV^2\cos^2\alpha}(\mathrm{e}^{2ks}-1)=\sec\alpha\tan\alpha-\sec\psi\tan\psi-\log\left(\frac{\sec\psi+\tan\psi}{\sec\alpha+\tan\alpha}\right)$$

Example 3.5

Consider the motion of a particle which is sliding on a fixed plane that is inclined at an angle α to the horizontal. The dominant forces which act on the particle in this case are the gravitational force and the reaction from the surface, which acts normal to the surface and will be labelled R. A resistance to the motion may also be included, which arises principally through the frictional interaction with the surface. This will be labelled F. It may be assumed that the gravitational force is constant, and that the plane is stationary relative to a frame of reference that is at least approximately inertial. All other forces may then be ignored (see figure 3.3).

It is first appropriate to consider the component of the equation of motion perpendicular to the plane. Since there is no motion in this direction, this gives

$$0=R-mg\cos\alpha$$

which effectively gives an expression for R. The reactional force R can alternatively be considered as the force of constraint which keeps the particle on the plane.

It is now possible to concentrate on the motion in the plane, and it is again appropriate to use intrinsic coordinates and tangential and normal components (see figure 3.4).

The equation of motion can now be expressed as

$$m\dot{v}\mathbf{T} + m\kappa v^2\mathbf{N} = mg \sin\alpha\,(\sin\psi\mathbf{T} + \cos\psi\mathbf{N}) - F\mathbf{T}$$

The frictional force must now be specified according to a well-established theory. It is found in practice that when a body slides over any surface, the frictional force is approximately proportional to the normal reaction with the surface, and that the constant of proportionality is a particular characteristic of the materials that compose the two surfaces in contact. The constant of proportionality is known as the *coefficient of friction.* Denoting it by μ, it is appropriate to assume that the magnitude of the frictional force is given by

$$F = \mu R$$

Fig. 3.3. The forces acting on a particle which is sliding on an inclined plane.

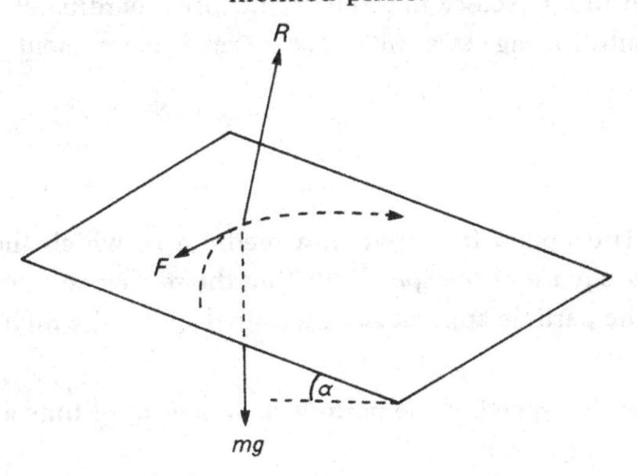

Fig. 3.4. The path of a particle which is sliding on an inclined plane. Only the components of force in the plane are shown.

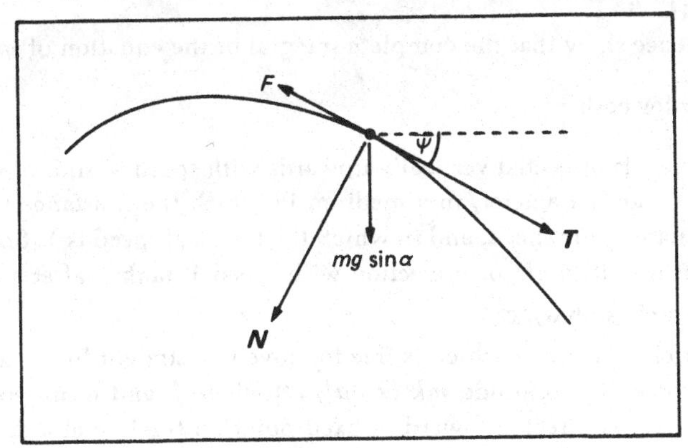

Substituting this into the equation of motion and using the fact that $\kappa = +\mathrm{d}\psi/\mathrm{d}s$ in this case, the tangential and normal components can be written, respectively, as

$$\frac{\mathrm{d}v}{\mathrm{d}t} = g \sin \alpha \sin \psi - \mu g \cos \alpha$$

$$v^2 \frac{\mathrm{d}\psi}{\mathrm{d}s} = g \sin \alpha \cos \psi$$

Using the identity $v = \mathrm{d}s/\mathrm{d}t$, the variables s and t can be eliminated between these components giving an equation which can be separated in the form

$$\int \frac{\mathrm{d}v}{v} = \int (\tan \psi - \mu \cot \alpha \sec \psi)\, \mathrm{d}\psi$$

Thus if the particle is originally projected horizontally with speed V, its speed when the velocity has turned through an angle ψ is given by

$$v = V \sec \psi\, (\sec \psi - \tan \psi)^{\mu \cot \alpha}$$

The equation of the trajectory in terms of intrinsic coordinates can now be obtained by substituting this into the normal component above and integrating.

Exercises

3.1 A particle is dropped from rest in a medium in which the resistance varies as the square of the speed. Writing the resistance term as $-mkv^2$, show that the particle approaches a terminal speed V which is given by $V^2 = g/k$

Hence show that speed of the particle as functions of time and distance travelled x, are given by

$$v = V \tanh \frac{gt}{V}$$

$$v = V(1 - \mathrm{e}^{-2gx/V^2})^{1/2}$$

and hence show that the complete integral of the equation of motion is

$$x = \frac{V^2}{g} \log \cosh \frac{gt}{V}$$

3.2 A particle is projected vertically upwards with speed $V \sinh \alpha$, where α is a constant parameter, in a medium in which the resistance varies as the square of the speed, and in which the terminal speed is V. Show that it returns to its point of projection with speed $V \tanh \alpha$ after a time $V(\alpha + \tan^{-1} \sinh \alpha)/g$.

3.3 A particle of mass m which is free to move in a straight line is acted on by a force of magnitude $mk^2(x + a^4/x^3)$ where k and a are constants, and which is directed towards a fixed point on the line at a distance x

from the particle. If it starts from rest at $x = a$, show that it reaches the fixed point after a time $\pi/4k$.

3.4 A particle moves in a straight line and is attracted towards a fixed point on that line with a force of magnitude μ/x^2 per unit mass, where μ is a constant and x is the distance of the particle from that point. If the particle starts from rest when $x = a$, show that the value of x at some subsequent time t is given by the equation

$$\sqrt{\left(\frac{x}{a}\right)}\sqrt{\left(1-\frac{x}{a}\right)}+\cos^{-1}\sqrt{\left(\frac{x}{a}\right)}=\sqrt{\left(\frac{2\mu}{a^3}\right)}t$$

3.5 A projectile is fired with speed V from a point O at a height h above a horizontal plane. Ignoring air resistance, show that the greatest distance from O at which the projectile can strike the plane is $(V^2/g)+h$, and that to achieve this the projectile must be fired at an angle α to the horizontal where

$$\sin^2\alpha=\frac{1}{2}\left(1+\frac{gh}{V^2}\right)^{-1}$$

3.6 If an explosion at a point on level ground hurls debris in all directions with equal speed given by $\sqrt{(2ag)}$, ignoring air resistance, show that a man standing a distance a away is in danger of being struck at two instants a time $2\sqrt{(a/g)}$ apart.

3.7 A particle is projected horizontally with speed V at a great height. Assuming that the only forces acting are a constant gravitational force mg and a resistance to the motion mkv^2, show that, when the velocity has turned through an angle ψ, the speed of the particle is given by

$$v=\sqrt{\left(\frac{g}{k}\right)}\left[\frac{g}{kV^2}\cos^2\psi-\sin\psi-\cos^2\psi\log(\sec\psi+\tan\psi)\right]^{-1/2}$$

3.8 A particle is projected horizontally with speed V along a fixed plane that is inclined at an angle α to the horizontal.

(a) If the frictional force is ignored, show that the particle follows a parabolic path.

(b) If $\mu<\tan\alpha$, show that the particle reaches a minimum speed when the velocity vector has turned through an angle $\psi=\sin^{-1}(\mu\cot\alpha)$, where μ is the coefficient of friction.

(c) if $\mu>\tan\alpha$, show that the particle comes to rest after a time

$$\frac{V}{g\sin\alpha}\int_0^{\pi/2}\sec^2\psi\,(\sec\psi-\tan\psi)^{\mu\cot\alpha}\,d\psi$$

By substituting $\sec\psi-\tan\psi=u$, or otherwise, show that this is equal to

$$\frac{\mu V}{g}\frac{\cot\alpha\,\mathrm{cosec}\,\alpha}{(\mu^2\cot^2\alpha-1)}$$

3.7. Comments on other axiomatic formulations

Before concluding this chapter it is appropriate to assess the relation of the formulation given above with the previous formulations of classical dynamics given by Newton, Mach, Hertz and others.

The central question in any formulation of classical dynamics is how to introduce the concept of force. This is a concept which has only been developed with great difficulty. Early workers in this subject soon came to understand what have been called impressed forces, but they were unwilling to accept any force whose mechanism they did not understand. Newton's great contribution at this point was to shift the emphasis from the question of why bodies move to the question of how they move. This permitted the introduction of forces that were not understood. The concept of gravitation which previously had been dismissed as an 'occult quantity' could then be included as a force. However, the idea of 'action at a distance', which was implied by the Newtonian theory of gravitation, has been regarded as unacceptable by many great men. Such ideas, however, are now rarely held, and the opposite view prevails. The development of relativity and quantum mechanics has led to the general acceptance of field theories. All forces are now frequently explained as being fundamentally the effects of force fields, and this has been reflected in the modern axiomatic formulations, even those of classical mechanics.

The formulation that has been given above follows very closely that originally given by Newton. The concept of force has been defined in exactly the same way. There is, however, a slight difference of terminology. Newton used the term force more generally, whereas it has been restricted here to what he called impressed forces. This latter term has been used in a more restricted sense, and a more general concept of fictitious forces has also been introduced. Newton was the first to introduce this concept, but he only considered the limited case of centripetal forces. By centripetal forces he meant centre-seeking forces such as gravitation and magnetism. A larger class of fictitious forces has been preferred here since an assumption about a frame of reference has been delayed until after stating the law of motion. It is therefore necessary to consider fictitious forces caused by relative accelerations and rotations. In this way the similarity between relative acceleration and gravitation can be demonstrated: a result which is developed into a basic postulate of the general theory of relativity. The classification of fictitious forces given here thus follows the spirit of Newton's formulation, although the concept is generalised.

There are, however, a number of points at which the above formulation differs from that of Newton. The most serious defect in Newton's formulation, as pointed out by Mach,[†] is that he has not given a proper definition of mass. His formal definition of mass as the product of density and volume clearly leaves density undefined. Following Mach's suggestion, the third law, stated as assumption 3.3, is used here to define mass ratio as the only possible objective definition of mass. Newton also to some extent confused his definition of force with his first two laws of motion. The attempt has therefore been made to separate out clearly the definition of force which is given here in definitions 3.2–4. Newton's first law has been rewritten here as assumption 3.2 in a form which permits the concept of force to be introduced. Once this concept has been developed, the law of motion, which is essentially Newton's second law, may be stated as proposition 3.2. This proposition, and particularly assumption 4.1, which is stated in the next chapter, include assumption 3.2 as a special case. Assumption 3.2 is therefore redundant in an axiomatic formulation of the subject. This feature is in fact also included in the Newtonian formulation since the first law is essentially contained in the second. Another difference is that Newton's assumptions concerning the nature of space and time have been replaced here. His assumptions of an absolute space and an absolute time are unnecessary for the theory and more general alternatives have been given here in assumptions 2.1, 2.2 and 3.2. The remaining assumptions, however, are also made by Newton. Assumption 3.4 is implied but not stated, and assumption 3.5 Newton gives as his corollary 1.

The influence of Mach's work can easily be seen above. Assumptions 3.3–5 are almost identical to Mach's three propositions, and his definition of mass ratio has been adopted. However, his philosophical approach has not been followed. One of Mach's main aims was to rid the whole of science of the idea of cause and effect. He was therefore entirely opposed to the concept of force described above. Using this approach classical dynamics ought to be a purely descriptive theory. Accordingly, he developed his theory of dynamics regarding acceleration as fundamental. He defined mass ratio as above, and concluded his system by defining force as the product of mass and acceleration. This, however, reduces the law of motion to a definition. He gave no rule to apply in practice. In a physical situation the acceleration of a particle can be measured. This, multiplied by its mass, Mach called

[†] E. Mach, *The Science of Mechanics* (English translation, Open Court, 1902).

the force. Since he did not consider the acceleration to be caused by a force, he was therefore formally unable to predict the motion of a body by considering the forces acting on it.

The attempt to remove the concept of force was taken a step further by Hertz,[1] who attempted to build the whole theory of classical dynamics from the three fundamental concepts of time, space and mass. Hertz's work is an attempt to present the whole subject in a formal and logical way, but in order to reformulate an approach to fictitious forces he introduced additional hypothetical invisible particles. He then regarded the traditionally understood forces as being determined by these hidden masses and their motions. However, in spite of the logical and mathematical beauty of this system, Hertz readily acknowledged that the classical formulation of the theory was easier to use in practical situations.

A more recent formal axiomatic formulation of classical particle mechanics has been presented by McKinsey, Sugar & Suppes[2] and repeated by Bunge.[3] However this assumes a working knowledge of classical particle dynamics and is not concerned with the physical foundations of the subject.

Another less formal axiomatic formulation of classical dynamics has been given by Kilminster & Reeve.[4] This has considerable merit. They follow Mach in regarding acceleration as fundamental, and the concept of force is developed in a sophisticated way using the concept of an acceleration field, a prominent place being given to gravitation. However, the concept of a force field can surely only be understood if the concept of force is understood first, at least intuitively. Kilminster & Reeve actually assume such an understanding of their readers. The field concept is more difficult. It has therefore seemed preferable here to develop the concept of a force from an intuitive idea of a 'push'. However, an advanced axiomatic formulation of classical dynamics for those already familiar with the subject could well concentrate on a field theoretic approach.

[1] H. Hertz, *The Principles of Mechanics* (English translation, London, 1899).
[2] J. C. C. McKinsey, A. C. Sugar & P. Suppes, 'Axiomatic foundations of classical particle mechanics'. *J. Rational Mech. Anal.*, **2**, 253 (1953).
[3] M. Bunge, *Foundations of Physics* (Springer, 1967).
[4] C. W. Kilminster & J. E. Reeve, *Rational Mechanics* (Longmans, 1966).

4

Newtonian relativity

It has been emphasised in previous chapters that position, and hence also velocity and acceleration, are relative concepts. Thus when stating or defining certain positions some frame of reference is always, at least implicitly, assumed. For example, a child may be told to sit still in a car, even though, together with the car, he may be moving at a great speed. In this case it is obviously implied that the child is required not to move relative to the car. Alternatively, the positions of the pieces in a game of chess are usually stated relative to the chessboard, and are therefore unambiguously defined, even if the board were to be moved from one room to another. In the same way, the items of furniture in a room can be described as being located in fixed positions relative to the floor and walls, in spite of the fact that they have a large velocity due to the rotation of the earth and its orbital velocity about the sun.

Clearly then, positions and dynamical properties such as velocities and accelerations can only be stated or determined relative to an assumed frame of reference. When an individual is required to state a position, he is free to choose an arbitrary frame of reference for convenience. However, when he attempts to measure a dynamical property such as the velocity or acceleration of a particle, in practice he would usually make the measurements relative to himself or to the instruments he is using. Thus when two individuals attempt to measure the same such quantity, their measurements may well differ. For example, the measured velocity of a bird will be different for an observer on the ground, an observer in a train, or one in an aircraft. What is measured in each case is the velocity relative to the observer. The observed object is the same, but the relative position and motion of each observer are different.

In cases such as this, in which the motion of the same point is being recorded by distinct observers, the results obtained are not totally independent but are found to be related. The actual relation between them can be modelled by a theory which must take into account the relative position and motion of each observer. Any theory that aims to relate the measurements and observations of different observers is known as a theory of relativity. Such a theory is, in fact, implicitly contained in the theory of classical particle dynamics that was outlined in the previous chapter. This theory, which is known as that of Newtonian relativity, will now be discussed in some detail.

4.1. Relative motion

In order to introduce a theory of relativity, it is appropriate initially to consider two observers, labelled O and O', who may be moving relative to each other, but not rotating. The position of O' from O can be denoted by the vector $\boldsymbol{r}_0$, and thus the velocity and acceleration of O' relative to the frame of reference of O are expressed by $\dot{\boldsymbol{r}}_0$ and $\ddot{\boldsymbol{r}}_0$ respectively.

We may now consider the situation in which both O and O' observe the motion of the same particle, which can be labelled P. The position of P relative to O and O' may be denoted by the vectors $\boldsymbol{r}$ and $\boldsymbol{r}'$ respectively. Now, since space is assumed to be Euclidean, the vectors $\boldsymbol{r}$, $\boldsymbol{r}'$ and $\boldsymbol{r}_0$ can respectively be interpreted as the line vectors from O to P, from O' to P and from O to O'. The situation can thus be represented as in figure 4.1. From this it can clearly be seen that these vectors can be related by the equation

$$\boldsymbol{r} = \boldsymbol{r}_0 + \boldsymbol{r}'$$

Now, since in this case O, O' and P can all be chosen arbitrarily, this equation can be interpreted as stating that *relative positions are related according to the vector law of addition.*

Fig. 4.1. The position of a particle P relative to two observers O and O'.

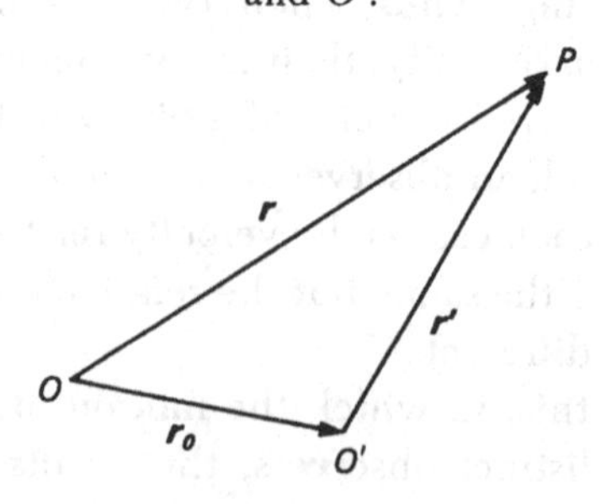

This result forms the first step in the theory of Newtonian relativity. It can be seen that this is an immediate consequence of assumption 2.2 which concerns the Euclidean nature of space. It involves only relative positions and is independent of any motion or time. It can thus be tested independently of any theory of dynamics, and it is found to be in complete agreement with observations, at least over the range in which tests can be made.

Now, according to assumption 2.1, time is assumed to be independent of space. It is therefore possible to differentiate the three vectors above with respect to the common time parameter. This assumption, therefore, together with the above identity, implies that

$$\dot{r} = \dot{r}_0 + \dot{r}'$$

This vector equation states that the velocity of P relative to O is equal to the velocity of P relative to O' plus the velocity of O' relative to O. Since, again, O, O' and P can be chosen arbitrarily, this implies that *relative velocities are related according to the vector law of addition.*

This result, which is the most widely known aspect of the theory of Newtonian relativity, is in complete accord with the intuition that is based on everyday experience. For example, if two cars approach each other on a road and both are travelling at a speed of 60 mph, it is understood that they would approach each other at a relative speed of 120 mph. Using the same approach, the velocity of an aircraft relative to the ground can be determined from a knowledge of its velocity through the air and the velocity of the wind.

In practice it is found that, for such everyday situations, observations are in very close agreement with the above theory. The complete theory has in fact been used with considerable success for over 300 years. However, every theory should be tested for as wide a range of parameters as is practically possible, and in this case it is found that the above theoretical result is not in agreement with observations in situations where the relative velocities are extremely large. A discrepancy was first found when attempts were made to measure differences in the speed of light which were expected to be caused by the relative motion of an observer. No such differences were detected. However, it was not the theory of light that was found to be in error, but the Newtonian theory of relative motion. In the following decades, particularly with the construction of particle accelerators, it has been found that the predictions of the theory of Newtonian relativity do not agree with observations in all situations where the magnitudes of relative velocities become a significant fraction of the speed of light.

It must therefore be concluded that the above theory describing the relation between relative velocities is wrong. However, the theory has been obtained simply as a logical consequence of assumptions 2.1 and 2.2, which state that space is Euclidean and that time is independent of space. Thus the error that appears in the theory of Newtonian relativity must lie in at least one of these statements. But it has already been pointed out that the Newtonian theory of relative positions, which is based only on the assumption that space is Euclidean, is in fact in agreement with all known observations. Assumption 2.2 is thus strongly corroborated. It therefore seems inevitable that the error has entered through the assumption that a unique time parameter can be introduced which is independent of position and motion in space.

In view of the ultimate failure of the Newtonian theory of relative velocities a new theory is required. Such a theory has in fact been provided by Einstein's special theory of relativity. In accordance with the above argument, this theory is based upon a new concept of time which is assumed not to be directly dependent on space, but more generally to depend on the motion of an observer in space. Thus a distinct time parameter is defined for each distinct observer, and the familiar concept of simultaneity has to be abandoned. It is then convenient to assume that space and time can be represented by a new four-dimensional entity, called space–time, in such a way that even the measurement of space depends upon the observer. It is found, however, that the special theory cannot be made to include a theory of gravitation. For this Einstein developed his general theory of relativity in which the four-dimensional manifold representing space–time is considered to be curved.

Unfortunately Einstein's theories of relativity are extremely difficult to use in most practical situations. However, as is clearly required by observations, its equations do approximate to those of Newtonian theory in situations in which relative velocities are small compared to the speed of light. In practice, therefore, the theory of classical dynamics may still be applied with confidence in such situations.

Returning now to the Newtonian theory and again using assumptions 2.1 and 2.2, the equation for relative velocities can again be differentiated with respect to the common time parameter. This gives the equation

$$\ddot{\boldsymbol{r}} = \ddot{\boldsymbol{r}}_0 + \ddot{\boldsymbol{r}}'$$

which expresses the relation between relative accelerations in Newtonian theory.

In deriving the above expressions for relative positions, velocities and accelerations, it was explicitly assumed that the observers O and O' were not rotating relative to each other. It is now convenient to relax this condition and accordingly to assume that, in addition to a possible relative motion, O' is also rotating with angular velocity $\boldsymbol{\omega}$ relative to O. The motion of the point P is thus now considered both relative to a frame of reference with origin at O, and to another frame of reference with origin at O' which is rotating relative to the frame at O. To obtain total generality, it may also be assumed that $\boldsymbol{\omega}$ is a vector function of time, so that O' is not necessarily rotating with constant angular velocity.

In this situation, the equation relating the relative positions of the particle P is formally identical to that given above: namely

$$\boldsymbol{r} = \boldsymbol{r}_0 + \boldsymbol{r}'$$

The reason for this identity is, basically, that vectors define real geometrical objects that are independent of any particular coordinate representation. In this case it is only the coordinate frame that is considered to rotate. However, in order to use this equation in practice, it is sometimes convenient to adopt some particular coordinate representation, in which case certain mathematical techniques, such as the introduction of matrix operators, sometimes have to be used to allow for the rotation of O'. It can also be pointed out here that a change in only the orientation of a reference frame has no effect upon the vector formulation of the position or motion of P.

However, when attempting to differentiate the above equation, or any other vector expression, with respect to time, care has to be taken. The point is that vectors represent real geometrical quantities. These may vary with time. In addition, relative to a rotating frame of reference, vectors also appear to vary simply as a consequence of that rotation. For example, a vector that is stationary with respect to one frame of reference, must appear to be rotating relative to a second frame that is itself rotating relative to the first. Thus there must be a difference between the time derivative of any vector with respect to frames that are rotating relative to each other. In fact, a relation between the derivatives can easily be obtained, and the result may be stated in the form of a lemma.

***Lemma* 4.1.** *The time derivative of a vector* **A** *relative to a frame of reference S is equal to its apparent time derivative evaluated in a frame of reference S′ which has the same origin as S, together with the vector product of the angular velocity* **ω** *of S′ relative to S with the vector* **A**.

$$\dot{\mathbf{A}}_S = \dot{\mathbf{A}}_{S'} + \boldsymbol{\omega} \times \mathbf{A}$$

This can easily be proved by considering the cartesian components of the vector in each frame.

This result may now be used to obtain an expression which relates the apparent velocities of the particle P to the two observers O and O'. The resulting equation is

$$\dot{\mathbf{r}} = \dot{\mathbf{r}}_0 + \dot{\mathbf{r}}' + \boldsymbol{\omega} \times \mathbf{r}'$$

This is interpreted as stating that the velocity of P relative to O is equal to the apparent velocity of P relative to O', plus the velocity of O' relative to O, plus the term $\boldsymbol{\omega} \times \mathbf{r}'$ which allows for the relative angular velocity of O'.

To obtain a general equation relating the apparent accelerations of the particle P relative to O and O', lemma 4.1 may be used a second time. The relation obtained is expressed by the equation

$$\ddot{\mathbf{r}} = \ddot{\mathbf{r}}_0 + \ddot{\mathbf{r}}' + \dot{\boldsymbol{\omega}} \times \mathbf{r}' + 2\boldsymbol{\omega} \times \dot{\mathbf{r}}' + \boldsymbol{\omega} \times (\boldsymbol{\omega} \times \mathbf{r}')$$

This equation will be required in the next section.

4.2. Inertial frames of reference

It is convenient to begin here with the preliminary assumption 3.2. Accordingly, it is assumed that there exists at least one inertial frame of reference, relative to which the natural motion of any particle is one of constant or zero velocity. When a particle accelerates relative to such a frame, it is assumed that there must be forces acting on it in such a way that the law of motion, as stated in proposition 3.2, is satisfied.

Since an inertial frame of reference has been assumed to exist, it is convenient to consider the motion of a particle relative to such a frame. In this case it is explicitly assumed that no fictitious forces arise simply as a result of the choice of reference frame. It was necessary to introduce the possibility of such forces in the previous chapter, but it was understood that they only arise when the frame of reference being used is accelerating or rotating relative to an inertial frame. The forces which arise in this way form an essential part of the theory of Newtonian relativity, and will be introduced in detail a little later. However, for the moment it is assumed that an inertial frame of reference has been adopted in which the law of motion is satisfied without the inclusion of these forces.

That is not to say, however, that no fictitious forces should be included. On the contrary, it is often found in practice to be necessary to include the fictitious force of gravitation. However, the forces of gravitation do appear to have real physical sources, and it is such forces, including the various types of impressed force, which are assumed in Newtonian dynamics to induce a particle to accelerate relative to an inertial frame of reference.

We may now consider the motion of a particle P relative to such a frame. According to proposition 3.2, the motion must satisfy the equation

$$m\ddot{\boldsymbol{r}} = \boldsymbol{F}$$

where m is the mass of the particle, and $\boldsymbol{F}$ is the resultant force which acts on it. In this case it is assumed that the components of $\boldsymbol{F}$ only include impressed forces and those fictitious forces, such as gravitation, which can be attributed to real physical sources.

We must now consider how this equation of motion can be stated relative to an arbitrary frame of reference. To do this it is convenient to imagine an observer O located at the origin of the inertial frame. Then, using the same notation as in the previous section, the motion of the particle P can also be considered relative to another observer O'.

To start with, it is convenient to consider only the case in which O' is moving, but not rotating, relative to the inertial frame. In this case the observed relative accelerations of P are related by the equation

$$\ddot{\boldsymbol{r}} = \ddot{\boldsymbol{r}}_0 + \ddot{\boldsymbol{r}}'$$

Thus, since the above equation of motion is assumed to be satisfied relative to O, the apparent acceleration of P relative to O' must be given by the equation

$$m\ddot{\boldsymbol{r}}' = \boldsymbol{F} - m\ddot{\boldsymbol{r}}_0$$

It can immediately be seen that this retains the form of the basic equation of motion, provided that the term $-m\ddot{\boldsymbol{r}}_0$ is somehow included as an additional fictitious force. This in fact is the exact expression for the fictitious force that is considered to be due to the relative acceleration of the observer. Such a force was alluded to in the previous chapter. With this single addition, Newton's law of motion retains the same form for the motion of any particle relative to any frame of reference that is not rotating relative to an inertial frame.

It should also be noticed at this stage that the equations of motion relative to O and O' are identical if the term $-m\ddot{r}_0$ is zero. This is the case if O' is moving with constant velocity relative to O. Now, the initial assumption of this section was that O is in an inertial frame of reference. But it has now been shown that O' can also be in such a frame, since the equation of motion relative to it clearly satisfies the appropriate condition. We are therefore led to the conclusion that there is no unique inertial frame. Rather, there is a class of such frames which move relative to each other with constant velocity. They are related to each other by what is known as the *group of Galilean transformations.* It is inappropriate to single out any one frame as a basic inertial frame, and so we assume the existence of the whole class of such frames.

The basic statement of the theory of classical dynamics may now be given.

***Assumption* 4.1.** *There exists a unique class of inertial frames of reference which move relative to each other with constant velocity and in which Newton's law of motion is always satisfied for all particles, without the addition of the fictitious forces that are considered to be due to the motion of the reference frame.*

This statement clearly contains several aspects. To start with, it postulates the existence of a whole class of inertial frames, and specifies the way in which these are related to each other. In addition, it states that, if a frame of reference is an inertial frame for the motion of one particle, then it is also an inertial frame for the motion of all particles. This is an important aspect of the theory which aims at a general application. Finally, also following the same spirit, it states that all particles satisfy the same equation of motion. The equation to be satisfied is that described by proposition 3.2, but with the clarification that certain fictitious forces should be omitted.

It can also immediately be seen that assumption 4.1 logically contains the preliminary assumption 3.2 as a particular special case. These statements are not independent, and therefore in any axiomatic formulation of the theory they should not both appear as basic axioms. In any such formulation, it is the essential part of assumption 4.1 that must be included in the axiomatic base. Assumption 3.2, however, can be used to provide the basic definition of the concept of an inertial frame of reference, provided it is appropriately rewritten.

In assumption 4.1 the idea is introduced of a class of equivalent frames of reference. Unfortunately this adds considerable complexity to the Newtonian concept of space–time. Had there existed a unique space, then assumptions 2.1 and 2.2 would naturally have led to space–time being considered in a mathematical form as the direct product $E^3 \times E^1$, where E^n denotes an n-dimensional Euclidean space. Such a space–time is referred to as Aristotelian space–time. Now, a point in space–time is referred to as an 'event', that is, a point having a unique position in space and occurring at a unique instant of time. It can be seen that in Aristotelian space–time, two events are separated by a unique distance and a unique time. However, it has now been shown that a unique space is not a concept that is useful in classical dynamics. Rather, we consider a class of equivalent spaces which move relative to each other with constant velocity. Thus two events are only separated by a unique distance if their time difference vanishes. This leads to a concept of space–time, referred to as Newtonian space–time, in which only the time difference between two events is well defined.

We must now return to the previous discussion, in which we obtained the equation of motion for a particle relative to a frame of reference that was moving with respect to an inertial frame. It was found in this case that Newton's law was satisfied provided an additional fictitious force was introduced to allow for the relative acceleration of the observer. It is now appropriate to attempt to generalise this result to include the case in which the observer is also rotating relative to an inertial frame.

Again the previous notation can be used. The motion of a particle P is considered relative to two observers O and O'. The observer O is assumed to be located at the origin of an inertial frame of reference, relative to which P has position vector $\boldsymbol{r}$. The frame of reference of O', which has position vector $\boldsymbol{r}_0$ relative to O, is assumed to be both moving and rotating with angular velocity $\boldsymbol{\omega}$ relative to the inertial frame. The position of P relative to O' is denoted by $\boldsymbol{r}'$. According to a result of the previous section, the relative accelerations of P are related by the formula

$$\ddot{\boldsymbol{r}} = \ddot{\boldsymbol{r}}_0 + \ddot{\boldsymbol{r}}' + \dot{\boldsymbol{\omega}} \times \boldsymbol{r}' + 2\boldsymbol{\omega} \times \dot{\boldsymbol{r}}' + \boldsymbol{\omega} \times (\boldsymbol{\omega} \times \boldsymbol{r}')$$

It may now be assumed that the motion of P relative to O satisfies Newton's equation

$$m\ddot{\boldsymbol{r}} = \boldsymbol{F}$$

where $\boldsymbol{F}$ is the sum of the impressed forces that act on P and those fictitious forces, such as gravitation, which can be associated with distinct physical sources. Using the above identity, this may now be rewritten to give the equation of motion of P relative to O' in the form

$$m\ddot{\boldsymbol{r}}' = \boldsymbol{F} - m\ddot{\boldsymbol{r}}_0 - m\dot{\boldsymbol{\omega}}\times\boldsymbol{r}' - 2m\boldsymbol{\omega}\times\dot{\boldsymbol{r}}' - m\boldsymbol{\omega}\times(\boldsymbol{\omega}\times\boldsymbol{r}')$$

This result, which clearly generalises that obtained above, is of considerable importance and can be stated in the following way.

***Proposition* 4.1.** *If a particle of mass m has a position vector* $\boldsymbol{r}$ *relative to some frame of reference, and if that frame has acceleration* $\ddot{\boldsymbol{r}}_0$ *and angular velocity* $\boldsymbol{\omega}$ *relative to an inertial frame, then its equation of motion is*

$$m\ddot{\boldsymbol{r}} = \boldsymbol{F} - m\ddot{\boldsymbol{r}}_0 - m\dot{\boldsymbol{\omega}}\times\boldsymbol{r} - 2m\boldsymbol{\omega}\times\dot{\boldsymbol{r}} - m\boldsymbol{\omega}\times(\boldsymbol{\omega}\times\boldsymbol{r})$$

where $\boldsymbol{F}$ *is the sum of the forces acting on it as measured in an inertial frame.*

This result effectively states the equation of motion for a particle relative to an arbitrary frame of reference. However, in order to use it in practice, the acceleration $\ddot{\boldsymbol{r}}_0$ and angular velocity $\boldsymbol{\omega}$ of the frame of reference relative to an inertial frame either have to be known explicitly, or must be postulated.

The additional terms that appear in the equation of motion, as stated in proposition 4.1, can be regarded as expressions for the fictitious forces that must necessarily be included whenever the frame of reference is noninertial. With use in applications, these terms have become very familiar. The term $-m\boldsymbol{\omega}\times(\boldsymbol{\omega}\times\boldsymbol{r})$ is known as the *centrifugal force* due to the rotation of the frame of reference. The term $-2m\boldsymbol{\omega}\times\dot{\boldsymbol{r}}$ is known as the *Coriolis force,* which is due to motion in a rotating frame of reference. The additional terms $-m\ddot{\boldsymbol{r}}_0$ and $-m\dot{\boldsymbol{\omega}}\times\boldsymbol{r}$ are due to the acceleration and angular acceleration of the reference frame respectively.

4.3. Motion relative to the earth: examples

To illustrate the theory described above, it is appropriate to consider here the motion of a particle relative to a frame of reference which is fixed at a point on the earth's surface. The purpose is to find the effect of the earth's rotation on the motion of particles near the surface, and to demonstrate that the earth does in fact rotate relative to an inertial frame in Newtonian theory.

Consider a point O located at an arbitrary point on the earth's surface, having latitude λ. A cartesian frame of reference can be chosen which has origin at O and axes which are fixed relative to the earth. It is convenient to choose the z axis to be vertically upwards. The x and y axes may then be chosen to be aligned respectively with the east and north directions at O, as illustrated in figure 4.2.

It may be assumed that the earth rotates with constant angular velocity $\boldsymbol{\Omega}$ about the north polar axis. Denoting the position vector of O relative to the earth's centre by $\boldsymbol{r}_0$, the relative acceleration of O can be expressed in the form

$$\ddot{\boldsymbol{r}}_0 = \boldsymbol{\Omega} \times (\boldsymbol{\Omega} \times \boldsymbol{r}_0)$$

since this is due only to a rigid rotation with angular velocity $\boldsymbol{\Omega}$.

It may also be assumed as a first approximation, although this can easily be generalised, that the earth is exactly spherical with radius R. In this case $\boldsymbol{r}_0$ is aligned with the vector $\boldsymbol{k}$ of the reference frame. Since the frame is fixed relative to the earth, it also must be rotating with angular velocity $\boldsymbol{\Omega}$ about an axis in the plane $x=0$ which is inclined at an angle λ to northerly direction (see figure 4.3). The angular velocity of the frame can therefore be expressed as

$$\boldsymbol{\Omega} = \Omega(\cos\lambda \boldsymbol{j} + \sin\lambda \boldsymbol{k})$$

Now consider the motion of a particle relative to this frame. It may be assumed that the particle is acted on by a gravitational force, which it is convenient here simply to denote by $m\boldsymbol{g}$, and a set of impressed forces whose resultant is denoted by $\boldsymbol{P}$. Then, according

Fig. 4.2. Cartesian axes chosen at a point O on the earth's surface.

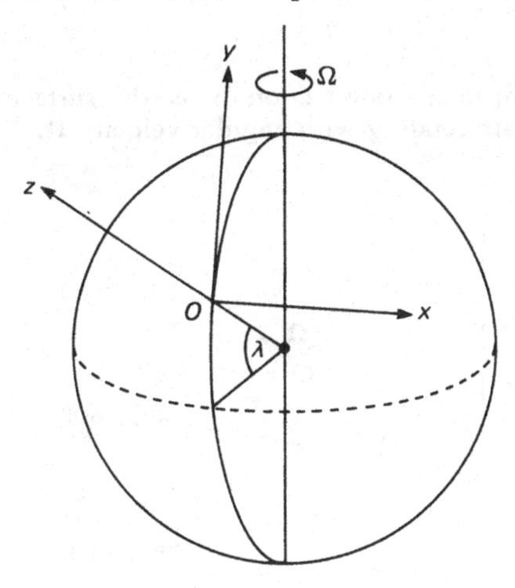

to proposition 4.1, and ignoring the acceleration of the earth's centre, the equation of motion of the particle relative to the frame is

$$m\ddot{\boldsymbol{r}} = m\boldsymbol{g} + \boldsymbol{P} - m\boldsymbol{\Omega}\times(\boldsymbol{\Omega}\times\boldsymbol{r}_0) - 2m\boldsymbol{\Omega}\times\dot{\boldsymbol{r}} - m\boldsymbol{\Omega}\times(\boldsymbol{\Omega}\times\boldsymbol{r})$$

Now, for the type of motion envisaged, the position vector of the particle $\boldsymbol{r}$ will be small compared with $\boldsymbol{r}_0$. The term $-m\boldsymbol{\Omega}\times(\boldsymbol{\Omega}\times\boldsymbol{r})$ may therefore be neglected.

At this point it is convenient to digress to consider the case when the particle is stationary relative to the frame. In this case

$$m\boldsymbol{g} + \boldsymbol{P} - m\boldsymbol{\Omega}\times(\boldsymbol{\Omega}\times\boldsymbol{r}_0) = 0$$

and therefore since $\boldsymbol{r}_0 = R\boldsymbol{k}$, the total impressed force required to keep the particle stationary is

$$\boldsymbol{P} = -m\boldsymbol{g} + m\Omega^2 R\cos\lambda(\sin\lambda\boldsymbol{j} - \cos\lambda\boldsymbol{k})$$

This effectively is the force that is measured when the particle is 'weighed'. It is therefore appropriate to introduce the concept of an 'apparent gravitational force' $m\boldsymbol{g}'$ where

$$m\boldsymbol{g}' = m\boldsymbol{g} - m\Omega^2 R\cos\lambda(\sin\lambda\boldsymbol{j} - \cos\lambda\boldsymbol{k})$$

It is this force which defines the familiar concept of a 'vertical', as can be seen by considering a stationary plumb-line. It is composed of a true gravitational force directed approximately towards the earth's centre, and a centrifugal force due to the earth's rotation. The earth's rotation therefore implies that the vertically downwards direction does not point exactly towards the earth's centre. This also accounts for its 'equatorial bulge', since mean sea level is everywhere horizontal which is normal to the vertical.

Fig. 4.3. Axes fixed at a point O on the earth's surface at latitude λ are rotating with angular velocity $\boldsymbol{\Omega}$.

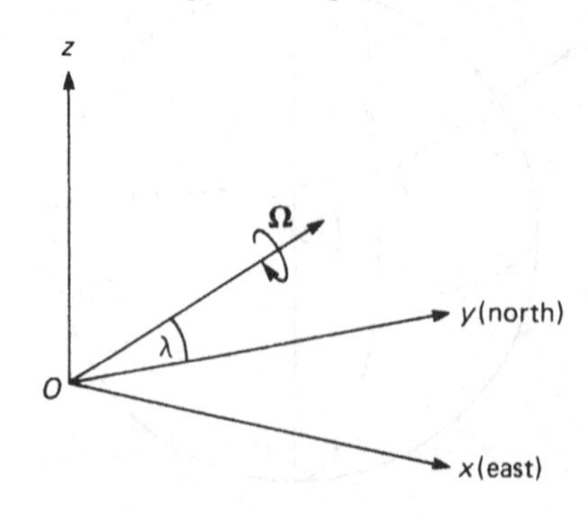

It can now be seen that the equation of motion of a particle relative to a frame which is fixed on the earth's surface is

$$m\ddot{\boldsymbol{r}} = m\boldsymbol{g}' + \boldsymbol{P} - 2m\boldsymbol{\Omega}\times\dot{\boldsymbol{r}}$$

Thus the perturbation of particle motion due to the earth's rotation effectively arises from the presence of the Coriolis force.

It should also be pointed out that the angular velocity of the earth is very small, being only 2π radius per day. It is therefore appropriate to consider only the first order effects, and to neglect terms of order Ω^2. At this level of approximation $m\boldsymbol{g}' = m\boldsymbol{g} = -mg\boldsymbol{k}$ and $\boldsymbol{r}_0 = R\boldsymbol{k}$.

Example 4.1: *projectiles*

Consider the effects of the earth's rotation on projectile motion. In order simply to estimate this effect, it is appropriate to ignore any resistance to the motion, and to assume that the particle is subject only to a constant vertical gravitational force and the Coriolis force due to the rotation of the reference frame. With these assumptions the equation of motion implies that

$$\ddot{\boldsymbol{r}} = \boldsymbol{g} - 2\boldsymbol{\Omega}\times\dot{\boldsymbol{r}}$$

This can immediately be integrated to give

$$\dot{\boldsymbol{r}} = \boldsymbol{g}t - 2\boldsymbol{\Omega}\times\boldsymbol{r} + \boldsymbol{A}$$

where $\boldsymbol{A}$ is an arbitrary constant of integration. If it is assumed that the particle is initially projected from the origin with velocity $\boldsymbol{V}$, then $\boldsymbol{A} = \boldsymbol{V}$. A second integral of the equation of motion cannot directly be obtained. However, since terms of order Ω^2 can be neglected, the first integral can be substituted back into the equation which then becomes

$$\ddot{\boldsymbol{r}} = \boldsymbol{g} - 2\boldsymbol{\Omega}\times\boldsymbol{g}t - 2\boldsymbol{\Omega}\times\boldsymbol{V}$$

This equation can now be integrated twice. For the initial conditions above, the complete integral of the equation of motion is

$$\boldsymbol{r} = \boldsymbol{V}t + \tfrac{1}{2}\boldsymbol{g}t^2 - \tfrac{1}{3}\boldsymbol{\Omega}\times\boldsymbol{g}t^3 - \boldsymbol{\Omega}\times\boldsymbol{V}t^2$$

Example 4.2

Consider a particle projected with speed V along a smooth horizontal plane. The purpose of this example is to investigate the effect on motion of the earth's rotation and, accordingly, any resistance to the motion may be neglected. The forces which act on the particle are thus assumed to be the gravitational force which acts vertically downwards, the Coriolis force, and a reaction from the surface denoted by $\boldsymbol{R}$ which acts vertically upwards. The equation of motion is thus

$$m\ddot{\boldsymbol{r}} = m\boldsymbol{g} + \boldsymbol{R} - 2m\boldsymbol{\Omega}\times\dot{\boldsymbol{r}}$$

Neglecting terms of order Ω^2 we may put $m\boldsymbol{g} = -mg\boldsymbol{k}$, $\boldsymbol{R} = R\boldsymbol{k}$ and $\boldsymbol{\Omega} = \Omega(\cos\lambda\boldsymbol{j} + \sin\lambda\boldsymbol{k})$. Now, since the particle slides over the plane, the vertical component of the equation of motion is

$$0 = -mg + R - 2m\Omega \cos \lambda \mathbf{j} \times \dot{\mathbf{r}}$$

This is simply an equation defining R. The equation for motion in the plane is thus

$$\ddot{\mathbf{r}} = -2\Omega \sin \lambda \mathbf{k} \times \dot{\mathbf{r}}$$

Since the acceleration is clearly perpendicular to the velocity, the speed remains constant. The acceleration therefore has magnitude $2\Omega V \sin \lambda$ and is directed towards the right of the velocity vector in the northern hemisphere. The particle therefore tends to move in a clockwise direction round a circular arc of radius $V/(2\Omega \sin \lambda)$.

It may generally be observed that any particle moving horizontally with speed V is acted on by a Coriolis force $2\Omega V \sin \lambda$ to the right. This force clearly increases with the speed of the particle. It is this force which induces winds to move in an anticlockwise direction round an anticyclone in the northern hemisphere. A region of low atmospheric pressure tends to attract the particles of air towards it. As these acquire velocity they are forced to the right by the Coriolis force, thus establishing the anticlockwise motion. Then, in a simple way, it may be considered that the force towards the centre induced by the pressure gradient is almost balanced by the opposing Coriolis force.

Example 4.3: *Foucault's pendulum*

This is essentially a spherical pendulum considered relative to a set of axes that are rotating with the earth. However, in practice it is composed of a very heavy bob suspended by a very long cord from a fixed point. The large

Fig. 4.4. Foucault's pendulum.

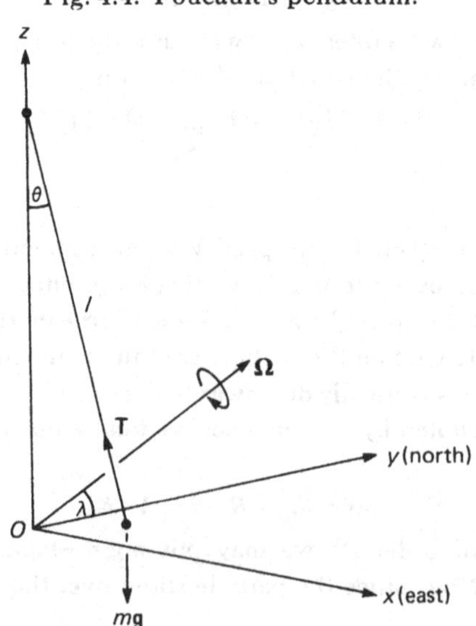

dimensions involved enable the bob to swing back and forth for a significant fraction of a day before its motion declines. The pendulum is set in motion by drawing it aside from its equilibrium position and releasing it from rest. As the pendulum swings, its plane of motion is found to rotate throughout the day. Such pendulums have been built in a number of cities, following an original experiment of Jean Foucault in Paris in 1851. They clearly demonstrate the rotation of the earth relative to an inertial frame according to Newtonian theory.

Foucault's pendulum can be modelled by considering a particle of mass m suspended from a fixed point by a light inelastic string of length l, as illustrated in figure 4.4.

Ignoring resistance and denoting the tension in the string by $\boldsymbol{T}$, the equation of motion is

$$m\ddot{\boldsymbol{r}} = m\boldsymbol{g} + \boldsymbol{T} - 2m\boldsymbol{\Omega}\times\dot{\boldsymbol{r}}$$

The vertical component of this is

$$m\ddot{z} = -mg + T\cos\theta - 2m\Omega\cos\lambda\boldsymbol{j}\times\dot{\boldsymbol{r}}$$

where θ is the inclination of the string to the vertical. It may now be assumed that the motion is such that θ remains small. The vertical component therefore implies that to first order T is approximately equal to mg. At this level of approximation, the equation for motion in a horizontal plane is

$$m\ddot{\boldsymbol{r}} = -T\sin\theta\hat{\boldsymbol{r}} - 2m\Omega\sin\lambda\boldsymbol{k}\times\dot{\boldsymbol{r}}$$

which can now be rewritten as

$$\ddot{\boldsymbol{r}} = -\frac{g}{l}\boldsymbol{r} - 2\Omega\sin\lambda\boldsymbol{k}\times\dot{\boldsymbol{r}}$$

This is now identical to the approximate equation of motion of a simple pendulum relative to axes which rotate about the vertical with angular velocity $\Omega\sin\lambda$. This is in excellent agreement with the rate at which the plane of Foucault's pendulum is found to rotate at the given latitude.

Exercises

In these exercises, assume that the gravitational force acting on each particle is constant, ignore resistive forces, but allow for the effect of the Coriolis force due to the rotation of the earth. Denote the latitude of the origin O of the reference frame by λ, and assume that terms of order Ω^2 can be neglected.

4.1 If a particle is projected vertically upwards from O with initial speed V, show that it returns to the same horizontal plane at a distance $4\Omega V^3\cos\lambda/3g^2$ to the west of O.

4.2 If a particle is projected in a northerly direction from O with initial speed V at an angle α to the horizontal, show that the projection of the trajectory on the original vertical plane of motion is a parabola, but that the particle deviates to the west of this plane by a distance given by

$\Omega V \sin(\alpha-\lambda)t^2 - \frac{1}{3}g\Omega \cos \lambda t^3$

4.3 If a particle is projected in an easterly direction from O with initial speed V at an angle α to the horizontal, show that it returns to the same horizontal plane after a time given approximately by

$$\frac{2V}{g}\sin\alpha\left(1+\frac{2\Omega V}{g}\cos\lambda\cos\alpha\right)$$

Hence show that the change in the horizontal range of the particle due to the earth's rotation, is given approximately by

$$\frac{4\Omega V^3}{g^2}\sin\lambda[\cos\lambda(1-\tfrac{4}{3}\sin^2\alpha)\boldsymbol{i}-\sin\lambda\cos\alpha\sin\alpha\boldsymbol{j}]$$

where $\boldsymbol{i}$ and $\boldsymbol{j}$ are directed towards the east and north respectively.

4.4 If water is constrained to move with constant velocity $\boldsymbol{V}$ along a straight horizontal canal of width a, show that the water level on the right-hand bank is higher than that on the left by an amount $2a(\Omega V/g)\sin\lambda$.

4.4. The search for an inertial frame

It has been shown that in the Newtonian theory of relativity certain fictitious forces need to be included in the equations of motion if the frame of reference being used is accelerating or rotating with respect to an inertial frame. In the application of classical dynamics it is therefore essential to determine initially an inertial frame of reference. In the past three centuries, during which classical dynamics has been applied, such an inertial frame has in fact been determined to a high degree of accuracy. The method of determining such a frame is now described.

In many simple applications a frame of reference is used that is fixed to a point on the earth's surface. The assumption that this is an inertial frame usually leads to good results for the motion of bodies near that point, but in some experiments, such as Foucault's pendulum experiment, the motion cannot be explained. These results can, however, be explained if the earth is rotating about its axis relative to an inertial frame. Foucault's pendulum manifests the effect of the Coriolis force due to the rotation of the earth. Also the shape of the earth with its equatorial bulge can be explained as a consequence of the centrifugal force due to rotation. It is therefore appropriate to change the frame of reference to one centred at the centre of the earth and with axes fixed relative to the near stars. The above effects are now explained, but not the motion of the planets, which appear to move about the sun. It is therefore appropriate to transfer the origin of the frame of reference to the sun. But we must also allow for the possibility that the sun is moving. In fact, in order to describe the motion of stars in our galaxy, it is convenient to choose a frame of reference which has its origin at the centre of the galaxy. And in order to describe the motion of the galaxies we must move our frame of reference again. In this way it can be seen that we can never ultimately find a basic inertial frame of reference. However, the method above ensures that if such an inertial frame exists we come closer and closer to it. Present experimental results indicate a convergence to a basic inertial frame whose axes are fixed relative to the directions of the most distant observed galaxies. Even though such galaxies are moving, their vast distance ensures that their transverse motion is unobservable.

Having thus determined a unique frame of reference that is, at least approximately, inertial, we can now determine the acceleration and rotation of any other frame of reference that we might find convenient to use. The fictitious forces associated with the relative acceleration and rotation can then be estimated. For example, it is found that for a frame of reference attached to a point on the earth's surface the forces caused by the earth's relative acceleration and rotation are very small and in many applications may be neglected.

It should be emphasised that the search for an inertial frame of reference involves only dynamical considerations. The method, essentially, is to look for the effects of the fictitious forces described in proposition 4.1 which are caused by relative accelerations and rotations. It does not in any sense involve metaphysical considerations. The idea that the earth, or the sun, or the galaxy, or anything else, is the centre of the universe is a metaphysical idea that has nothing to do with inertial frames of reference. An inertial frame, if such a concept really exists, can be determined by dynamical principles alone. Even if the earth were permanently covered with a dark cloud, the earth's relative acceleration and rotation could still be determined by experiments such as Foucault's pendulum. The fact that we can also observe large-scale motions, such as that of the stars, merely enables us to determine an inertial frame with greater accuracy.

4.5. Absolute rotation or Mach's principle

In Newton's original formulation of classical dynamics, he introduced the concept of an absolute space. However, it has been argued above that classical dynamics does not single out such a unique space. Rather, it singles out a class of spaces that move relative to each other with constant velocity. There is no need to distinguish one absolute space. It is therefore meaningless to introduce the concept of an absolute velocity. Any constant velocity relative to an inertial frame can be transformed away by making a Galilean transformation to the inertial frame in which that velocity is zero. However, it has also been concluded that any frame of reference which is rotating relative to an inertial frame is a noninertial frame. Thus the angular velocity of a space or frame is uniquely defined. It therefore appears to be meaningful to speak of the absolute angular velocity of a given space or frame.

Newton's argument for the reality of absolute rotations centres on his bucket experiment. Newton considered a bucket of water which was made to rotate about a vertical axis. Initially, the surface of the water is flat and the water is rotating relative to the bucket, but as the motion of the bucket is communicated to the water the surface begins to dip at the centre and to rise at the edge of the bucket. This is clearly an effect of the rotation of the water. Newton argued that the water is not affected by the relative motion of the bucket, but only by its own absolute rotational motion. Berkeley was the first to criticise this argument, pointing out that it merely indicates the effect of the water's rotation relative to the earth. Mach later generalised this criticism by regarding the rotation as relative to the universe as a whole.

Mach's argument at this point is worth considering in greater detail. By considering purely dynamical ideas, we have deduced that our basic inertial frame is one which is fixed relative to the most distant galaxies. Mach suggested that this is not a pure coincidence but that, somehow, distant mass may determine the inertial properties of local mass. Thus Newton's bucket experiment may demonstrate the effect of a rotation relative to the distant distribution of mass. In Mach's view, rotation is relative in this sense and is not necessarily absolute. He pointed out that an appropriate test of this suggestion would be to fix a bucket of water and rotate the Universe! If the surface of the water remains flat this would be consistent with Newtonian classical dynamics, but if the surface became concave it would prove that all rotations are in some sense relative.

Nowadays we refer to Mach's principle as the principle that the inertial properties of particles are related to the distribution of matter at large distances. This principle can be regarded as letting the universe define the properties of space rather than considering the universe merely to move in space. Mach's ideas had a considerable influence on Einstein and the general theory of relativity manifests some aspects of Mach's principle. The classical theory of dynamics, however, does not satisfy Mach's principle and rotation is always regarded as absolute. This is not a serious criticism though, since the principle obviously can not be tested experimentally.

5

Newtonian gravitation

The theory of gravitation really starts from a consideration of the motion of bodies near the earth's surface. It is observed that all bodies have a tendency to accelerate vertically downwards unless they are prevented from doing so by impressed forces acting on them. This acceleration is explained in classical dynamics as the effect of a hypothetical gravitational field which exerts a force on any body that is placed in the field. The general theory of gravitation has, however, been deduced largely from a consideration of the motion of heavenly bodies. This is the approach that is followed here.

5.1. Kepler's laws

The motion of the planets seems to have fascinated many past civilisations. Over many centuries the wanderings of the planets across the sky have been recorded. However, as far as the subject of this book is concerned it is appropriate to start with the work of Johannes Kepler. Kepler had at his disposal the mass of observations recorded by previous generations. In particular, he had access to the detailed observations of Tycho Brahe. Tycho's work was remarkable in the degree of accuracy to which he attained, though his observations like those of his predecessors were all made with the naked eye. It is interesting to notice that Kepler published his first two laws in 1609, the same year as Galileo developed the telescope.

Kepler's work is significant in that he introduced a new theory relating to planetary motion. The ancient Ptolemaic system of epicycles could have been further developed to fit the observations of Tycho Brahe. However, Kepler preferred to introduce a completely new theory of his own, and it is this theory which paved the way for the Newtonian theory of gravitation.

Kepler's theory of planetary motion is usually presented in the form of three laws. However, to be consistent with the approach of the rest of this book we must describe Kepler's laws as assumptions. It must be emphasised that they are not foundational assumptions of classical dynamics. They are foundational assumptions of Kepler's theory of planetary motion. They can be stated as follows.

***Kepler's assumption* 1.** *The planets follow elliptic orbits about the sun as focus.*

***Kepler's assumption* 2.** *The radius vector drawn from the sun to a planet sweeps out equal areas in equal times.*

***Kepler's assumption* 3.** *The squares of the periodic times of the different planets are proportional to the cubes of their major axes.*

Following the standard terminology, these will be referred to respectively as Kepler's laws.

The theoretical nature of these laws is obvious. They are hypotheses put forward in order to be tested. Kepler himself had tested them using Tycho's observations and had found them to be approximately correct. Subsequently, however, more accurate observations, which have been made possible by the development of the telescope, have falsified this theory. But it is still of considerable use as an approximate theory.

It is remarkable that Kepler's laws are just of sufficient accuracy for the development of the theory of gravitation. A less accurate Ptolemaic theory, or even a more accurate theory involving further perturbations, would have considerably hindered the development of any such theory. It is the identification of basic elliptic orbits and the later discovery of the third law which historically form the starting point for the dynamical theory of orbital motion and gravitation.

It should first be pointed out that, as far as a planet's motion about the sun is concerned, the planet can be represented as a particle. Although planets are very large their dimensions are very small when compared with the dimensions of their orbits. It is therefore appropriate to represent them as particles of equivalent mass. Further justification of this assumption is given in sections 7.3 and 6.5.

According to the classical theory of dynamics, planetary motion must satisfy Newton's law of motion for a particle. Now, planets do not move with uniform motion in a straight line. They must therefore be subject to forces. There are, however, no impressed forces that appear to be acting. There is no string attaching them to the sun. Nor are heavenly creatures observed pushing the planets along. Nor is it possible to find a frame of reference in which all planets move with constant velocity, so that the forces that accelerate planets are not due to relative accelerations or rotations. The planets, however, are continuously accelerating, and in order to explain this in terms of classical dynamics it is necessary to invent a new type of force which is called gravitation. This fictitious gravitational force is invented purely to satisfy the equations of motion. It is defined therefore as the product of the mass and the acceleration of each planet.

Kepler's first law implies that the motion of a planet is in a plane. The gravitational force therefore has no component perpendicular to the plane of motion.

It is now convenient to introduce polar coordinates in the orbital plane with origin at the sun. The gravitational force acting on a planet is now given by

$$\boldsymbol{F} = m\ddot{\boldsymbol{r}} = m(\ddot{r} - r\dot{\theta}^2)\boldsymbol{e}_r + \frac{m}{r}\frac{\mathrm{d}}{\mathrm{d}t}(r^2\dot{\theta})\boldsymbol{e}_\theta$$

where $\boldsymbol{r} = r\boldsymbol{e}_r$ is the position vector of the planet.

Now Kepler's second law states that the radius vector sweeps out equal areas in equal times, or that area is swept out at a constant rate. In the small time δt the area swept out is given by $\frac{1}{2}r^2\delta\theta$, and the second law therefore implies that $r^2\dot{\theta}$ is a constant. It follows immediately from this that the gravitational force has no angular component in the plane. This is exactly equivalent to stating that the angular momentum of the planets about the sun is constant. Gravitational forces must therefore be purely radial in direction.

It is convenient to put

$$r^2\dot{\theta} = h$$

which is the angular momentum per unit mass of a planet, or twice the rate at which area is swept out by its radius vector. The gravitational force acting on a planet can now be reduced to the form

$$\boldsymbol{F} = m\left(\ddot{r} - \frac{h^2}{r^3}\right)\boldsymbol{e}_r$$

Further use can now be made of Kepler's first law. In order to determine the gravitational force, we must substitute for r in the above expression, and for this we use the assumption that the planets move in elliptic orbits. Now the polar equation for an ellipse is

$$\frac{l}{r} = 1 + e\cos\theta$$

where e and $l = a(1-e^2)$ are constants determining its size and shape. The parameter e is called the eccentricity of the ellipse, l the semilatus rectum and a the semimajor axis. The polar angle θ is referred to by astronomers as the true anomaly of the orbit. Using the polar equation with $r^2\dot{\theta} = h$ we obtain that

$$\ddot{r} = \frac{h^2 e\cos\theta}{lr^2}$$

and substituting this into the above expression gives

$$\boldsymbol{F} = -m\frac{\mu}{r^2}\boldsymbol{e}_r$$

where

$$\mu = h^2/l$$

The coefficient μ is clearly a constant for any planetary orbit since h and l are constants. The gravitational force on a planet has thus been shown to be a force of attraction towards the sun which varies inversely as the square of the distance from it.

It has been shown above that the rate at which the radius vector sweeps out area is $h/2$, and in terms of μ this is given by $\frac{1}{2}\sqrt{(\mu l)}$ or $\frac{1}{2}\sqrt{(\mu a(1-e^2))}$. Now the total area of an ellipse is given by $\pi a^2\sqrt{(1-e^2)}$. From these expressions the periodic time τ of a planetary orbit can be written as

$$\tau = 2\pi\sqrt{\left(\frac{a^3}{\mu}\right)}$$

It is now appropriate to introduce Kepler's third law, which states that τ^2 is proportional to a^3 for all planetary orbits. This therefore implies that μ is a constant for all planetary orbits. It is therefore concluded that all planets are subject to the same law of attraction to the sun expressed by

$$\boldsymbol{F} = -\frac{m\mu}{r^2}\hat{\boldsymbol{r}}$$

where m is the mass of the planet and μ is a constant. This conclusion has been reached purely on the assumption of Kepler's laws, with the requirement that the motion of planets satisfies Newton's equation of motion for a particle.

The constant μ has been defined above in terms of the angular momentum per unit mass of the planet and the semilatus rectum of the planet's orbit. It may appear surprising that μ is a constant, but this is clearly required by the theory. It therefore follows that the angular momentum of the planet is related to the semilatus rectum of its orbit. This is in fact just one example of the fact that dynamical properties are related to the geometrical properties of orbits. Another example that can easily be demonstrated is that the energy of a planet determines not only the type of orbit but also the length of its major axis.

5.2. An intermediate theory of planetary motion

We must now consider the above law of planetary motion in greater detail. It has been concluded that all planets accelerate towards the sun with an acceleration which is inversely proportional to the square of their distance from the sun. We must now ask what causes this acceleration. The answer given by the theory of classical dynamics is that it must be a fictitious force. But the mechanism of such a force has not been explained. The classical theory is that somehow it is the sun itself that causes this acceleration. As to how the sun does this, no conclusive answer is given. But somehow it is understood that the sun generates a force field such that if any mass is placed in the field it is caused to accelerate towards the sun with an acceleration which is inversely proportional to the square of its distance from it.

The theory of planetary motion suggested above is a very useful theory. It is approximately in agreement with observations to within the accuracy achieved by Tycho Brahe. It also compares favourably with the Ptolemaic theory that Tycho preferred. But it must now be considered in a larger context. The theory must be shown to be consistent with classical dynamics. It is also appropriate to try to generalise this theory to include other systems involving orbital motion.

First, the above theory of planetary motion must be made compatible with assumption 3.3. This assumption, which is Newton's third law, states that where two particles interact the mutual forces acting upon each other are always equal in magnitude and act in opposite directions. In the model of the solar system used above, the planets and the sun were represented as particles. It was then suggested that there is some kind of interaction between the sun and the planets in which the planets are subject to forces induced by the sun. Clearly, this hypothesis can only be made consistent with assumption 3.3 if it is also assumed that the sun is acted on by a set of equal and opposite forces that are directed towards the various planets.

It has now been assumed that each planet is attracted to the sun with a force $\boldsymbol{F}_{ps}$ given by

$$\boldsymbol{F}_{ps} = -\frac{m_p\mu}{r_{ps}^3}\,\boldsymbol{r}_{ps}$$

where m_p is the mass of the planet which has position vector $\boldsymbol{r}_{ps}$ relative to the sun. To make this consistent with assumption 3.3 it must also be assumed that the sun is attracted to each planet with a force $\boldsymbol{F}_{sp}$ given by

$$\boldsymbol{F}_{sp} = -\frac{m_p\mu}{r_{sp}^3}\,\boldsymbol{r}_{sp}$$

where $\boldsymbol{r}_{sp}$ is the position of the sun relative to the planet.

Now it is appropriate to introduce a degree of symmetry between these formulae by putting

$$\mu = Gm_s$$

where G is a new constant and m_s is the mass of the sun. The theory now developed states that the planets and the sun attract each other with a force whose magnitude is equal to Gm_sm_p/r_{sp}^2.

It is necessary to consider how closely this modified theory of planetary motion is in agreement with observations. It can be seen that if the mass of the sun were very much larger than that of the planets, then it would accelerate very little compared to the accelerations of the planets. In fact this theory is almost the same as that of Kepler, provided the sun is regarded as having a sufficiently large mass. It is therefore supported by the same observational evidence. However, further modifications are required before the theory can be made consistent with accurate modern observations.

5.3. Newton's theory of universal gravitation

It is now appropriate to consider the question as to what it is in the sun that gives rise to its associated force field, which is of such a character that any particle situated in the field is caused to accelerate towards the sun. No definitive answer needs to be given, but a consideration of the question does lead to some interesting suggestions that may be used to generalise the theory.

Whatever it is that causes the force field is clearly not unique to the sun. It is also common to the planets since it has been assumed that the sun is also attracted to each planet. But if it is assumed that a planet also has an associated force field, then this implies that any particle situated in such a field must be caused to accelerate towards the planet. In fact, such accelerations are clearly observed. Several planets have moons orbiting about them. Each moon accelerates towards its particular planet. Also, it is well known that all particles near the earth accelerate towards it unless they are prevented from doing so by impressed forces. It therefore seems reasonable to conclude that the gravitational forces which attract bodies to the earth are similar in essence to the forces that attract planets to the sun.

The classical argument at this point is that given by Newton. It is known that the squares of the periodic times of the moons of Jupiter are proportional to the cubes of their major axes. Jupiter's moons therefore satisfy Kepler's third law but with a different constant of proportionality. This therefore lends support to the suggestion that the forces keeping Jupiter's satellites in orbit are qualitatively the same as the forces causing the planets to orbit about the sun. Exactly the same argument applies to the moons of Saturn. It is observed that the squares of the periodic times are proportional to the cubes of the major axes. Unfortunately, the earth only has one moon. However, when its motion is compared with the motion of a projectile near the earth's surface the same law is found to hold approximately.

It has now been argued that the sun somehow induces a force field that causes the planets to accelerate towards it. In order to be consistent with assumption 3.3, the planets must also be considered to have associated force fields through which they attract the sun. These force fields associated with planets are also required to explain the motion of their moons and of particles moving near their surface. Further, in order to be consistent with assumption 3.3, such moons and particles must also be considered to induce force fields and thus to attract their respective planet, and of course any other particles in their vicinity.

We may now return to the question of what it is that gives rise to the force fields. The answer given classically is that it is a new quantity known as the *gravitational mass.* Stars, planets, moons and smaller bodies are said to have gravitational mass. Gravitational masses are said to induce force fields that attract other bodies that have gravitational mass. However, the formula derived above for the gravitational attraction between the sun and the planets indicates that the mutually induced gravitational force is proportional to the masses of the sun and the planets. The greater the mass of the planet the greater the force it exerts on the sun. The force field is proportional to the mass of the body generating it. The term 'gravitational mass' was in fact specifically coined to suggest a connection between it and mass, as defined in chapter 3. The exact relation between gravitational and inertial masses will be considered later. At this stage it may simply be conjectured that the force field exerted by a body is directly related to its mass.

If this suggestion that the gravitational field exerted by a body is related to its mass is correct, then all bodies that have mass must induce a gravitational force field. It also follows that the gravitational force between any two massive bodies is qualitatively similar to the gravitational force between the sun and a planet. The term massive is used here in a technical sense to describe any body that has mass, however small. The conclusion of the above argument may now be summarised.

Assumption 5.1. *Any two particles attract each other with a force which is proportional to their masses, and inversely proportional to the square of the distance between them.*

This is known as Newton's law of universal gravitation. It can be represented by the formula

$$\boldsymbol{F}_{12} = -\frac{Gm_1m_2}{r_{12}^2}\hat{\boldsymbol{r}}_{12}$$

where $\boldsymbol{F}_{12}$ is the force on one mass m_1 due to the second m_2, $\boldsymbol{r}_{12}$ is the position of the first mass relative to the second and G is known as the universal constant of gravitation.

It is important to notice that this law was not obtained purely by a flash of inspiration as an apple fell from a tree. Rather, it was the result of a considerable amount of thought. The thought processes that gave rise to it were partly speculative and partly logical. They were not purely logical in the sense that the theory follows logically and directly from observations. It does not do so. On the other hand, the theory is not the result of pure speculation that ignores observations and related theories.

The nature of the law of gravitation, however, is ultimately speculative. It is put forward as a hypothesis to be tested. The above argument that preceded the statement of the law is therefore logically irrelevant. Assumption 5.1 is an axiom of classical dynamics that stands or falls only according to its agreement with observations in the real world. However, the argument that preceded its statement is useful in that it illustrates the method by which fictitious forces are considered in the theory of classical dynamics. It also introduces the new concept of gravitational mass, and it is a consideration of this concept that suggests further experiments that can be used to test the theory.

It should also be pointed out here that Newton's theory of gravitation is a generalisation of previously suggested theories. In fact, it is a generalisation both of Kepler's theory of planetary motion and of Galileo's theory of free fall. It has been developed here from the former, but it can easily be shown that Newton's original development of the theory was also considerably influenced by the latter.

5.4. Observational evidence

Newton's law of universal gravitation has been very extensively tested during the last 300 years, and its predictions have been found to be in very close agreement with physical observations. On small-scale experiments the predictions of the theory are well within the estimated errors of the experiments. Also, on a large scale, predictions have been confirmed to an amazingly high degree of accuracy.

It is found that any two bodies appear to attract each other with a force given by Newton's formula in which G is always the same constant. This constant has now been evaluated in the laboratory as

$$G = (6.674 \pm 0.004) \times 10^{-11}\ \mathrm{m^2/(s^2\,kg)}$$

According to the theory there exists a mutual attractive force between any two massive bodies. The mutually attractive forces between the sun and each planet have already been considered. Such forces induce the planets to follow elliptic orbits about the sun. However the theory also predicts the existence of mutually attractive forces between the planets themselves. Such forces would cause each planet to deviate from its basic elliptic orbit about the sun. Now, such deviations are in fact observed. Modern observations are far more accurate than those considered by Kepler, and these observations are in excellent agreement with the predictions of Newton's theory of universal gravitation in which each planet is considered as being attracted to the sun and also to all the other planets.

There is, however, one apparent difficulty that occurs in celestial mechanics. This arises from the fact that it is not possible to directly measure the mass of the sun or the planets, or even the earth. Methods exist for measuring the distances between the planets, so that from their apparent size their volume may be estimated. But there is no accurate direct way of measuring their mass. In practice, the masses of the sun and planets are determined entirely by their gravitational effects, and this of course involves the assumption of Newton's law of universal gravitation. It is therefore not possible to directly verify Newton's law at this scale. There is no way of measuring the constant G for such interactions. All that is possible is to estimate the consistency of the Newtonian theory with observations.

This apparent difficulty, however, is compensated for in a remarkable way. The masses of celestial bodies are determined by measuring the acceleration of other bodies moving in their gravitational fields. It is therefore G times the mass of the attracting body that is determined. This quantity can in fact be determined with considerable accuracy. For example, G times the mass of the sun (μ) is known to eight significant figures, whereas G is only known to three significant figures. The mass of the sun can therefore only be estimated to the accuracy with which G can be measured in the laboratory. However, this does not matter since it is always the product of G with mass that is required in calculations of orbital motions. Thus, although their masses may not be known very accurately, predictions of the motions of celestial bodies can still be made with very great theoretical accuracy.

The predicted perturbations of the planetary orbits are in almost exact agreement with observations. The remarkable degree of accuracy achieved by this theory can be seen by considering the case in which the greatest apparent discrepancy occurs, namely the orbit of the planet Mercury.

The elliptic orbit of Mercury appears to be rotating about the sun. The perihelion of the orbit appears to advance at a rate of 5599.74 ± 0.41 seconds of arc per century. Using Newtonian theory 5025.64 ± 0.50 seconds of this can be explained as the effect of using a noninertial frame of reference based on the earth. An advance of a further 531.54 ± 0.68 seconds is predicted as the perturbations on the orbit due to the presence of the other planets, chiefly Venus, earth and Jupiter. This, however, leaves an advance of 42.56 ± 0.94 seconds of arc per century still to be explained. Although this residue is extremely small it cannot be explained by Newtonian theory. The oblateness of the sun for example is not sufficient to explain an advance of this magnitude. It can therefore be concluded that the Newtonian theory of gravitation is incorrect. The remaining advance in the perihelion of Mercury can in fact be explained by Einstein's general theory of relativity. The orbits of other planets and asteroids that pass relatively close to the sun also exhibit a similar, though smaller, residual advance of perihelion which can be explained in the same way. However, it should be emphasised that the discrepancy between Newtonian theory and observations is extremely small.

In the early nineteenth century, the most distant planet known was Uranus. However it was also known that Uranus was not following its predicted orbit very accurately. This naturally led to the suggestion that the Newtonian theory was not accurate over such large distances. However, there was also the possibility that the orbit of Uranus might be perturbed by the gravitational field of a further undiscovered planet. Adams and Leverrier independently made this latter assumption and, from the known positions of Uranus and Newton's theory of gravitation, were able to predict the rough position of such a planet. The planet Neptune was thus discovered in 1846 at very nearly its predicted position.

The perturbation of the orbit of Neptune, in its turn, was observed to be greater than that predicted by Newtonian theory. This led to the prediction of a further planet whose position was roughly estimated. Such a planet, now called Pluto, was discovered in 1930 in very nearly the exact position predicted. It should be pointed out though that, at the time, the perturbations of Neptune were not known with sufficient accuracy to predict the orbit of Pluto. In addition, it has subsequently been found that, unless it is peculiarly nonreflective, or is abnormally dense, Pluto does not have sufficient mass to have caused the supposed perturbations on the basis of which its existence and position were originally predicted. Nowadays most astronomers appear to believe that the discovery of Pluto was somewhat fortuitous.

The discovery of these new planets was a spectacular success for Newton's theory of universal gravitation. It was therefore natural to suggest that the unexplained perturbations of the orbit of Mercury might be due to the presence of another planet orbiting closer to the sun. This possible new planet was called Vulcan. However, no such planet has ever been observed, in spite of a considerable search for it. It has therefore had to be concluded that the perturbations of the orbit of Mercury are unexplainable in terms of the Newtonian theory of gravitation.

The situation may now be summarised by stating that Newton's theory of universal gravitation provides an extremely accurate model for the world as it is observed. In most areas in which it is applicable, its predictions are well within experimental accuracy, even for the highest precision that is at present attainable. However, for planetary orbits that are closest to the sun there is observed a slight deviation from the predictions of Newtonian theory.

5.5. Gravitational and inertial mass

It is clear that the definition of mass given in chapter 3, which may be described as a measure of a body's natural resistance to being accelerated and may be called inertial mass, is very different from the concept of gravitational mass defined in this chapter. The gravitational mass of a body is a measure of its power to attract other bodies and appears to have no logical connection with its power to resist being accelerated. Gravitational mass and inertial mass are unrelated concepts.

In classical dynamics, however, the concepts of gravitational and inertial masses are directly related. This is a direct consequence of applying assumption 3.3 or Newton's third law to gravitational forces. This can be seen by considering two isolated particles moving only under the action of their mutual gravitational fields. Denote their inertial masses by m_1 and m_2 and their gravitational masses by M_1 and M_2 respectively. The first particle is assumed to induce in the second an acceleration proportional to M_1, since its gravitational mass M_1 is a measure of its power to accelerate other bodies towards it. Also the force acting on the second particle is assumed to be equal to m_2 times its acceleration, since its inertial mass is a measure of its resistance to being accelerated by a force. Thus the force acting on the second particle must be proportional to m_2M_1. Similarly, the force acting on the first particle must be proportional to m_1M_2. Now, if the forces on the two particles are equal and opposite in accordance with assumption 3.3, then m_1M_2 must be equal to m_2M_1 and so

$$\frac{m_1}{m_2}=\frac{M_1}{M_2}$$

This states that the ratio of inertial masses is the same as the ratio of gravitational masses. It has already been shown that mass ratio is the only objective definition for mass, and so gravitational mass must be proportional to inertial mass. The constant of proportionality can be absorbed into the constant G, and in classical dynamics the gravitational and inertial masses of a body are equated.

The above argument is purely theoretical and based on the foundational assumptions of classical dynamics. The theoretical model is consistent if the magnitudes of the gravitational and inertial masses of a body are equal. This result is, however, also found to be consistent with observations and physical experiments within the experimental errors.

Essentially, the method of testing the equivalence of gravitational and inertial mass is to observe how bodies with different inertial mass accelerate in the same gravitational field. The theory predicts that different bodies should fall with the same acceleration. This is what is said to have been observed by Galileo in the apocryphal Leaning Tower experiment. Such experiments have actually been carried out with great precision by Eötvös and more recently by Dicke. In order to fully test the theory different materials are used as well as different masses, and since different materials have different proportions of protons and neutrons any difference in the acceleration of elementary particles could also have been deduced. In fact, no difference in acceleration has ever been observed down to an accuracy of 1 part in 10^{11}.

5.6. The general relativistic correction

Einstein's general theory of relativity is a mathematically complicated theory which describes gravitation in terms of the curvature of space–time. The theory predicts that a small particle moving in the spherically symmetric gravitational field of a large mass would have an acceleration given by

$$\ddot{\mathbf{r}} = -\frac{\mu}{r^2}\left(1+\frac{3h^2}{c^2r^2}\right)\hat{\mathbf{r}}$$

where μ is G times the mass of the attracting body, h is the angular momentum per unit mass of the particle relative to the attracting body, and c is the speed of light. It can be seen that general relativity predicts an inverse square law of acceleration and an additional inverse fourth law of acceleration which is related to the particle's angular momentum.

The first term in the above expression is in exact agreement with Newtonian theory and the second term is negligibly small in most observable situations. The predictions of Einstein's theory for orbital motion are therefore in almost exact agreement with the predictions of classical Newtonian theory. There are in fact very few observed situations in which the additional term is significant. However, it is significant for planetary orbits that pass close to the sun, and can be used to explain the advance of the perihelion of Mercury which is observed to be in contradiction to the Newtonian theory.

The above expression for gravitational attraction is in agreement with all known observations. Although it has in fact been derived from Einstein's theory of relativity there is no reason why it should not be incorporated into the classical theory of dynamics. It is perfectly reasonable to include the factor $(1+3h^2/c^2r^2)$ in the statement of the universal law of gravitation given here as assumption 5.1. The reason why it was initially excluded is purely historical. The original theory was based on Kepler's laws, which are consistent with the inverse square law term only. However, now that more accurate observations are available the inverse fourth law term may well be included. It is therefore appropriate to include this term in the law of gravitation.

In practice, however, the additional term is so small that it can almost always be neglected. It is therefore more appropriate to leave the statement of assumption 5.1 as it stands, as this is the form in which it is always used. On the rare occasions when this term becomes significant it may always be included as an additional fictitious force. The two terms are treated separately, and so it is convenient to describe the second term in an additional assumption.

***Assumption* 5.2.** *In addition to the attractive forces mentioned in assumption* 5.1, *any two bodies with masses m_1 and m_2 attract each other with a force given by*

$$\boldsymbol{F}_{12} = -\frac{3Gm_1m_2h_{12}^2}{c^2r_{12}^4}\hat{\boldsymbol{r}}_{12}$$

where G is the universal constant of gravitation, $\boldsymbol{r}_{12}$ is the position of m_1 relative to m_2, h_{12} is the angular momentum per unit mass of one mass about the other with respect to an inertial frame of reference, and c is the speed of light.

This additional force is referred to as the *general relativistic correction.*

The most appropriate way to include the general relativistic correction is to include it as suggested as an additional fictitious force. It could alternatively have been included as a modification of the gravitational constant. In this form G could simply have been replaced by $G(1+3h^2/c^2r^2)$ or, equivalently, the gravitational mass of a body could be regarded as the factor $(1+3h^2/c^2r^2)$ times its inertial mass.

It is now well known that the theory of classical dynamics has been falsified. But it is important to notice that it is not the theory of gravitation which has proved the point of failure. The theory of universal gravitation can be rewritten, as suggested, with the additional term. In this form it is consistent with all currently known observations. The points of failure for classical dynamics are only its theory of relativity and its inappropriateness on the atomic scale.

5.7. Weight

The term weight has been defined in so many different ways and used with such confusion that it is probably best never to use the term in a scientific context. However, since the aim of this book is to explain the foundational concepts of classical dynamics it is appropriate to include a discussion of the various concepts that are sometimes covered by the vague term weight.

This term was in common use long before the theories of dynamics were developed. It presumably arose initially as a quantitative measure of bulk. As trade developed, items came to be valued in terms of an equivalent weight of gold or silver. Thus weight was a measure of what we now call mass. In fact, the noun form of the word still frequently carries this meaning. For example, the weights of a balance are more properly called masses.

Difficulties occur, however, when it is found necessary to measure a weight, or rather to assign a weight value to an object. The old method of using the balance type of scales is simply a method of comparing and equating masses. It relies on the fact that the earth's gravitational field is virtually constant in the region of the balance, and then compares the gravitational forces on the two masses. From this it is a simple step to the spring-balance method of determining weight. It is found by suspending different masses from a spring that its extension is related to the mass suspended. The spring can therefore be calibrated, and an unknown mass can be estimated by the extension it induces in the spring. It is essentially the gravitational force acting on the mass that is measured.

Unfortunately, these two methods of determining weight are not exactly equivalent, at least not as they are commonly used. A set of scales with its associated set of standard weights can always be used to compare masses anywhere in the world. But if a spring balance is transported to a different latitude or altitude it may come to assign different weights to the same mass. The reason for this is that, whereas the scales merely compare the gravitational forces acting on two masses, the spring balance directly measures the gravitational force acting on a mass, and this may vary slightly over the surface of the earth. The variation can of course be overcome by recalibrating the scale at every place, but this is never done in practice.

Over the centuries the term weight has come to be regarded as the gravitational force that acts on a body, rather than its mass. But it is regarded as the gravitational force as it might be measured. This is the apparent gravitational force, rather than that which is actually described by the theory of gravitation. A difference here may be due, for example, to the earth's rotation and its related centrifugal forces. One important effect of this is that the earth is not basically spherical, but has what is called an equatorial bulge. At any point on the earth's surface the concept of a vertical is defined in terms of a stationary plumb-line. This is found to be perpendicular to the plane defined by the horizon or equivalent sea level, which is called the horizontal plane. But because of the earth's oblateness caused by its rotation, the vertical at any point on the surface is not necessarily in the direction of the earth's centre. Thus, the apparent gravitational force which can be measured as acting on a body is in the direction of the downward vertical, whereas the actual gravitational force is directed more towards the earth's centre, even after allowing for the possible nonspherical distribution of mass.

It can therefore be seen that if, as according to some textbooks, the weight of an object is defined as the gravitational force acting on it, then this is not consistent with the common usage of the word. The common usage is that weight is the observed fictitious force acting on a mass as would be measured by a spring balance in the understood frame of reference. Thus an astronaut in an earth satellite is said to be 'weightless'. Now, he is certainly not outside the earth's gravitational field. The gravitational force acting on him is only slightly less than it would be on the earth's surface. The point is that both the satellite and the astronaut are subject to gravitation, and that both are in a state of 'free fall'. In fact, they are continually accelerating towards the earth, and since they have a large orbital velocity the acceleration curves the orbit around the earth. The term weightless is used because the astronaut would register zero weight on a spring balance inside the satellite. It is therefore inappropriate to define weight scientifically as the gravitational force acting on a body unless it is clearly emphasised that this is contrary to the common usage of the term.

The confusion over the term weight has been further compounded by the attempts of some scientists in the past to define it as the force supporting the body against gravitation, instead of the apparent gravitational force itself. They would thus define weight as the force exerted by a spring balance on the body, rather than the apparent force of gravitation acting on it. For them, weight is a force acting upwards!

Now, it has been known since the time of Galileo that all objects would accelerate towards the earth's centre with an equal and approximately constant acceleration unless they are prevented from doing so by impressed forces. This can also be deduced as an approximation from Newton's theory of gravitation. This acceleration can be measured quite accurately at any point on the earth's surface and is usually denoted by the symbol $\mathbf{g}$. It should be noticed that this is a vector quantity whose direction is vertically downwards. The theory then states that any particle in this vicinity having mass m is acted on by a gravitational force given by $m\mathbf{g}$. This quantity is then frequently defined as the particle's weight. This defines weight as the apparent gravitational force, since $\mathbf{g}$ has actually been physically measured, or at least estimated, near the location of the motion. However, even this definition as the apparent gravitational force acting on a body is not always consistent with common usage and therefore invites confusion. For example, many people think that, because an astronaut in orbit is said to be weightless, he is necessarily outside the effect of the earth's gravitational field.

In view of the different meanings of the term weight which are used, and the confusion caused by them, it is probably wiser not to use the term at all in a scientific way. The terms gravitational force, apparent gravitational force, or simply $m\mathbf{g}$, can all be unambiguously defined and it is these concepts that are required in the theory of dynamics.

6

Particle dynamics

In this chapter a theoretical framework is developed within which the motion of a particle may be discussed. Such motion is assumed to satisfy the law of motion, as stated in proposition 3.2 and expressed by the equation

$$m\ddot{\boldsymbol{r}} = \boldsymbol{F}$$

It is assumed that a frame of reference is specified, and that if this is not an inertial frame, then the fictitious forces mentioned in proposition 4.1 have to be included. The theoretical results obtained in this chapter are consequences of the law of motion for a stated frame of reference.

When the theory is required to make predictions of the motion of a particle, the method used is based on a guess or estimate of the forces that would be acting on it. These forces are substituted into the law of motion, which then becomes a second-order differential equation. Since this is a vector equation it is equivalent to a set of ordinary differential equations of order six. Now, in order to make predictions of the motion, it is necessary to solve these equations subject to appropriate boundary conditions. However, only rarely is it easy to solve these equations analytically. Of course, numerical solutions may be obtained, but analytic solutions are usually preferred. If complete analytic solutions cannot be obtained, even a knowledge of certain first integrals can be of great value. In the abstract study of particle dynamics, therefore, considerable attention is given to the derivation of possible first integrals of the equation of motion. This emphasis is reflected in this and subsequent chapters.

It is also appropriate here to introduce and define a number of terms and concepts which are not only of use in particle dynamics, but which are also useful in an analysis of more complicated systems.

6.1. Kinetic energy, work and the activity equation

Let us start with the definitions of work and power.

***Definition* 6.1.** *The work done by a force acting on a particle is the product of the force and the distance moved in the direction of that force.*

Denoting work by the symbol W, the small amount of work δW done by a force $\boldsymbol{F}$ when a particle moves through a small distance $\delta\boldsymbol{r}$ is given by the scalar product

$$\delta W = \boldsymbol{F}\cdot\delta\boldsymbol{r}$$

When a particle moves along a path C, the work done by the force $\boldsymbol{F}$ is the sum of the work done by that force in each small step along the path. This is given, in the limit, by the line integral

$$W = \int_C \boldsymbol{F}\cdot \mathrm{d}\boldsymbol{r}$$

The value of this quantity depends of course on the particular path C, as well as on the position of its end points.

***Definition* 6.2.** *The power exerted by a force is the rate at which work is done by that force.*

This can be described by the equation

$$\text{Power} = \frac{\mathrm{d}W}{\mathrm{d}t} = \boldsymbol{F}\cdot\boldsymbol{v}$$

The concept of power is particularly useful in the application of mechanics to engineering problems, but does not form an integral part of the theoretical discussions described in this book.

When a particle is acted on by a number of forces, the power and work done by each force can be calculated separately. However, it is instructive also to consider the total work done by all the forces which are acting on the particle. In this case the equation of the motion can be applied.

Consider the total work done when a particle moves between any two points $\boldsymbol{r}_1$ and $\boldsymbol{r}_2$ along a particular path. In this case the motion is actually caused by the net force $\boldsymbol{F}$. Using the equation of motion, the total work done can be expressed as

$$W = \int_{\boldsymbol{r}_1}^{\boldsymbol{r}_2} \boldsymbol{F}\cdot\mathrm{d}\boldsymbol{r} = \int_{\boldsymbol{r}_1}^{\boldsymbol{r}_2} m\ddot{\boldsymbol{r}}\cdot\dot{\boldsymbol{r}}\,\mathrm{d}t$$
$$= [\tfrac{1}{2}mv^2]_{\boldsymbol{r}_1}^{\boldsymbol{r}_2}$$

The total work done is therefore equal to the change in the quantity $\frac{1}{2}mv^2$, which is called the kinetic energy.

***Definition* 6.3.** *The kinetic energy of a particle is one-half of the product of its mass and the square of its velocity.*

Kinetic energy is denoted by the symbol T. It is a nonnegative scalar quantity given by

$$T = \tfrac{1}{2}mv^2$$

The above result can now be stated as a proposition.

***Proposition* 6.1.** *The increase in the kinetic energy of a particle in any interval is equal to the total work done by the forces acting on it in that interval.*

The central step in the proof of this proposition is the use of the equation of motion in the integrand. The result can therefore be considered as a first integral of the equation of motion. The existence of this integral is of great importance in classical dynamics, as will be shown later. However, it should be noticed at this point that if the motion of a particle is such that a component force does no work, then that component does not contribute to the change in the kinetic energy.

The above result may also be differentiated with respect to time to give the interesting result that power is equal to the rate of change of kinetic energy

$$\boldsymbol{F} \cdot \boldsymbol{v} = \frac{\mathrm{d}}{\mathrm{d}t} T$$

This is known as the *activity equation.*

6.2. Irrotational fields

At this point it is appropriate to consider in greater detail the general concept of a vector field. It is also necessary to state a few of the basic results of the mathematical subject of vector field theory that will be required in the following section.

The general concept of a vector field is one which assigns vectors to all points in a region of space. A vector field is thus simply a vector function of position. Such a concept has many applications. For example, in the motion of a liquid the velocity of small elements of the liquid at any time can be represented by a vector field. At any point the velocity is uniquely determined. In steady flow the vector field of velocity is a function of position only. In unsteady flow it is also a function of time.

The mathematics of vector fields is also particularly applicable in an analysis of fictitious forces and of the motion of bodies being acted on by them. It has been shown in the previous chapter that the concept of gravitation gives rise to a field theory. It has been postulated that any massive body generates a gravitational field. This field is such that, if another mass were situated in it, then that mass would be attracted to the first by a gravitational force. At all points in an appropriate region a unique vector can be defined which is equal to the force that would be exerted on a particle of unit mass if it were placed at that point. Gravitational forces can thus be analysed by introducing vector fields defined with the aid of such hypothetical test particles.

In the previous section, the work done in moving a particle from one point to another was expressed as a line integral involving the forces that act on the particle. Without going into the methods for evaluating line integrals, certain properties of line integrals will be required later and it is convenient to introduce them separately here.

Consider a general vector field and denote it by the symbol $\boldsymbol{A}$. In order to include the most general case the vector field may at this stage also be assumed to be time dependent. Thus

$$\boldsymbol{A} = \boldsymbol{A}(\boldsymbol{r}, t)$$

Now consider the line integral of $\boldsymbol{A}$ between any two arbitrary points $\boldsymbol{r}_1$ and $\boldsymbol{r}_2$ in the field, along a particular path between them. This can be expressed as

$$\int_{\boldsymbol{r}_1}^{\boldsymbol{r}_2} \boldsymbol{A} \cdot \mathrm{d}\boldsymbol{r}$$

where it is assumed that the line integral is evaluated along the particular path. In this case it is also assumed that the line integral is evaluated at an instant of time.

In general, the value of this line integral depends on the position of the end points $\boldsymbol{r}_1$ and $\boldsymbol{r}_2$, and also on the particular path taken between them. In this general case it may also depend on time. However, for certain types of vector field it is found that the value of the line integral is always independent of the path chosen. In this case the same value of the line integral is obtained along any arbitrarily chosen path between the same end points. Now, it turns out that many of the vector fields that are considered in practical applications are of this type. In particular, these include some of the most common force fields that are considered in classical dynamics. These will be introduced in the following sections.

It is appropriate to consider this particular class of vector fields in greater detail. It is therefore now assumed that $\boldsymbol{A}$ is such that its line integral is independent of the path between any two arbitrarily chosen points in the field. By considering two different paths between any two points, it can be seen that in this case the line integral around any closed path must necessarily be zero. This result can be stated in the form

$$\oint_C \boldsymbol{A} \cdot \mathrm{d}\boldsymbol{r} = 0$$

for all closed curves C, and at any instant of time.

Since the above argument can also be reversed, it is possible to use the above property to characterise the particular type of vector field that is being considered. Fields of this type are described as irrotational, since a closed line integral in some applications is related to some kind of rotation. This term may now be defined.

***Definition* 6.4.** *If the line integral of a vector field around a closed curve is zero at all times and for all curves in the field, then the field is said to be irrotational.*

The result used in the above argument is well known in vector field theory and may now be stated in the following way.

***Lemma* 6.1.** *The line integral between any two points in a vector field is independent of the path between them if, and only if, the field is irrotational.*

It can also be shown that, if the value of the line integral between any two points is independent of the path taken, then this value can be expressed as the difference between the values of some scalar function evaluated at the end points. Denoting such a function by ϕ, the result referred to can be stated in the form that, for any irrotational field $\boldsymbol{A}(\boldsymbol{r}, t)$, there exists a scalar field $\phi(\boldsymbol{r}, t)$ such that

$$\int_{\boldsymbol{r}_1}^{\boldsymbol{r}_2} \boldsymbol{A} \cdot \mathrm{d}\boldsymbol{r} = \phi(\boldsymbol{r}_2, t) - \phi(\boldsymbol{r}_1, t)$$

for any two points $\boldsymbol{r}_1$ and $\boldsymbol{r}_2$ in the field.

Since in this case the line integral is evaluated at an instant of time, it follows that the integrand is a perfect differential.

$$\boldsymbol{A} \cdot \mathrm{d}\boldsymbol{r} = \mathrm{d}\phi$$

From this it also follows that the vector field is the gradient of the scalar field $\phi(r, t)$.

$$A(r, t) = \text{grad } \phi$$

The gradient can be considered in some ways as the vector derivative $d\phi/dr$. In cartesian coordinates it can be expressed in the form

$$\text{grad } \phi(x, y, z, t) = i\frac{\partial\phi}{\partial x} + j\frac{\partial\phi}{\partial y} + k\frac{\partial\phi}{\partial z}$$

The argument outlined above is well known in vector field theory, and can also easily be reversed. The result is summarised in the following statement.

***Lemma* 6.2.** *A vector field is irrotational if, and only if, it can be expressed as the gradient of a scalar field.*

It can immediately be seen that an irrotational vector field can be determined uniquely from its associated scalar field. Thus if the scalar field only is given, the vector field can be deduced. The scalar field is therefore referred to as the *potential function* for the vector field.

If a constant is added to a scalar field, it has no effect on its gradient. The potential function introduced above is thus defined only up to an arbitrary additive constant. Its magnitude at any point can therefore have no objective meaning apart from some convention by which this constant is determined.

The results described here are general and can be used in any application of vector field theory. In the following sections they will be applied particularly to the classical analysis of particle dynamics.

6.3. Conservative fields and potential energy

In the subject of classical dynamics, a number of the vector fields that are considered are found to be irrotational. For example, all gravitational fields are irrotational in the sense defined in the previous section.

It is convenient here to restrict attention to force fields that do not depend on time. If such a field is also irrotational, then it is called conservative. The reason for this term will be explained in the next section. Conservative fields are of great importance in classical dynamics and it is appropriate to give them a clear definition. With applications in view, this definition is best given in the following way.

Definition 6.5. *If the work done by a force on a particle when it is moved between any two points is independent of the path taken, then the force field is said to be conservative.*

The term conservative is applied only to force fields. It can immediately be seen from lemma 6.1, and the fact that the work done is evaluated by a line integral, that conservative fields are necessarily irrotational force fields.

There is, however, another more subtle difference between definitions 6.4 and 6.5. Conservative fields are defined in terms of the work done by a force on a particle which moves. The motion of the particle necessarily takes time. In contrast to this, irrotational fields are defined in terms of line integrals which are evaluated at an instant. Irrotational fields can therefore be time dependent. On the other hand, the motion of a particle along different paths may take different times. The particle may also be moved slowly or quickly along the same path. Thus, in order for the work done to be the same in each case, conservative fields must necessarily be independent of time.

Having clarified this point, lemma 6.2 can be used to give the important result that, *a force field is conservative if, and only if, it can be expressed as the gradient of a time independent scalar field.* In other words, a necessary and sufficient condition that a force field $\boldsymbol{F}(\boldsymbol{r})$ is conservative is that there should exist a scalar field $\phi(\boldsymbol{r})$ such that $\boldsymbol{F} = \text{grad}\ \phi$.

The function $\phi(\boldsymbol{r})$ is a potential function for the conservative field. However, in practice it is found to be more convenient to introduce the concept of a potential energy instead. This can be defined as follows.

Definition 6.6. *The potential energy of a particle situated at a point in a conservative field is the work that would be done by the force on the particle if it were moved from that point to some standard position.*

Denoting the potential energy of a particle by V and the standard position by $\boldsymbol{r}_0$, the definition can be expressed in the mathematical form

$$V(\boldsymbol{r}) = \int_{\boldsymbol{r}}^{\boldsymbol{r}_0} \boldsymbol{F}(\boldsymbol{r}') \cdot \mathrm{d}\boldsymbol{r}'$$

To investigate the properties of this potential energy, it is convenient to consider the work done by a conservative force $\boldsymbol{F}(\boldsymbol{r})$ in moving a particle between two arbitrary points $\boldsymbol{r}_1$ and $\boldsymbol{r}_2$.

$$W = \int_{\boldsymbol{r}_1}^{\boldsymbol{r}_2} \boldsymbol{F} \cdot \mathrm{d}\boldsymbol{r}$$

Since the force is conservative, the path can be chosen arbitrarily, and it is convenient to choose a path through the standard point $\boldsymbol{r}_0$. Thus,

$$W = \int_{\boldsymbol{r}_1}^{\boldsymbol{r}_0} \boldsymbol{F} \cdot \mathrm{d}\boldsymbol{r} - \int_{\boldsymbol{r}_2}^{\boldsymbol{r}_0} \boldsymbol{F} \cdot \mathrm{d}\boldsymbol{r}$$

Using the above definition of potential energy, this can be expressed in the form

$$\int_{\boldsymbol{r}_1}^{\boldsymbol{r}_2} \boldsymbol{F} \cdot \mathrm{d}\boldsymbol{r} = V(\boldsymbol{r}_1) - V(\boldsymbol{r}_2)$$

Since this result applies to any two arbitrary points $\boldsymbol{r}_1$ and $\boldsymbol{r}_2$, an argument of the previous section can be repeated to show that

$$\boldsymbol{F} = -\text{grad } V$$

This is an important result which applies to any conservative field. It clearly indicates that a suitable potential function $\phi(\boldsymbol{r})$ for the force field can be taken to be $-V(\boldsymbol{r})$.

***Proposition* 6.2.** *In any conservative field the force is equal to minus the gradient of its associated potential energy.*

The negative sign that appears here was introduced implicitly through definition 6.6. The reason for it is purely a matter of convenience, as will be shown in the next section.

Another point to notice in passing is that the potential function for any irrotational field is only defined up to an arbitrary additive constant. This arbitrariness is also reflected in the definition of potential energy where it appears through the arbitrary choice of some standard position. Once this position has been agreed, the constant is specified. However, it is often convenient in practice not to specify the standard position, in which case an arbitrary additive constant is included in the potential energy. This constant, or the standard position, can always be chosen for convenience at a later stage if that is necessary.

This section started with a consideration of conservative forces which are necessarily time independent. However, it is sometimes necessary to consider time-dependent force fields. These are not conservative in the usual sense, but it is sometimes still possible to express them in terms of the gradients of appropriate scalar functions.

For a time-dependent force field, it can be seen from lemma 6.2 that the vanishing of the line integral

$$\oint_C \boldsymbol{F} \cdot \mathrm{d}\boldsymbol{r} = 0$$

for all closed curves C, and at any instant of time, is a necessary and sufficient condition for there to exist a scalar field $V(\boldsymbol{r}, t)$ such that

$$\boldsymbol{F} = -\mathrm{grad}\ V(\boldsymbol{r}, t)$$

It must, however, be stressed that, in spite of obvious similarities, irrotational force fields are not necessarily conservative. This can be seen since the evaluation of a line integral at an instant of time bears no relation to the evaluation of the work done in moving a particle along a path. The latter process necessarily takes a finite time during which the field may change. This can also be seen by noticing that the differential of a time dependent potential function can be expanded in the form

$$\mathrm{d}V = -\boldsymbol{F} \cdot \mathrm{d}\boldsymbol{r} + \frac{\partial V}{\partial t}\mathrm{d}t$$

From this it can be seen that the integrand in the expression for the work done on a particle is only a perfect differential if the force and its associated potential function are independent of time.

Force fields that are explicitly time dependent are not conservative. However, if they are irrotational they can still be expressed in terms of a potential function $V(r, t)$, and this can sometimes still be interpreted as the potential energy of a particle situated in the field.

A further extension of this approach is required when classical dynamics is generalised to include electrodynamics. In this case it is convenient to introduce the concept of a velocity dependent potential function from which the force acting on a charged particle can be determined. Thus it is sometimes necessary to consider force fields whose properties can be described by a single scalar function of position, time and velocity. The term *monogenic* has been introduced by Lanczos to describe such forces. The associated scalar potential may simply be described as a *generalised potential function.* Within the traditional subject area of classical dynamics, however, this generalisation is not usually required. In this book, therefore, potential functions are usually assumed to be at most functions of relative positions. However, an additional time dependence is sometimes included.

6.4. The energy integral

Bearing the results of the previous sections in mind, it is now appropriate to reconsider the motion of a particle in a conservative force field $\boldsymbol{F}(\boldsymbol{r})$. It may initially be assumed that this is the only force that acts on the particle, so that its possible motions are largely specified.

Consider any particular path of motion, and two arbitrary points $\boldsymbol{r}_1$ and $\boldsymbol{r}_2$ on it. According to proposition 6.1, the work done by the force on the particle in moving it between these points is equal to the kinetic energy that the particle gains in the process.

$$\int_{\boldsymbol{r}_1}^{\boldsymbol{r}_2} \boldsymbol{F}\cdot \mathrm{d}\boldsymbol{r} = T(\boldsymbol{r}_2) - T(\boldsymbol{r}_1)$$

But, since the field is conservative, the work done can also be expressed in terms of the difference in the potential energy of the particle at the two points

$$\int_{\boldsymbol{r}_1}^{\boldsymbol{r}_2} \boldsymbol{F}\cdot \mathrm{d}\boldsymbol{r} = V(\boldsymbol{r}_1) - V(\boldsymbol{r}_2)$$

Thus $T(\boldsymbol{r}_1)+V(\boldsymbol{r}_1)=T(\boldsymbol{r}_2)+V(\boldsymbol{r}_2)$ and, since the two points were chosen arbitrarily along a particular motion, this sum must be constant at all points on that motion. In this case the general relation has been obtained that along any particular motion

$$T+V=\text{const}$$

This is a very important result. It is referred to as the energy integral. However, before stating it formally, it must be considered in greater detail so that it can be expressed in its most general form.

The assumptions of the previous paragraph may now be relaxed, and it is convenient to consider the motion of a particle which is acted on by both conservative and nonconservative forces. The conservative forces can be denoted by the field $\boldsymbol{F}_1(\boldsymbol{r})$, and the sum of the nonconservative forces by $\boldsymbol{F}_2$. The motion of the particle is determined by both of these forces through the equation

$$m\ddot{\boldsymbol{r}} = \boldsymbol{F}_1 + \boldsymbol{F}_2$$

Again consider a particular motion and two arbitrary points $\boldsymbol{r}_1$ and $\boldsymbol{r}_2$ on it. Then

$$\int_{\boldsymbol{r}_1}^{\boldsymbol{r}_2} m\ddot{\boldsymbol{r}}\cdot \mathrm{d}\boldsymbol{r} = \int_{\boldsymbol{r}_1}^{\boldsymbol{r}_2} \boldsymbol{F}_1\cdot \mathrm{d}\boldsymbol{r} + \int_{\boldsymbol{r}_1}^{\boldsymbol{r}_2} \boldsymbol{F}_2\cdot \mathrm{d}\boldsymbol{r}$$

The work done by the conservative forces can of course be expressed in terms of a potential function $V(\mathbf{r})$. It is conventional to regard this as the potential energy of the particle, even though it is also acted on by other forces. The above integrals can thus be evaluated in the form

$$T(\mathbf{r}_2)-T(\mathbf{r}_1)=V(\mathbf{r}_1)-V(\mathbf{r}_2)+\int_{\mathbf{r}_1}^{\mathbf{r}_2}\mathbf{F}_2\cdot\mathrm{d}\mathbf{r}$$

Therefore, along any particular motion, *the gain in kinetic energy is equal to the loss in potential energy plus the work done by the nonconservative forces.*

It can immediately be deduced from this result that, in all situations in which the nonconservative forces do no work, the sum of the kinetic and potential energies is constant for any particular motion. This approach therefore leads to a more general statement of the energy integral than that originally presented. This result may now be stated formally.

***Proposition* 6.3.** *The sum of the kinetic energy and potential energy of a particle is constant throughout a motion if, and only if, it moves in such a way that the only forces which do work are conservative.*

Because of the importance of this proposition, and in order to clarify the above point, it is appropriate to consider some examples. For instance, in the motion of a simple pendulum the tension in the string does no work since the motion of the bob is always perpendicular to the string. The only other force acting in this case is the gravitational force, which is conservative. A similar situation occurs when a body rolls over a surface. In this case the rolling condition implies that the reactional and frictional forces that may be acting do no work, and the system again may be conservative. In both of these examples the sum of the kinetic and potential energies is a constant in any particular motion.

It may have been noticed that the proof of proposition 6.3 has involved an integration of the equation of motion. The equation $T+V=\text{constant}$ is therefore a first integral of the equation of motion under these conditions. It is for this reason that it is known as the energy integral.

If the motion of a particle is such that the sum of its kinetic and potential energies is constant, it is said that energy is conserved. It is for this reason that the forces which do work in this case are described as being 'conservative'. Indeed, any dynamical system in which the total kinetic and potential energy is conserved is referred to as being 'conservative'.

It can now also be seen why the negative sign was included in the definition of potential energy. It seems more natural to regard the sum of two kinds of energy as being constant, rather than their difference. The potential function is used with the negative sign in this case in order that $T+V$ may be a constant.

In the broader development of science, a more general principle of conservation of energy has proved to be of great use. According to such a principle, energy cannot be created or destroyed, but can be transformed from one type to another. It is then necessary to consider numerous types of energy – nuclear, chemical, thermodynamic, electromagnetic, etc. Kinetic and potential energies, which arise purely from a body's motion and position, are added to this list. This approach enables an intuitive understanding of the concept of a conservative system to be developed. A conservative system is one in which the kinetic and potential energies are not being transformed into other types of energy. For example, when a stable particle moves in a vacuum it may be that the only transfer of energy is between kinetic and potential. Such a system would be conservative. On the other hand, a body sliding over the surface of another body generates heat in the surfaces in contact. In this case, kinetic or potential energy is being transformed into heat and $T+V$ is not a constant. In fact, systems in which any form of resistance occurs are the most common nonconservative systems that are considered in classical dynamics.

Although proposition 6.3 can thus be related to a more general concept of conservation of energy, this has little value in the study of classical dynamics, other than to give an intuitive idea of whether a given system is conservative or not. In classical dynamics the main significance of proposition 6.3 is as a first integral of the equation of motion for cases in which the forces that do work are conservative. It is this approach that is used in this book, and therefore the philosophical problem of ascribing objective meanings to the concepts of kinetic and potential energy is avoided.

So far in this section, force fields have usually been assumed to be conservative. However, it is convenient also to consider the slightly more general case that arises when the forces that do work can be expressed as the gradient of a time-dependent potential function $V(\boldsymbol{r}, t)$. In this case the activity equation can be written in the form

$$\frac{\mathrm{d}}{\mathrm{d}t}(T+V)=\frac{\partial V}{\partial t}$$

for which no simple integral exists. This result again emphasises the point that a time-dependent force field is not conservative.

Throughout this chapter it is assumed that a frame of reference is specified and that if this is not an inertial frame, then the additional fictitious forces mentioned in proposition 4.1 have to be included in the equation of motion. It is necessary therefore to indicate whether or not these forces are conservative.

The fictitious force $-m\ddot{\boldsymbol{r}}_0$, which appears when the origin is accelerating relative to an inertial frame, may or may not be conservative according to the application. In any particular situation it is not difficult to determine whether or not this force is conservative, and if so then a suitable potential function may be determined for it.

The centrifugal force can always be derived from a potential function, since

$$-m\boldsymbol{\omega}\times(\boldsymbol{\omega}\times\boldsymbol{r})=-\text{grad}\,\{-\tfrac{1}{2}m(\boldsymbol{\omega}\times\boldsymbol{r})^2\}$$

The term $-\frac{1}{2}m(\boldsymbol{\omega}\times\boldsymbol{r})^2$ may therefore be regarded as the potential energy of a particle associated with the centrifugal force. This force, however, is only strictly conservative when the angular velocity of the frame $\boldsymbol{\omega}$, is a constant.

The Coriolis force, on the other hand, is not conservative. However, since it always acts perpendicular to the instantaneous displacement, it does no work and therefore does not affect the possible existence of an energy integral.

Finally, the fictitious force $-m\dot{\boldsymbol{\omega}}\times\boldsymbol{r}$, which arises from the angular acceleration of the frame of reference relative to an inertial frame, is never conservative.

It can therefore be seen that, for the motion of a particle in a conservative force field, relative to a frame of reference which is rotating with constant angular velocity relative to an inertial frame, an energy integral can still be obtained, provided the potential energy is modified to take into account the centrifugal force. For example, in the circular restricted three-body problem that is considered in celestial mechanics, such a term is included in Jacobi's integral, which is in fact just the energy integral. However, if the angular velocity of the reference frame is not constant, then the fictitious forces are time dependent and an energy integral does not exist.

6.5. Gravitational fields

In the previous sections we have been developing a theoretical framework for the study of particle dynamics, and the concepts of conservative fields and potential energy have been introduced in their most general form. However, in practice, the most common conservative fields that are considered are those of gravitation. In fact, in the applications of the theory of classical dynamics, it is only very rarely that conservative forces other than those of gravitation need to be included, whereas gravitational forces appear more often than not. It is therefore appropriate to include here a number of results associated with the concept of gravitation.

To start with, consider the force acting on a particle of mass m situated at a position defined by the vector $\boldsymbol{r}$, due to the gravitational field of a particle of mass m_1 and position $\boldsymbol{r}_1$. According to the law of gravitation, stated here as assumption 5.1, this force is given by the expression

$$\boldsymbol{F} = -\frac{Gmm_1}{|\boldsymbol{r}-\boldsymbol{r}_1|^3}(\boldsymbol{r}-\boldsymbol{r}_1)$$

It can easily be seen that this can be written as

$$\boldsymbol{F} = -\text{grad}\left\{-\frac{Gmm_1}{|\boldsymbol{r}-\boldsymbol{r}_1|}\right\}$$

and therefore a suitable potential function for this force field may be taken to be

$$V = -\frac{Gmm_1}{|\boldsymbol{r}-\boldsymbol{r}_1|}$$

This may be interpreted as the potential energy of the mass m due to the presence of the mass m_1.

Before proceeding further, two comments need to be made on the above calculation. Firstly, an arbitrary constant can be added to the potential energy without altering the value of the force field. However, it may intuitively be felt that the potential energy of the mass m is somehow due to the presence of the mass m_1, and that if this mass was not present or if it was removed to infinity, then m would have no potential energy due to m_1. Such an assumption is convenient, though not mathematically necessary. It is therefore conventional to omit the additional arbitrary constant, and to adopt the above expression for the potential energy, since this can be seen to vanish in limit as $|\boldsymbol{r}-\boldsymbol{r}_1|\to\infty$.

It should be noticed that the potential energy defined above is a function of both $\boldsymbol{r}$ and $\boldsymbol{r}_1$. If the vector $\boldsymbol{r}_1$ is constant, then it can clearly be seen that the gravitational field is conservative. On the other hand, if the mass m_1 is moving and its position vector is a known function of time, then the potential function defined above depends on time explicitly. In such a case the force field $\boldsymbol{F}$ is not conservative in the sense defined above. However, it is irrotational at any instant of time, so that the potential function still exists and can still be interpreted as the potential energy of the mass m.

Let us now move on to consider the gravitational field induced by a number of masses. Following the above notation, consider the force on a particle of mass m situated at a point with position vector $\boldsymbol{r}$, due to the gravitational fields of n particles of mass m_i $(i = 1, \ldots, n)$, each respectively having position vectors $\boldsymbol{r}_i$. According to assumption 3.5 the forces induced by each particle are independent of each other, and the total gravitational force is obtained by summing over all particles.

$$\boldsymbol{F} = -\sum_{i=1}^{n} \frac{Gmm_i}{|\boldsymbol{r}-\boldsymbol{r}_i|^3}(\boldsymbol{r}-\boldsymbol{r}_i)$$

Since each component of this sum is irrotational, it follows that the resultant force field is also irrotational, and the total potential energy of the mass m is simply the sum of the potential energies due to all the other particles. This can also be seen, since it is possible to write

$$\boldsymbol{F} = -\text{grad } V$$

where

$$V = -\sum_{i=1}^{n} \frac{Gmm_i}{|\boldsymbol{r}-\boldsymbol{r}_i|}$$

Having obtained these general results, let us now consider the special case of the gravitational field of a spherically symmetric mass distribution. Let the position vector $\boldsymbol{r}'$ denote the centre of the distribution, let m' denote its total mass, and let $\boldsymbol{F}$ be the force it induces on a particle of mass m outside the distribution. Then, using techniques of integration, it can be shown that

$$\boldsymbol{F} = -\text{grad } V$$

where

$$V = -\frac{Gmm'}{|\boldsymbol{r}-\boldsymbol{r}'|}$$

The procedure that is used to obtain this result is simply to consider the distribution of mass to be composed of a set of thin concentric spherical shells. It can then easily be shown that, if mass is uniformly distributed over any such shell, it would induce a force of the type stated above on a particle of mass m outside of the shell.

It can immediately be seen from the discussion above that the gravitational field induced by a spherically symmetric mass distribution is identical to that of a single particle of equal mass. This result is of far-reaching importance. For example, in celestial mechanics it provides further justification for the representation of the sun and planets as classical particles. Not only are these bodies small compared to the size of their orbits, but since they are approximately spherical their gravitational fields are almost identical to those of their particle representation.

Also, the above result is valid however close the particle comes to the spherical mass distribution. Thus in terrestrial mechanics, since the earth is approximately spherical, the gravitational field acting on a particle near the earth's surface is approximately the same as that of a single particle situated at the earth's centre and with mass equal to that of the earth. This result is particularly important since it means that the gravitational force acting on such a particle may be considered to be dominated by the gravitational field of the earth as a whole rather than by the local distribution of mass in its vicinity. This therefore introduces considerable simplification.

As a further consequence, it can be seen that at any point near the earth's surface the magnitude of the earth's gravitational field is approximately constant, and is directed approximately towards the centre of the earth. Thus, for a particle which moves such that its distance from a point on the earth's surface is small compared to the radius of the earth, the total gravitational force acting on it is approximately constant. And, since this force is also proportional to the mass of the particle itself, it is convenient to write this force as

$$\boldsymbol{F} = m\boldsymbol{g}$$

where $\boldsymbol{g}$ may be interpreted as the acceleration induced in the mass by the gravitational field. According to the above discussion, $\boldsymbol{g}$ must be approximately a constant vector, and must be aligned with the downward vertical. This result is in agreement with observations, and it is found that the magnitude of $\boldsymbol{g}$ is approximately $9.813\ \mathrm{ms}^{-2}$.

In the study of motion near a point on the earth's surface it is often convenient to introduce cartesian coordinates at that point, defined such that the z axis is in the vertically upwards direction. Relative to such a frame, the gravitational force acting on a particle situated near the origin is given by

$$\boldsymbol{F} = -mg\boldsymbol{k}$$

Such a force is conservative, and is consistent with the potential function

$$V = mgz + c$$

where c is an arbitrary constant. This expression may be interpreted as the gravitational potential energy of the particle.

6.6. Momentum and angular momentum

Let us start here with a definition

***Definition* 6.7.** *The linear momentum of a particle is the product of its mass and velocity.*

Linear momentum is thus the vector quantity $m\boldsymbol{v}$. It can easily be seen by differentiating this quantity with respect to time that the law of motion is equivalent to the statement that *the rate of change of linear momentum of a particle is equal to the sum of the forces which act on it.*

If no forces act on a particle, then its linear momentum is a constant of the motion. No undue significance should be attached to this result. It is merely a consequence of the definition and of assumption 3.2.

If the forces acting on a particle are such that the net force in any particular direction is zero, then the component of linear momentum in that direction is a constant. In such a case the linear momentum in that direction is said to be conserved. This result is a first integral of the equation of motion and may be very useful in some applications. Again, however, this result is not unexpected. It merely states that if a particle is not subjected to a force in a particular direction, then it does not accelerate in that direction.

Another concept that arises in classical dynamics is that of angular momentum. This arises principally in a consideration of the rotational motion of a rigid body, but it also arises in particle dynamics in central force problems. It is therefore convenient to introduce this concept at this stage. However, it is first necessary to define the moment of a vector.

***Definition* 6.8.** *The moment about a point O of a vector* $\boldsymbol{A}$ *which acts at a point with position vector r relative to O is the vector product* $\boldsymbol{r}\times\boldsymbol{A}$.

***Definition* 6.9.** *The angular momentum of a particle about a point is the moment of its linear momentum about that point.*

If we denote the position of a particle at some time relative to a given point by $\boldsymbol{r}$, its velocity by $\dot{\boldsymbol{r}}$ and its angular momentum by $\boldsymbol{L}$, then

$$\boldsymbol{L}=\boldsymbol{r}\times m\dot{\boldsymbol{r}}$$

With this definition it appears that the concept of angular momentum is in some ways parallel to that of linear momentum when the moments of forces, or couples, are considered instead of the forces themselves. This similarity is demonstrated in the following proposition.

***Proposition* 6.4.** *The rate of change of angular momentum of a particle about a point O is equal to the sum of the moments about O of the forces which act on it.*

This proposition can be written in the form

$$\begin{aligned}\frac{\mathrm{d}}{\mathrm{d}t}\boldsymbol{L}&=\boldsymbol{r}\times\boldsymbol{F}\\&=\boldsymbol{M}\end{aligned}$$

where $\boldsymbol{M}$ is the sum of the moments of the forces which act on the particle. This result can be obtained by differentiating the above expression for $\boldsymbol{L}$ and using the equation of motion. Formally it is nothing more than the moment of the equation of motion. As such it is a direct consequence of that equation, but contains less information since the component in the direction of $\boldsymbol{r}$ is removed. It is therefore not surprising that the concept of angular momentum is not of great importance in particle dynamics. It becomes of great significance, however, when systems of more than one particle are considered.

It can be deduced from the above proposition in the cases in which no forces act, or in which the net force is always directed towards or away from the centre O, that the angular momentum about O is a constant of the motion. Angular momentum is then said to be conserved. As for linear momentum, this also is merely a consequence of the definition and the equation of motion.

It can also be shown by taking the scalar product of the above equation with a constant vector that, if there is no couple about a particular axis through O, then the angular momentum of the particle about that axis is a constant of the motion. In such a situation this result is a first integral of the equation of motion, and may be of great use in some applications.

6.7. Impulsive motion

One further topic needs to be introduced in this chapter, namely that of impulsive motion. This is a concept that arises naturally; for example, when a ball is kicked, or an object falls on the floor, or when one particle collides with another. In such situations the motion of a body is drastically changed by the action of a large force acting for a very short time. During that time interval the laws of motion are assumed to hold, but the detailed analysis is very complicated. The dominating force is usually not conservative, and it usually causes the body to deform so that elastic forces also come into play. In such situations it is convenient to introduce the concept of an impulse. We say that an impulsive force, or simply an impulse, occurs when a force acts for a very small time interval.

Consider first the equation of motion $m\ddot{\boldsymbol{r}} = \boldsymbol{F}$, and integrate this with respect to time over a time interval $t_1 \leq t \leq t_2$ to obtain

$$m\boldsymbol{v}(t_2) - m\boldsymbol{v}(t_1) = \int_{t_1}^{t_2} \boldsymbol{F}\, \mathrm{d}t$$

The left-hand side of this equation is the change in linear momentum over the time interval, and the right-hand side is the time integral of the forces acting over that interval. This integral, however, can only be evaluated if the total force is given explicitly as a function of time only. In the general case the force would also be expected to be a function of the position of the particle, and in this case the value of this integral could not be obtained.

Now, return to the problem of impulsive motion, and consider the motion of a particle subjected to a large force for a very short time. According to the above equation, it experiences a change in momentum equal to the integral of the force over the time interval. It is this time integral of the force which is called its impulse.

Definition* 6.10.** ***The impulse of a force is its integral with respect to time over the interval in which it acts.

The above equation, which is the basic equation for the impulsive motion of a particle, may now be stated in the form of a proposition.

***Proposition* 6.5.** *When a particle is acted on by a large force for a short time interval, then it experiences a change in its linear momentum equal to the impulse of that force.*

It should be pointed out that there may well be other forces acting on the particle during the instant of the impulse. However, if the time interval of the impulse is sufficiently short, then these other forces will have negligible effect during this interval and so may be omitted. After the impulse, however, they must again be included in order to determine the subsequent motion of the particle.

Although an impulse has been defined in terms of the time integral of a force, it is not usually practical to evaluate it in this way. However, this does not matter since an impulse, like a force, is estimated by its effects. If a force acts on a particle for a long time interval, then it is measured by the acceleration of the particle, and in applications the motion of a particle can be predicted on the basis of an assumed force. If, however, the force only acts for a small time interval, then it can be measured by the change of momentum caused, and in applications the motion of a particle can be predicted on the basis of an assumed impulse.

In situations, therefore, in which a particle is acted on by an impulse, it is not necessary to know the details of how the force acts; it is only necessary to know its net effect. In fact, the actual magnitude of the force at any instant is irrelevant. It is therefore possible in a mathematical model to replace the actual force by any other that has the same net effect. The actual impulse can even be replaced in a model by the mathematical idealisation of an infinite force acting for an infinitesimal time period, provided that its time integral is bounded and equal to the impulse of the actual force.

This result is very important. It means that in a mathematical model for the motion of a particle subjected to an impulsive force the actual impulse can be replaced by an infinite force acting at an instant. When a particle is acted on by such a force, it experiences a change in momentum equal to the required impulse. However, since the time during which the force acts is infinitesimal, the distance the particle moves during this instant is also infinitesimal. It can therefore be seen that the position of the particle $\boldsymbol{r}(t)$ is continuous across the instant of the impulse, whereas the momentum, and hence the velocity $\dot{\boldsymbol{r}}(t)$, is discontinuous. In classical dynamics an impulse is always considered in this idealised sense.

Clearly, the replacement of impulsive forces by infinite instantaneous forces in a theoretical model is only an approximation to what actually happens in the real world. However, it seems to be a reasonable approximation in situations in which the motion of a body is considered over a time interval which is considerably larger than the time interval of the actual impulse.

Finally, consider the moment of the equation of motion for particle about a fixed point and integrate it over a short time interval $t_1 \leqslant t \leqslant t_2$,

$$\int_{t_1}^{t_2} \boldsymbol{r} \times m\ddot{\boldsymbol{r}}\, \mathrm{d}t = \int_{t_1}^{t_2} \boldsymbol{r} \times \boldsymbol{F}\, \mathrm{d}t$$

It can be seen that the left-hand side is just the change in the angular momentum about the origin. The change in angular momentum of a particle is therefore equal to the time integral of the moment of the force acting on it.

Now, in the case of impulsive motion it is appropriate to consider the above idealisation in which the time interval is considered to be infinitesimal. The position of the particle is considered to be continuous across the instant of the impulse. So integrating across this instant gives

$$\boldsymbol{L}(t_2) - \boldsymbol{L}(t_1) = \boldsymbol{r} \times \int_{t_1}^{t_2} \boldsymbol{F}\, \mathrm{d}t$$

This result can be stated as a proposition.

***Proposition* 6.6.** *When a particle is subject to an impulse, its change in angular momentum about a given point is equal to the moment of the impulse about that point.*

In this section the results have been obtained that for impulsive motion the changes in the linear momentum and the angular momentum of a particle are equal to the impulse and the moment of the impulse respectively. It is perhaps appropriate to conclude here by pointing out that these results are parallel to those of the previous section. There it was shown that the rate of change of the linear momentum and the angular momentum of a particle are equal to the force and the moment of the force respectively.

6.8. Standard examples

The purpose of this section is to illustrate how the theory of particle dynamics developed above can be applied to a number of typical problems.

Example* 6.1: *the simple pendulum

A simple pendulum can be represented as a particle of mass m suspended from a fixed point O by a light inelastic string of length l. If the Coriolis force due to the earth's rotation is ignored, it can be shown that the motion of the particle remains in a fixed vertical plane. It is also convenient initially to ignore any resistance to the motion.

While the string is taut, the particle is constrained to move along the arc of a circle with centre at O. Its position at any time can then be uniquely defined by the parameter θ, which is the inclination of the string to the downward vertical. This is illustrated in figure 6.1.

It has now been assumed that the only forces which act on the particle are the gravitational force, and the tension in the string. The gravitational force is clearly conservative. The tension, however, is not conservative, but since this force does no work in the motion of the particle, the model as a whole is conservative and an energy integral can be obtained. Using the notation defined in example 2.2, the unit tangent and normal to the motion are $\boldsymbol{e}_\theta$ and $-\boldsymbol{e}_r$ respectively. The velocity of the particle can thus be written as

$$\boldsymbol{v} = l\dot{\theta}\boldsymbol{e}_\theta$$

Then, taking the potential energy relative to the horizontal plane through O, the energy integral can be expressed as

$$\tfrac{1}{2}ml^2\dot{\theta}^2 - mgl\cos\theta = \text{const}$$

The derivative of this equation is also important, and can be written in the form

$$\ddot{\theta} = -\frac{g}{l}\sin\theta$$

Fig. 6.1. The simple pendulum.

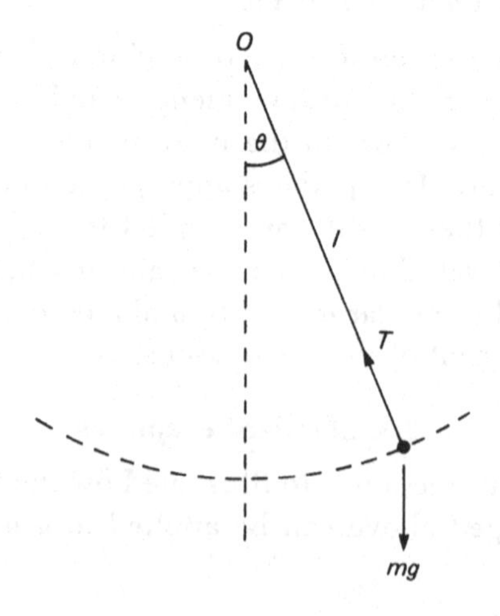

It can immediately be seen from the latter equation that, if θ remains small, the motion is periodic with frequency approximately equal to $\sqrt{(g/l)}$.

It is also instructive to consider the full equation of motion, which can be written in the form

$$-ml\dot{\theta}^2\mathbf{e}_r + ml\ddot{\theta}\mathbf{e}_\theta = mg(\cos\theta\mathbf{e}_r - \sin\theta\mathbf{e}_\theta) - T\mathbf{e}_r$$

The tangential component of this equation is the derivative of the energy integral. It can therefore be seen that the energy equation used above, in this case, is the first integral of this component. The normal component, on the other hand, simply gives an equation for the tension in the string

$$T = mg\cos\theta + ml\dot{\theta}^2$$

This can now be expressed in terms of θ only, using the energy integral.

Appropriate boundary conditions, however, have to be defined first. For example, if the pendulum is set in motion by initially drawing it aside through an angle θ_0 and releasing it from rest, then the energy integral implies that

$$l\dot{\theta}^2 = 2g(\cos\theta - \cos\theta_0)$$

On the other hand, if it were set in motion by projecting it from the position of stable equilibrium with speed V, then the energy integral would imply that

$$l\dot{\theta}^2 = V^2/l - 2g(1 - \cos\theta)$$

Using the latter condition, the tension in the string is given by

$$T = mV^2/l + mg(3\cos\theta - 2)$$

It may immediately be noticed from this expression that the tension becomes zero if a value of θ is reached at which

$$\cos\theta = -\tfrac{1}{3}\left(\frac{V^2}{gl} - 2\right)$$

At such a point the string would become slack and the particle would fall away from its otherwise circular path. Such a point could only be reached, however, if the initial velocity were sufficiently great that the bob would rise above the level of O. The energy integral implies that this only occurs if $V^2/l > 2g$. On the other hand, the critical value of θ given by the above expression is only real if $V^2/gl < 5$. Initial velocities greater than those satisfying this inequality would generate complete circular motion, and the tension would remain positive, even at the top of the circle. Thus it may be concluded that the string would become slack at a value of θ between $\pi/2$ and π if $2gl < V^2 < 5gl$.

Example 6.2: inverse square law orbits

Consider the motion of a particle of mass m which is acted on only by an inverse square law of attraction to a point O. Using an inertial frame of reference with origin at O the equation of motion can be written as

$$m\ddot{\mathbf{r}} = -\frac{m\mu}{r^2}\hat{\mathbf{r}}$$

where μ is a constant. It can be seen from chapter 5 that this model can be used to describe the approximate motion of a planet or asteroid about the sun, or of a satellite or moon about a planet. In such applications μ is equal to the constant of gravitation G times the mass of the attracting body.

Since the force acting on the particle is parallel to its position vector, it can immediately be seen using proposition 6.4 that the angular momentum $\boldsymbol{L}$ of the particle about O is constant. But, since $\boldsymbol{L}=\boldsymbol{r}\times m\dot{\boldsymbol{r}}$, the direction of $\boldsymbol{L}$ is perpendicular to the instantaneous plane of motion. It follows that the particle remains in a fixed plane through O, which is perpendicular to the direction of the angular momentum vector.

It is now convenient to use polar coordinates in the plane of motion. The equation of motion can then be written as

$$m(\ddot{r}-r\dot{\theta}^2)\boldsymbol{e}_r+\frac{m}{r}\frac{\mathrm{d}}{\mathrm{d}t}(r^2\dot{\theta})\boldsymbol{e}_\theta=-\frac{m\mu}{r^2}\boldsymbol{e}_r$$

The angular component of this equation immediately implies that $r^2\dot{\theta}$ is constant. As in section 5.1, this can be written as

$$r^2\dot{\theta}=h$$

and h can be interpreted as the angular momentum per unit mass L/m, thus also confirming that the magnitude of the angular momentum vector is a constant of the motion.

Using this first integral to eliminate $\dot{\theta}$, the equation of motion now reduces to the single equation

$$\ddot{r}-\frac{h^2}{r^3}=-\frac{\mu}{r_2}$$

Although this equation is nonlinear, it can be linearised by introducing a reciprocal coordinate $u=1/r$, and by changing the independent variable to θ using the above first integral. It then takes the form

$$\frac{\mathrm{d}^2u}{\mathrm{d}\theta^2}+u=\frac{\mu}{h^2}$$

whose general solution can be written as

$$u=\frac{\mu}{h_2}(1+e\cos(\theta-\omega))$$

where e and ω are arbitrary constants of integration.

It is clearly possible to measure θ from that point on the orbit which is closest to O. With this choice $\omega=O$, and θ is known as the true anomaly. It is also convenient to put $h^2/\mu=l$, and the polar equation of the orbit of the particle takes the form

$$\frac{l}{r}=1+e\cos\theta$$

This is in fact the polar equation of a family of curves known as 'conic sections', where l is called the semilatus rectum and e the eccentricity. When

$e=0$ the orbit is a circle with centre at O. When $O<e<1$ the orbit is an ellipse with O located at a focus. When $e=1$ the orbit is a parabola with focus O. And finally, when $e>1$ the orbit is a hyperbola with focus O. It has now been shown that these are the only possible orbits of a particle acted on only by an inverse square law central force.

As first pointed out in section 5.1, it should again be noticed that the identity $h^2/\mu = l$ directly relates the angular momentum of the particle to the semilatus rectum of the orbit. A similar result in which a geometrical property of the orbit is related to a dynamical quantity can be found by considering the energy integral.

The potential energy of the particle is $-m\mu/r$, and therefore the total energy per unit mass of the particle is a constant E given by

$$\tfrac{1}{2}v^2 - \frac{\mu}{r} = E$$

The velocity of the particle, however, can be expressed using the above results in the form

$$\mathbf{v} = \frac{h}{l} e \sin\theta \mathbf{e}_r + \frac{h}{r}\mathbf{e}_\theta$$

and, again using the above results, this implies that the energy per unit mass is given by

$$E = \frac{\mu}{2l}(e^2-1)$$

There are now three cases to consider. If $e<1$ the orbit is an ellipse and, since $l=a(1-e^2)$ in this case, $E=-\mu/2a$. If $e=1$ the orbit is a parabola and $E=0$. Finally, if $e>1$ the orbit is a hyperbola, $l=a(e^2-1)$, and $E=\mu/2a$.

It can therefore be seen that the sign of the energy determines the type of orbit, whether an ellipse, parabola or hyperbola, and its magnitude determines the semimajor axis $a=\mu/2|E|$. This, together with the above result that the semilatus rectum is determined by the angular momentum, enables the eccentricity of the orbit to be expressed in terms of dynamical quantities as

$$e = \left[1 + \frac{2h^2E}{\mu^2}\right]^{1/2}$$

The above energy integral can also be used to give the useful expressions for the speed of the particle as

$$v^2 = \mu\left(\frac{2}{r} - \frac{1}{a}\right) \quad \text{for an ellipse}$$

or

$$v^2 = \mu\left(\frac{2}{r} + \frac{1}{a}\right) \quad \text{for a hyperbola}$$

Example 6.3

Consider the situation in which a particle moving in a circular orbit of radius a under the action of an inverse square law of attraction to the centre, suddenly has its speed increased by a fraction λ without changing its direction of motion.

Using the results and notation of example 6.2, the speed of the particle in its original circular motion must be $\sqrt{(\mu/a)}$. Thus, after the impulsive increase in velocity, the energy per unit mass of the particle is

$$E = \tfrac{1}{2}(1+\lambda)^2\frac{\mu}{a} - \frac{\mu}{a}$$

Thus, providing $\lambda < \sqrt{2} - 1$, the particle subsequently follows an elliptic orbit with semimajor axis a' given by

$$a' = a(1 - 2\lambda - \lambda^2)^{-1}$$

Also, the angular momentum of the particle per unit mass implies that the semilatus rectum of the new orbit is

$$l = a(1+\lambda)^2$$

and hence its eccentricity is

$$e = \lambda(2+\lambda)$$

These parameters uniquely determine the new orbit.

Example 6.4: the Hohmann transfer orbit

Consider a satellite launched via a shuttle into an initial circular orbit of radius a_1, but required finally to be in a coplanar circular orbit of greater radius a_2. It may be assumed that the satellite is effectively outside the earth's atmosphere and that it is only acted on by an inverse square law force due

Fig. 6.2. Cotangential transfer between two circular coplanar orbits.

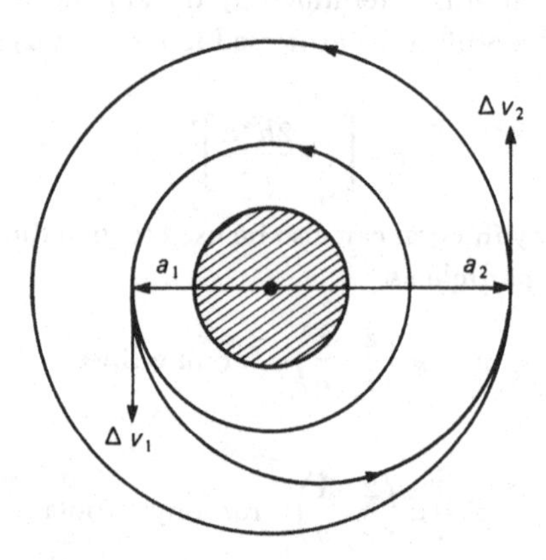

to the earth's gravitational field. The results of example 6.2 may therefore be used where μ is now the constant of gravitation times the mass of the earth.

An appropriate way to transfer the satellite to the required orbit is to use its rocket motor to increase its velocity, putting it into a cotangential elliptic orbit which also approaches the required orbit tangentially. The rocket motor must again be used to achieve the required orbit as this is approached. If the rocket is only fired for a short time in each case its effects may be considered as impulses. The resultant tangential velocity increments can then be denoted by Δv_1 and Δv_2 (see figure 6.2).

Now, the speed of a satellite at any point in an elliptic orbit is given by the expression

$$v^2 = \mu\left(\frac{2}{r} - \frac{1}{a}\right)$$

where a is its semimajor axis. Thus the speed in the initial circular orbit is $\surd(\mu/a_1)$, and that required in the final orbit is $\surd(\mu/a_2)$. In addition the semimajor axis of the cotangential transfer ellipse is $(a_1 + a_2)/2$. Thus the required expressions for the speed of the satellite at the required points can be obtained by substituting a_1 or a_2 for r above. The velocity increments required are then found to be

$$\Delta v_1 = \sqrt{\left(\frac{\mu}{a_1}\right)}\left[\sqrt{\left(\frac{2a_2}{a_1 + a_2}\right)} - 1\right]$$

$$\Delta v_2 = \sqrt{\left(\frac{\mu}{a_2}\right)}\left[1 - \sqrt{\left(\frac{2a_1}{a_1 + a_2}\right)}\right]$$

This particular orbit is the optimum transfer between two coplanar circular orbits, in the sense that it uses the minimum amount of fuel.

Example 6.5: *bouncing ball*

As a further example of impulsive motion consider a ball bouncing vertically on a horizontal plane. It is appropriate to ignore any resistance to the motion and to assume that the only forces acting are a constant gravitational force and the occasional impulsive forces as the ball bounces.

With these assumptions, the equation of motion of the ball while not in contact with the ground is simply

$$\ddot{z} = -g$$

where z is its height above the plane. This can immediately be integrated to give

$$\dot{z} = V_0 - gt$$

$$z = h_0 + V_0 t - \tfrac{1}{2}gt$$

where h_0 and V_0 are the initial values of z and $\dot{z}$ respectively.

Consider now the case in which the ball is simply dropped, or released from rest, from a height h above the plane. According to the above equations it is expected to strike the plane after a time $\surd(2h/g)$, and with speed $\surd(2gh)$.

When the ball lands on the plane, it is acted on by a large impulsive force which effectively reverses the sign of its velocity so that the ball rises again. A detailed analysis of this force would be very complicated, but this is not required provided the time during which the ball is in contact with the plane is sufficiently small to be ignored. In this case it can be replaced by the idealised concept of an impulse.

In practice it is found that when a ball bounces the speed at which it rebounds from a surface is approximately proportional to the speed at which it approached it. This enables a theoretical model to be constructed by introducing a *coefficient of restitution* e defined such that

$$e = \frac{\text{speed of separation}}{\text{speed of approach}}$$

and by assuming that this is constant in all bounces of the ball. This approach is found to work very well in practice with each type of object having its own characteristic value of e. It can be seen that if e were equal to 1, no kinetic energy would be absorbed in the impact, but for smaller values of e energy is absorbed.

Returning now to the problem being considered, it can be seen that if the ball lands with speed $\sqrt{(2gh)}$, it will rebound with initial speed $e\sqrt{(2gh)}$, and therefore the impulse which acts at the bounce must be $(1+e)m\sqrt{(2gh)}$.

It can also be concluded from this that the ball will bounce again after a further time $2e\sqrt{(2h/g)}$. It will then land with speed $e\sqrt{(2gh)}$ and rebound with speed $e^2\sqrt{(2gh)}$, landing again for a third bounce after the further time $2e^2\sqrt{(2h/g)}$. It can thus be seen that the total time before the ball finally comes to rest after many bounces is

$$(1+2e+2e^2+2e^3+\cdots)\sqrt{\left(\frac{2g}{h}\right)} = \sqrt{\left(\frac{2g}{h}\right)}\left[\frac{1+e}{1-e}\right]$$

Exercises

6.1 A particle is displaced from rest at the top of a sphere of radius a. Neglecting friction, show that the particle leaves the surface of the sphere after it has fallen through a vertical distance of $a/3$.

6.2 A toy car moves on a track which is fixed in a vertical plane, and contains a circular loop of radius a. Neglecting friction, show that the minimum height above the bottom of the loop from which the car may start from rest and not leave the track at any point is $5a/2$.

6.3 Show that in an elliptic orbit of semimajor axis a and eccentricity e the minimum and maximum distances from the focus are given, respectively, by $a(1-e)$ and $a(1+e)$.

6.4 A satellite is observed at a distance 16R from the earth's centre to have speed $\frac{1}{4}\sqrt{(\mu/R)}$, where R is the earth's mean radius and μ is the constant of gravitation times the mass of the earth. If the velocity of the satellite is inclined at an angle $\cos^{-1}(11/12)$ to the direction towards the earth, by evaluating its energy and angular momentum per unit mass and using the result of exercise 6.3, show that it would miss the earth's surface by a distance $R/3$ if it were acted on only by the inverse square law of attraction to the earth.

6.5 Consider the motion of a particle which is acted on only by an inverse square law of attraction to a fixed point which can be taken as the origin of an inertial frame of reference. Writing the equation of motion as

$$\ddot{\boldsymbol{r}} = -\frac{\mu}{r^2}\hat{\boldsymbol{r}}$$

show that the vector

$$\boldsymbol{h} = \boldsymbol{r} \times \dot{\boldsymbol{r}}$$

is constant throughout the motion, and interpret it physically.

Derive also the identity

$$\ddot{\boldsymbol{r}} \times \boldsymbol{h} = \mu \frac{\mathrm{d}}{\mathrm{d}t}\hat{\boldsymbol{r}}$$

and hence show that the vector

$$\boldsymbol{e} = \frac{1}{\mu}(\dot{\boldsymbol{r}} \times \boldsymbol{h}) - \hat{\boldsymbol{r}}$$

is also constant throughout the motion.

By taking the scalar product of this with the position vector $\boldsymbol{r}$, show that $\boldsymbol{e}$ may be interpreted as a vector with magnitude equal to the eccentricity of the orbit and directed towards that point on the orbit which is nearest to the focus.

6.6 Hailstones moving at an angle of 30° to the vertical strike the frozen surface of a lake and rebound at an angle of 60° to the vertical and to a height h. Ignoring any horizontal impulsive forces, show that the speed with which the hailstones strike the surface is approximately $2\sqrt{(6gh)}$, and that the coefficient of restitution is $\frac{1}{3}$.

7

Systems of several particles

7.1. Systems of several particles as a model

Up to this point we have been considering mainly the motion of a single particle relative to a frame of reference that has been somehow determined. However, some concepts associated with the possible interaction of one particle with another have been introduced. We are now in a position to consider systems of several particles relative to some frame of reference. The particles may be considered to be interacting with each other and each particle may also be acted on by other external forces. Each particle may still be considered individually, but, in addition, certain general properties of the motion of the whole system may also be considered, without necessarily determining the motion of every individual particle.

Such an approach may be used, for example, to analyse the motion of stars in a galaxy. Each star may be represented as a particle since a galactic scale of distance is being used, and the motion of each star relative to an inertial frame is affected by the gravitational forces induced by all the other stars. In such a case the dominant forces are the mutual interactions between the particles.

Of course, it is possible also to consider just a cluster of stars. In this case the mutual interaction between the individual stars is important, but so also is the effect of the gravitational field caused by the other stars in the galaxy which are not part of the particular cluster under consideration. This illustrates the distinction between the internal forces between the particles of the system, and the external forces that also act on the individual particles but whose source is external to the system being considered.

In this chapter some general results for systems of several particles are obtained. However, the results have a far wider application than simply to systems that can be represented by several, clearly isolated, particles.

Almost every physical object that can be discerned by our natural senses appears to be composed of parts in which matter is continuously distributed. Gases, liquids, the flexible bodies of animals, as well as apparently rigid objects such as chairs or mountains, look and feel like continuous distributions of matter of different composition and of various states. Of course it is now known that all matter is composed of atoms, and that each atom has a distinct structure. But atoms are so small that, on the scale of every-day objects, it is only the combined effect of many atoms that can be discerned. It is therefore appropriate to initially consider the possibility that such objects can be represented as being composed of continuous distributions of matter. This representation is a characteristic of the classical level of approximation. Of course the representation is not ultimately correct, but it is a useful one. It enables a model of a physical object to be theoretically divided into component parts. The various properties of the body can then be estimated by appropriately summing over the component parts. And since a continuous distribution can be divided into an arbitrarily large number of parts, techniques of integration may be used.

If a body is represented in terms of continuous distributions of matter, then it is always possible to divide it up into a sufficiently large number of component parts for each part to be representable as a particle. The body may thus be regarded as an aggregate of particles, each of which must satisfy the laws of particle dynamics. In this way any fluid or solid body may be represented as a system of several, usually a large number of, particles.

In particular applications of classical dynamics, it is usual practice, if it is not appropriate to represent a body by a single particle, to represent it in terms of continuous distributions of mass. However, such a representation is not essential to the theory. All that is required of a body is a knowledge of its mass distribution. Whether this is piecewise continuous or discrete, the distribution is still representable as an aggregate of particles. In practice though, the distribution of mass through a body is always represented by an idealised approximation in order to simplify the calculations involved.

The consideration of systems of several particles is therefore not only appropriate for the analysis of physical systems having a finite number of distinct isolated parts, but also forms the basis for the analysis of the motion of fluid and solid bodies. The various component parts of such bodies, however, are not usually free to move arbitrarily, but are constrained in certain ways. These constraints can be represented in terms of internal forces acting between the constituent particles of the system. For example, for an elastic body, the elasticity is modelled in terms of appropriate internal forces acting between neighbouring particles. Similarly, in analysing the flow of a liquid, it is usually appropriate to consider pressure forces which act between neighbouring particles in such a way that, in the motion of the particles of the fluid, the density remains constant.

The theory is now developed by considering a system of n particles, where n is an arbitrary number. Each particle may be distinctly labelled, and its properties denoted with the appropriate suffix. Thus each particle has mass m_i and position $\mathbf{r}_i$ relative to some defined origin, where $i = 1, \ldots, n$. The total mass of the system is given by

$$m = \sum_{i=1}^{n} m_i$$

Of great importance is the centroid or centre of mass of the system. This is the weighted average position of the particles with respect to their mass. It is defined as follows.

***Definition* 7.1.** *The centroid of a system of particles is the point denoted by the position vector $\bar{\mathbf{r}}$ relative to the origin where*

$$\bar{\mathbf{r}} = \frac{\sum_{i=1}^{n} m_i \mathbf{r}_i}{\sum_{i=1}^{n} m_i}$$

This equation defining the centroid can be written in the form

$$m\bar{\mathbf{r}} = \sum_{i=1}^{n} m_i \mathbf{r}_i$$

and its time derivative is

$$m\dot{\bar{\mathbf{r}}} = \sum_{i=1}^{n} m_i \dot{\mathbf{r}}_i$$

The right-hand side of this equation is the total linear momentum of the system, and the following may be deduced.

***Proposition* 7.1.** *The total linear momentum of a system of particles is equal to the product of the total mass of the system and the velocity of its centroid.*

It can thus be seen that the total linear momentum could be interpreted in terms of a single hypothetical particle, having mass equal to the total mass of the system, situated at the centroid and moving with it. Such an artificial representation is often particularly useful in classical dynamics and will be discussed later in this chapter.

7.2. Centroids: examples

At this point it is convenient to illustrate how the centroid of a body having a continuous distribution of mass may be determined. However, it is only appropriate to consider a few very simple examples.

Example 7.1. *a uniform rod*

Consider the idealised situation in which mass is uniformly distributed along a thin straight line of length l. Coordinates may be chosen with the x axis aligned with the rod, and the origin at one end.

The standard method of approach is now to theoretically divide the rod into n short pieces each of length δx_i and each a distance x_i from the origin, as illustrated in figure 7.1. Provided n is a sufficiently large number, each piece may now be represented by a particle.

Since mass is assumed to be uniformly distributed along the rod, the mass of each piece is

$$m_i = \frac{m}{l}\,\delta x_i$$

where m is the total mass of the rod. The position of the centroid can now be determined directly from definition 7.1. It is immediately clear that $\bar{y}=0$ and $\bar{z}=0$, so that the centroid is located in the line of the rod, and

$$\bar{x} = \frac{\sum_{i=1}^{n} \frac{m}{l}\,\delta x_i x_i}{\sum_{i=1}^{n} \frac{m}{l}\,\delta x_i}$$

It is now appropriate to theoretically let n become arbitrarily large, and each length δx_i arbitrarily small, so that the above sums can be expressed as integrals. The denominator becomes

$$\int_0^l \frac{m}{l}\,\mathrm{d}x = \frac{m}{l}[x]_0^l = m$$

Fig. 7.1. Consider a rod to be composed of n short sections each of length δx_i a distance x_i from one end.

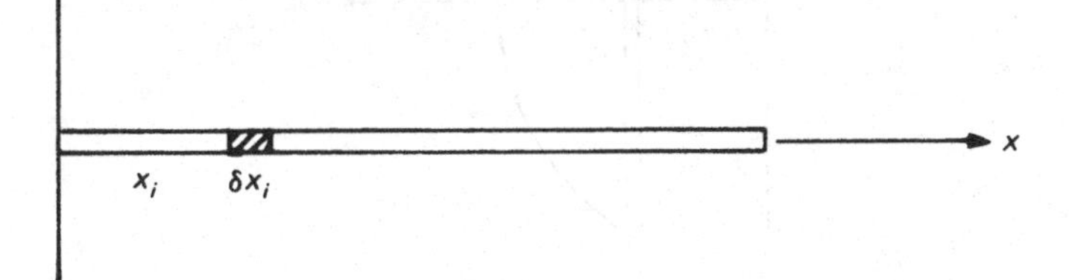

which is the total mass as required, and

$$\bar{x} = \frac{1}{m}\int_0^l \frac{m}{l} x \, dx = \frac{1}{l}\left[\frac{x^2}{2}\right]_0^l = \frac{l}{2}$$

which indicates that the centroid is the midpoint of the rod as expected.

Example 7.2: a uniform semicircular plate

Consider now a flat semicircular plate on which mass is uniformly distributed. The total mass of the plate can be denoted by m and its radius by a.

A convenient frame of reference can be chosen in which the plate is in the plane $z=0$ and its straight edge is taken to define the y axis, as in figure 7.2.

It is immediately clear that $\bar{z}=0$, and the symmetry of the body about the x axis can be used to show that $\bar{y}=0$.

To determine the position of the centroid along the x axis, it is necessary to divide the plate theoretically into a very large number of small pieces. It is then convenient to collect together all the pieces that are the same distance from the y axis. This effectively results in a division of the plate into n thin strips. If the thickness of each strip is denoted by δx_i, its distance from the y axis is x_i, and its mass is

$$m_i = \frac{m}{\frac{1}{2}\pi a^2} 2\sqrt{(a^2 - x_i^2)}\delta x_i$$

Fig. 7.2. A uniform semicircular plate can be considered in terms of n thin strips.

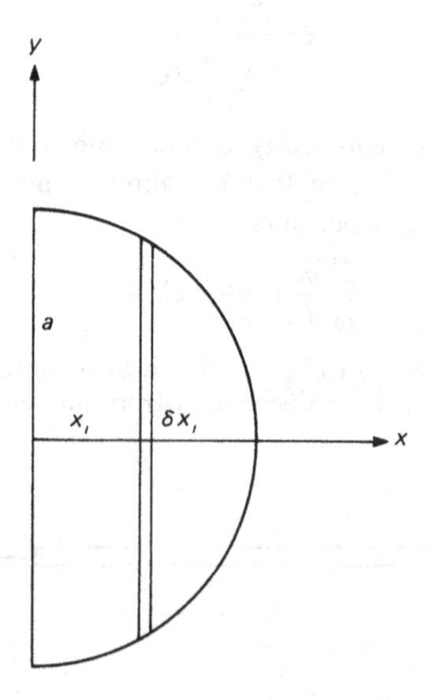

The position of the centroid is therefore given by

$$\bar{x} = \frac{1}{m}\sum_{i=1}^{n} \frac{4m}{\pi a^2}\sqrt{(a^2 - x_i^2)}\delta x_i x_i$$

which can be evaluated in the limit as

$$\bar{x} = \frac{4}{\pi a^2}\int_0^a \sqrt{(a^2 - x^2)}x \, dx$$

$$= \frac{4a}{3\pi}$$

Exercises

7.1 Show that the centroid of a uniform solid cone of height h is a distance $h/4$ from the base.

7.2 Show that the centroid of a uniform solid hemisphere of radius a is a distance $3a/8$ from the base.

7.3 A uniform piece of wire of length l is bent into the shape of an arc of a circle of radius a. Show that the centroid is a distance $2a^2/l \sin l/2a$ from the centre of curvature.

7.4 Show that the centroid of a lamina in the shape of a sector of a circle of radius a, which subtends an angle 2α at the centre, is a distance $2a/3\alpha \sin \alpha$ from the vertex.

7.3. Energy and angular momentum

For a system of n particles the total kinetic energy is

$$T = \tfrac{1}{2}\sum_{i=1}^{n} m_i \dot{r}_i^2$$

This is defined relative to an arbitrary frame of reference. It is often convenient, however, to consider the motion of each particle relative to a frame of reference with origin at the centroid, at least for purposes of calculation. It is assumed that this frame has axes parallel to those of the original arbitrary frame. Relative to the arbitrary frame, each particle has the position vector

$$\boldsymbol{r}_i = \boldsymbol{\rho}_i + \bar{\boldsymbol{r}}$$

where $\boldsymbol{\rho}_i$ is the position of the particle relative to the centroid. Substituting this into the above expression for the total kinetic energy of the system gives

$$T = \tfrac{1}{2}\sum_{i=1}^{n} m_i \dot{\boldsymbol{\rho}}_i^2 + \dot{\bar{\boldsymbol{r}}} \cdot \sum_{i=1}^{n} m_i \dot{\boldsymbol{\rho}}_i + \tfrac{1}{2}\sum_{i=1}^{n} m_i \dot{\bar{\boldsymbol{r}}}^2$$

and from the definition of the centroid it is clear that $\sum_{i=1}^{n} m_i \boldsymbol{\rho}_i = 0$

so that the middle term vanishes. This establishes a result known as König's theorem.

***Proposition* 7.2.** *The kinetic energy of a system relative to a given frame of reference is equal to the sum of two parts, (a) the kinetic energy of the system calculated relative to a frame with origin at the centroid and axes parallel to the given frame, and (b) one-half the product of the total mass of the system and the square of the velocity of the centroid.*

$$T=\tfrac{1}{2}\sum_{i=1}^{n} m_i\dot{\rho}_i^2+\tfrac{1}{2}m\dot{\bar{r}}^2$$

The last term in the above equation may be interpreted in a similar way to the total linear momentum of the system described above. It is the kinetic energy of a single hypothetical particle of mass m located at the centroid of the system. But in this case the kinetic energy relative to the centroid has to be added to give the total kinetic energy of the system.

There is nothing in proposition 7.2 that is of foundational importance. Its great importance, however, lies in the fact that in many practical applications it considerably simplifies the calculation of kinetic energies. The simplification arises from the particular properties of the centroid. It does not work for any other point of the system.

Similar simplifications also occur in the calculation of angular momentum. The total angular momentum of the system relative to an arbitrary frame of reference is given by

$$\boldsymbol{L}=\sum_{i=1}^{n} \boldsymbol{r}_i\times m_i\dot{\boldsymbol{r}}_i$$

which may be expanded in the form

$$\boldsymbol{L}=\sum_{i=1}^{n} \boldsymbol{\rho}_i\times m_i\dot{\boldsymbol{\rho}}_i+\left(\sum_{i=1}^{n} m_i\boldsymbol{\rho}_i\right)\times\dot{\bar{\boldsymbol{r}}}$$
$$+\bar{\boldsymbol{r}}\times\left(\sum_{i=1}^{n} m_i\dot{\boldsymbol{\rho}}_i\right)+\bar{\boldsymbol{r}}\times m\dot{\bar{\boldsymbol{r}}}$$

Again, because of the properties of the centroid, the central terms vanish and the following is obtained.

***Proposition* 7.3.** *The total angular momentum of a system relative to a given frame of reference is the sum of two parts, (a) the total angular momentum about its centroid using axes parallel to those of the given frame, and (b) the angular momentum of a* *hypothetical particle with mass equal to that of the system moving with the centroid relative to the origin.*

$$\boldsymbol{L} = \sum_{i=1}^{n} \boldsymbol{\rho}_i \times m_i \dot{\boldsymbol{\rho}}_i + \bar{\boldsymbol{r}} \times m \dot{\bar{\boldsymbol{r}}}$$

It is also appropriate at this point to consider the potential energy of a system of n particles in a constant external gravitational field, since this particular situation is frequently required in applications. In such a field, the gravitational force which acts on each particle is equal to its mass times the constant vector $\boldsymbol{g}$, which is the acceleration which the gravitational field would induce on the particles in the absence of other forces. Thus, for a system of n particles, the total gravitational force is given by

$$\boldsymbol{F} = \sum_{i=1}^{n} m_i \boldsymbol{g}$$
$$= m\boldsymbol{g}$$

and the sum of the moments of the gravitational forces is

$$\boldsymbol{M} = \sum_{i=1}^{n} \boldsymbol{r}_i \times m_i \boldsymbol{g}$$
$$= \bar{\boldsymbol{r}} \times m\boldsymbol{g}$$

From these two expressions it can be seen that the gravitational forces which act on all the particles are together equivalent to a single force $m\boldsymbol{g}$ which may be considered to act at the centroid. This single force is identical to the gravitational force which would act on a single particle of mass m. Thus, in a constant gravitational field, the gravitational force which is considered to act on a system of particles is the same as that which may be considered to act on its single particle representation.

***Proposition* 7.4.** *The effect of a constant gravitational field on a system of particles is equivalent to that of a single force which acts at its centroid and is equal to that which the gravitational field would induce on a single particle with the same total mass.*

It is perhaps worth pointing out in passing that the effect of a nonconstant gravitational field on a system of particles can also be replaced by that of a single force acting a point at any instant. However, both the magnitude of the force and the point of its application, which may be referred to as the instantaneous centre of gravity, may change as the system moves. Thus the general concept of a centre of gravity is not a useful one in classical mechanics. It is only a unique point of a system if the gravitational field is constant, and in this case it is identical to the centroid or centre of mass, which is always the most appropriate centre to consider.

Having obtained a simplification for the effect of a constant gravitational force field on a system of particles, an expression for the associated potential energy can be trivially obtained. The single equivalent force is clearly conservative and, using the approach of section 6.4, the potential energy due to a constant gravitational field which acts vertically downwards is given by

$$V = mg\bar{z} + c$$

where c is an arbitrary constant, and $\bar{z}$ is the vertical height of the centroid above some horizontal plane. It can be seen that this could alternatively have been obtained as the sum of the potential energies of all the component particles.

7.4. Equations of motion

In a system of n particles, each particle is required to satisfy the law of motion as stated in proposition 3.2 and expressed by the equation

$$m_i\ddot{\boldsymbol{r}}_i = \boldsymbol{F}_i$$

where $\boldsymbol{F}_i$ is the sum of the forces acting on the ith particle. It is assumed that a frame of reference is specified, and that if this is not an inertial frame, then it is necessary to include in $\boldsymbol{F}_i$ the fictitious forces mentioned in proposition 4.1. However, in practice it is almost always convenient to adopt a frame of reference that can at least be assumed to be inertial to some level of approximation.

It is now convenient to distinguish between forces that are internal and external to the system.

The internal forces arise as interactions between the particles. They may, for example, be due to the mutual gravitational interactions between them, or to the forces of constraint that are assumed to operate between neighbouring particles in order to build a model for a particular system. Such forces are denoted in vector form by the symbol $\boldsymbol{R}_{ij}$, where the first subscript indicates the particle on which the force acts, and the second indicates the particle which induces the force. It is of course assumed that no particle induces a force on itself, so that $\boldsymbol{R}_{ii} = 0$, and that according to assumption 3.3, which is Newton's third law, the interaction between distinct particles is mutually opposite, so that

$$\boldsymbol{R}_{ij} = -\boldsymbol{R}_{ji}$$

The external forces that act on a particle of the system are those caused by particles and fields that are not included in the n particles of the system. They include all the forces that are not internal, including those arising from the possible use of a noninertial frame. The sum of these forces is denoted by $\boldsymbol{P}_i$ and it is possible to put

$$\boldsymbol{F}_i = \boldsymbol{P}_i + \sum_{j=1}^{n} \boldsymbol{R}_{ij}$$

The equation of motion for each particle is now

$$m_i \ddot{\boldsymbol{r}}_i = \boldsymbol{P}_i + \sum_{j=1}^{n} \boldsymbol{R}_{ij}$$

Summing this over all particles and using the definition of the centroid and the antisymmetry of $\boldsymbol{R}_{ij}$ gives

$$m\ddot{\bar{\boldsymbol{r}}} = \sum_{i=1}^{n} \boldsymbol{P}_i$$

It is now convenient to denote the sum of the external forces acting on all the particles by the single vector $\boldsymbol{F}$. The result obtained is that *the sum of the external forces acting on the particles of a system is equal to the total mass times the acceleration of the centroid.*

$$m\ddot{\bar{\boldsymbol{r}}} = \boldsymbol{F}$$

This is one of the basic equations for the motion of a system. Using proposition 7.1 it may be stated as follows.

Proposition 7.5. *The rate of change of the total linear momentum of a system of particles is equal to the sum of the external forces which act on it.*

Proposition 7.5 is particularly significant since it clarifies the use of the concept of a particle, and also provides further justification for it. Initially, it was assumed that a body could be represented by an idealised particle if its physical size could be neglected in a given situation. The theory of particle dynamics was then developed, culminating in the law of motion that its mass times its acceleration is equal to the sum of the forces acting on it. On this foundation, it has now been shown that a physical body whose size is not negligible also satisfies a similar equation of motion. Assuming that a body can always be considered as an aggregate of particles, it has now been found to be convenient to also consider a single hypothetical particle situated at the centroid of the body, having mass equal to that of the whole body. The sum of the external forces acting on the various parts of the body is equal to the mass times the acceleration of this single hypothetical particle, which may here be taken to represent the body. The size and internal structure of the body plays no part in such an equation. This is consistent with the approach that a finite body can sometimes be represented by the concept of a particle. It also clarifies the representation in that it specifies that the single representative particle must be located at the centroid of the body or system.

The previous propositions also clarify the situation in which the single particle representation of a system is appropriate. According to proposition 7.1, the total linear momentum of a system is always identical to that of its single representative particle. According to propositions 7.2 and 7.3 on the other hand, the total kinetic energy of a system, and its angular momentum about a point, are each composed of two parts. One part is the kinetic energy or angular momentum of the single representative particle. The other part is the kinetic energy of all the individual particles, or their angular momentum, evaluated relative to the centroid. Thus, if these latter components are negligible in any situation compared with the former, then it may be appropriate to consider only the single particle representation.

To proceed further it is necessary to reconsider the internal forces of the system. Some further constraints have to be introduced concerning the mutual interactions between the individual particles of the system. These amount to a stronger form of Newton's third law which was given earlier as assumption 3.3.

Assumption 3.3 simply states that when two particles interact with each other, the mutually induced forces are equal in magnitude and opposite in direction. However, the direction of these forces is not specified. Now, in almost all applications the forces that are induced between the particles of a system act in the direction of the straight lines between the particles. They may be attractive or repulsive, but they usually satisfy this condition. In addition, the forces of constraint that are sometimes required in a system may also usually be chosen to satisfy the same condition. It is therefore appropriate to introduce the additional assumption.

***Assumption* 7.1.** *When two particles interact with each other the mutual forces induced act in the direction of the line joining the particles.*

With this assumption we may now reconsider the equation of motion of each particle

$$m_i\ddot{\boldsymbol{r}}_i = \boldsymbol{P}_i + \sum_{j=1}^{n} \boldsymbol{R}_{ij}$$

Taking moments about the origin and summing over all particles gives

$$\sum_{i=1}^{n} \boldsymbol{r}_i \times m_i\ddot{\boldsymbol{r}}_i = \sum_{i=1}^{n} \boldsymbol{r}_i \times \boldsymbol{P}_i + \sum_{i=1}^{n}\sum_{j=1}^{n} \boldsymbol{r}_i \times \boldsymbol{R}_{ij}$$

As a consequence of assumption 7.1, the final term vanishes since the forces $\boldsymbol{R}_{ij}$ and $-\boldsymbol{R}_{ji}$ are equal and have the same line of action, and hence the same moments. The expression on the left-hand side is simply the rate of change of the total angular momentum of the system about the origin. The resulting equation indicates that the sum of the moments of the external forces about the origin is equal to the rate of change of angular momentum.

$$\frac{\mathrm{d}}{\mathrm{d}t}\boldsymbol{L} = \sum_{i=1}^{n} \boldsymbol{r}_i \times \boldsymbol{P}_i$$

It is, however, convenient to use the single vector $\boldsymbol{M}$ to denote the sum of the moments of the external forces which act on all the particles.

So far no particular choice of origin or coordinate directions of a frame of reference have been made. The above equation therefore applies with reference to an arbitrary point. However, if the frame of reference adopted is noninertial, all the associated fictitious forces induced on all the individual particles would have to be included in the external forces of the system, and this would considerably complicate the evaluation of this equation. In practice, therefore, it is appropriate to consider only inertial frames of reference, and the result may be stated in the following form.

***Proposition* 7.6.** *The rate of change of the angular momentum of a system of particles about the origin of an inertial frame of reference is equal to the sum of the moments about the origin of the external forces which act on the particles of the system.*

$$\frac{\mathrm{d}}{\mathrm{d}t}\boldsymbol{L} = \boldsymbol{M}$$

In the case of a single particle, this result has already been stated as proposition 6.3. But in that case, the result contained less information than the original equation of motion. In the present case, however, there is additional information in this result which is not contained in proposition 7.5. Proposition 7.5 relates to the motion of the centroid of a system, whereas proposition 7.6 also contains information about the angular motion about the centroid. These two results therefore give independent equations for the motion of a system.

It is helpful at this point to consider motion relative to a frame of reference with origin at the centroid, and axes parallel to those of an inertial frame. Using a suffix c to denote the value of a quantity relative to the frame with origin at the centroid, proposition 7.3 may be stated in the form

$$\boldsymbol{L} = \boldsymbol{L}_c + \bar{\boldsymbol{r}} \times m\dot{\bar{\boldsymbol{r}}}$$

This, with proposition 7.6, gives

$$\begin{aligned} \frac{\mathrm{d}}{\mathrm{d}t}\boldsymbol{L}_c &= \sum_{i=1}^{n} (\boldsymbol{\rho}_i + \bar{\boldsymbol{r}}) \times \boldsymbol{P}_i - \bar{\boldsymbol{r}} \times m\ddot{\bar{\boldsymbol{r}}} \\ &= \sum_{i=1}^{n} \boldsymbol{\rho}_i \times \boldsymbol{P}_i \\ &= \boldsymbol{M}_c \end{aligned}$$

The step in the argument here has been made with the aid of proposition 7.5. The result can be stated as follows.

***Proposition* 7.7.** *The rate of change of the angular momentum of a system of particles about its centroid, using axes parallel to those of an inertial frame, is equal to the sum of the moments of the external forces about the centroid.*

It can be seen from the derivation of this proposition that its content differs mathematically from that of proposition 7.6 only by components of terms that appear in proposition 7.5. Only two independent vector equations for the motion of a system of particles have been obtained. In practice it is convenient to use proposition 7.5 to give an equation for the centroid of a system, and then to use either proposition 7.6 or 7.7, depending on which is the most convenient, to give an equation for the angular motion of the system.

It has been shown that the rate of change of angular momentum is equal to the sum of the moments of the external forces when the point of reference is either the origin of an inertial frame or the centroid of the system. It is now worth considering whether this result applies to any other point.

Consider an arbitrary point P having position vector $\boldsymbol{r}_{\mathrm{p}}$ relative to an inertial frame of reference, and consider a frame of reference with origin at P having axes parallel to those of the inertial frame. Denote with a prime quantities taken relative to the frame at P. Thus the position vectors of the centroid and the ith particle of the system relative to P are $\bar{\boldsymbol{r}}'$ and $\boldsymbol{r}_i'$ respectively, as illustrated in figure 7.3. Thus

$$\boldsymbol{r}_i = \boldsymbol{r}_{\mathrm{p}} + \boldsymbol{r}_i' \qquad \bar{\boldsymbol{r}} = \boldsymbol{r}_{\mathrm{p}} + \bar{\boldsymbol{r}}'$$

From proposition 7.3, the angular momentum relative to the frame at P is

$$\boldsymbol{L}' = \boldsymbol{L}_{\mathrm{c}} + \bar{\boldsymbol{r}}' \times m\dot{\bar{\boldsymbol{r}}}'$$

and therefore

$$\boldsymbol{L}' = \boldsymbol{L} - \bar{\boldsymbol{r}} \times m\dot{\bar{\boldsymbol{r}}} + \bar{\boldsymbol{r}}' \times m\dot{\bar{\boldsymbol{r}}}'$$

Substituting this into proposition 7.6 and expanding some terms gives

$$\frac{\mathrm{d}}{\mathrm{d}t}\boldsymbol{L}' = \sum_{i=1}^{n} (\boldsymbol{r}_{\mathrm{p}} + \boldsymbol{r}_i') \times \boldsymbol{P}_i - (\boldsymbol{r}_p + \bar{\boldsymbol{r}}') \times m\ddot{\bar{\boldsymbol{r}}} + \bar{\boldsymbol{r}}' \times m(\ddot{\bar{\boldsymbol{r}}} - \ddot{\boldsymbol{r}}_{\mathrm{p}})$$

Simplifying and using proposition 7.5, this finally gives

$$\frac{\mathrm{d}}{\mathrm{d}t}\boldsymbol{L}' = \sum_{i=1}^{n} \boldsymbol{r}_i' \times \boldsymbol{P}_i - \bar{\boldsymbol{r}}' \times m\ddot{\boldsymbol{r}}_{\mathrm{p}}$$

Thus, relative to a nonrotating frame of reference defined at an arbitrary point P, the rate of change of angular momentum about P is equal to the moments of the external forces about P, together with

Fig. 7.3. Position vectors for a system of particles. O is the origin of an inertial frame of reference, C is the centroid of the system, and P is an arbitrary point.

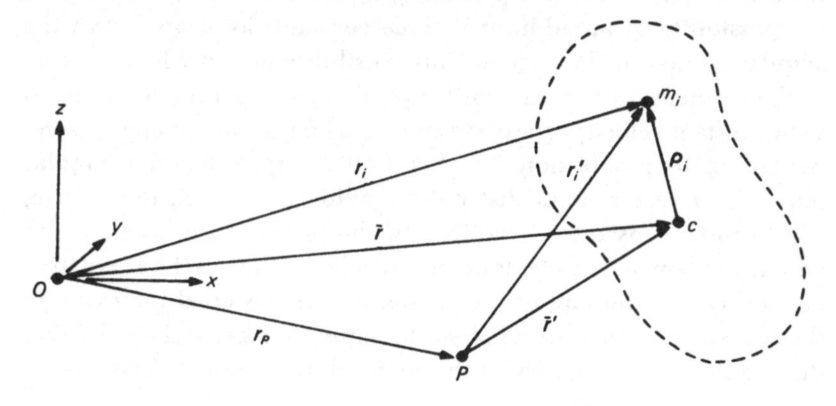

an additional term $-\bar{r}' \times m\ddot{r}_p$. This additional term clearly vanishes when either $\ddot{r}_p$ or $\bar{r}'$ are zero. In these cases either P is an inertial point, or it is at the centroid of the system. Both cases are therefore already covered by propositions 7.6 and 7.7 respectively. The only remaining case in which the additional term vanishes is that in which the acceleration of P and the position vector of the centroid relative to P are parallel.

Situations in which it is helpful to consider this case do not arise very often in applications. Even in situations in which it could be used, it is probably easier to consider the frame of reference as being noninertial, so that additional fictitious forces $-m_i\ddot{r}_p$ have to be included in the external forces $\boldsymbol{P}_i$ acting on each particle. The sum of these fictitious forces over all the particles is equivalent to a force $-m\ddot{r}_p$ acting on the centroid, and this can be seen to give the extra term in the above equation. This approach is simpler and can be used with greater confidence, since it places no restrictions on the relative directions of the acceleration of the origin and the position vector of the centroid.

Returning now to the equations of motion, it can be seen immediately from proposition 7.5 that when the sum of the external forces acting on all the particles is zero, the centroid moves with constant velocity. In this case there would exist six constants of the motion that could be interpreted in terms of the velocity of the centroid and its position at some time. In the less restrictive case, if the sum of the external forces had zero component in one or two independent directions, then this would give rise to two or four constants of motion which would be interpreted in the same way in terms of the general theory. The possibility of such constants has already been described for the motion of a single particle and, as in that case, there is also the possibility of an additional three constants associated with the angular motion of the system. This possibility occurs when the sum of the moments of the external forces about any point which moves with constant velocity relative to an inertial frame of reference is zero. According to proposition 7.6 this would imply that the angular momentum vector about that point would be constant, thus giving rise to three constants of motion. In the present case, however, in which a system of particles is being considered, there is the alternative possibility that the sum of the moments of the external forces about the centroid may be zero. This would imply from proposition 7.7 that the angular momentum about the centroid was constant. Less restrictive cases also sometimes occur, in which the sum of the moments of the external forces about either the centroid or an inertial point has a zero component in a particular direction. This would lead to a constant of angular momentum about that direction. These possible constants of motion are sometimes of importance in the analysis of systems of particles.

7.5. *n*-body problems: examples

Example* 7.3: *the two-body problem

Consider the motion of two isolated bodies under the action only of their mutual gravitational fields. Although it is strictly not possible to isolate two bodies in this way, this idealisation does have many applications in celestial mechanics.

It may be assumed that each body can be represented as a particle. Then, according to Newton's theory of universal gravitation, which is stated here as assumption 5.1, the equation of motion of each particle relative to an inertial frame of reference is

$$m_1\ddot{\boldsymbol{r}}_1 = -\frac{Gm_1m_2}{|\boldsymbol{r}_1-\boldsymbol{r}_2|^3}(\boldsymbol{r}_1-\boldsymbol{r}_2)$$

$$m_2\ddot{\boldsymbol{r}}_2 = -\frac{Gm_1m_2}{|\boldsymbol{r}_1-\boldsymbol{r}_2|^3}(\boldsymbol{r}_2-\boldsymbol{r}_1)$$

It can immediately be seen that the sum of these equations is zero, as is required by proposition 7.5, since it is assumed that there are no external forces. It follows that the centroid moves with constant velocity, and therefore may appropriately be chosen as the origin of an inertial frame of reference.

Having completely integrated the sum of the above equations, it is appropriate also to divide each by its mass, and their difference then takes the form

$$\ddot{\boldsymbol{r}}_1-\ddot{\boldsymbol{r}}_2 = -\frac{G(m_1+m_2)}{|\boldsymbol{r}_1-\boldsymbol{r}_2|^3}(\boldsymbol{r}_1-\boldsymbol{r}_2)$$

Denoting the position of one body relative to the other $(\boldsymbol{r}_1-\boldsymbol{r}_2)$ by $\boldsymbol{r}$, this equation can be written in the form

$$\ddot{\boldsymbol{r}} = -\frac{\mu}{r^2}\hat{\boldsymbol{r}}$$

where

$$\mu = G(m_1+m_2)$$

This equation is now identical to the equation of motion of a particle under an inverse square law of attraction to a fixed point, as considered in example 6.2, but with a different value for μ. Therefore all the properties of inverse square law orbits described in example 6.2 can immediately be applied to the two-body problem, provided the value of μ is adapted and the centre of attraction at the origin is not regarded as a fixed point.

This result also helps to clarify the intermediate theory of planetary motion described in section 5.2. If the motion of the sun and any one planet is considered in isolation, the above results imply that the planet would be attracted towards the sun according to the above inverse square law, where $\mu = G(m_s + m_p)$. The mass of every planet, however, is very small compared to that of the sun. Thus μ is approximately equal to Gm_s, and each planet approximately follows an elliptic orbit as described earlier.

Example 7.4: The restricted three-body problem

Although, as shown in the previous example, a complete solution can be obtained for the two-body problem, no general analytic solutions exist for the motion of three particles under the action of their mutual gravitational fields. First integrals of linear momentum, angular momentum and energy can all be obtained, but these only specify ten constants for a system of equations of order 18, and the remaining equations are intractable analytically.

In many applications, however, it is found that the mass of one body is very small compared to the others. For example, the dominant force on a satellite in a high earth orbit is the gravitational force towards the earth, but its orbit is perturbed by the gravitational field of the moon. The satellite, however, would have a negligible effect on the motion of the earth and moon, since its mass, and hence its gravitational field, is comparatively very small. In such situations it is appropriate to ignore the mass of one body. The resulting situation is known as the restricted three-body problem, and the two bodies with larger mass are known as the primaries.

Regarding the mass m_3 as negligibly small compared with m_1 and m_2, the equation of motion of each particle is

$$m_1\ddot{\boldsymbol{r}}_1 = -\frac{Gm_1m_2}{|\boldsymbol{r}_1-\boldsymbol{r}_2|^3}(\boldsymbol{r}_1-\boldsymbol{r}_2)$$

$$m_2\ddot{\boldsymbol{r}}_2 = -\frac{Gm_1m_2}{|\boldsymbol{r}_2-\boldsymbol{r}_1|^3}(\boldsymbol{r}_2-\boldsymbol{r}_1)$$

$$m_3\ddot{\boldsymbol{r}}_3 = -\frac{Gm_1m_3}{|\boldsymbol{r}_3-\boldsymbol{r}_1|^3}(\boldsymbol{r}_3-\boldsymbol{r}_1) - \frac{Gm_2m_3}{|\boldsymbol{r}_3-\boldsymbol{r}_2|^3}(\boldsymbol{r}_3-\boldsymbol{r}_2)$$

In this case the centroid of the system is located on the line between the primaries which orbit about each other in standard inverse square law orbits. The equation of motion for the small mass, however, still cannot be completely integrated.

Consider now making the further simplifying assumption that the primaries follow circular orbits about each other. This situation is known as the circular restricted three-body problem. Denoting the constant distance between the primaries by a, the periodic time of their orbits is

$$2\pi\sqrt{\left(\frac{a^3}{G(m_1+m_2)}\right)}$$

At this point it is convenient to introduce a cartesian frame of reference that has origin at the centroid and is rotating with the primaries. It is possible to choose the x axis to be the line containing the primaries, and the z axis to be perpendicular to their plane of motion. This frame, which is illustrated in figure 7.4, therefore rotates with angular velocity

$$\boldsymbol{\omega}=\sqrt{\left(\frac{G(m_1+m_2)}{a^3}\right)}\boldsymbol{k}$$

Relative to such a frame the primaries have position vectors

$$\boldsymbol{r}_1=-\frac{m_2}{m_1+m_2}a\boldsymbol{i} \qquad \boldsymbol{r}_2=\frac{m_1}{m_1+m_2}a\boldsymbol{i}$$

and, allowing for the rotation of the frame, the equation of motion for the small mass implies that

$$\ddot{\boldsymbol{r}}_3=-2\boldsymbol{\omega}\times\dot{\boldsymbol{r}}_3-\boldsymbol{\omega}\times(\boldsymbol{\omega}\times\boldsymbol{r}_3)-\frac{Gm_1}{|\boldsymbol{r}_3-\boldsymbol{r}_1|^3}(\boldsymbol{r}_3-\boldsymbol{r}_1)-\frac{Gm_2}{|\boldsymbol{r}_3-\boldsymbol{r}_2|^3}(\boldsymbol{r}_3-\boldsymbol{r}_2)$$

Now, as described in section 6.4, the Coriolis force does no work and the centrifugal force is conservative. An energy integral can therefore be obtained which, having divided by the mass, takes the form

$$\tfrac{1}{2}\dot{\boldsymbol{r}}_3^2-\tfrac{1}{2}(\boldsymbol{\omega}\times\boldsymbol{r}_3)^2-\frac{Gm_1}{|\boldsymbol{r}_3-\boldsymbol{r}_1|}-\frac{Gm_2}{|\boldsymbol{r}_3-\boldsymbol{r}_2|}=c$$

Fig. 7.4. The restricted three-body problem.

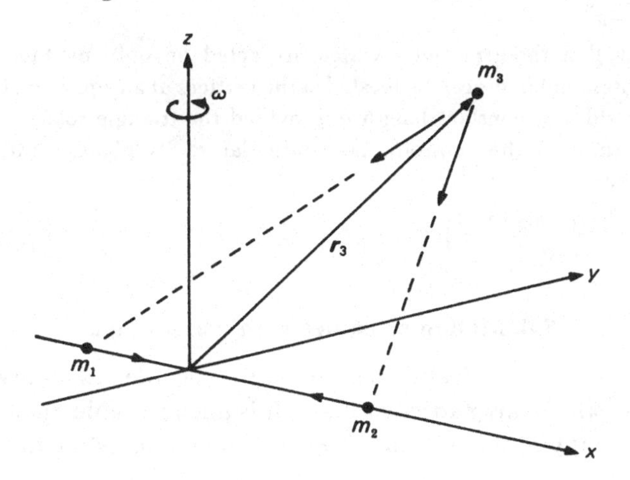

where c is a constant. This can be rewritten in terms of its components as

$$\dot{x}^2+\dot{y}^2+\dot{z}^2=\frac{G(m_1+m_2)}{a^3}(x^2+y^2)+\frac{2Gm_1}{|r_3-r_1|}+\frac{2Gm_2}{|r_3-r_2|}+2c$$

where

$$|r_3-r_1|=\left[\left(x+\frac{m_2a}{m_1+m_2}\right)^2+y^2+z^2\right]^{1/2}$$

$$|r_3-r_2|=\left[\left(x-\frac{m_1a}{m_1+m_2}\right)^2+y^2+z^2\right]^{1/2}$$

This result is often referred to as Jacobi's integral in celestial mechanics, where the constant can be used to classify orbits.

Exercises

7.5 Two particles of total mass m move under the action of their mutual gravitational fields. If they are initially a distance a apart, denoting their masses by εm and $(1-\varepsilon)m$, show that their centroid is initially at a distance $(1-\varepsilon)a$ from the particle of mass εm. If the particles initially have speeds v_1 and v_2 in opposite nonaligned directions, show also that they each subsequently follow elliptic orbits with the centroid at a focus provided $a(v_1+v_2)^2<2Gm$.

7.6 For a system of n particles, show that if the total angular momentum about the centroid is nonzero at some time, then it is not possible at any time for all the particles to come together at a single point.

7.7 Show that three particles of equal mass m, which are acted on only by their mutual gravitational fields, can be located at three points on a straight line an equal distance a apart, provided that line rotates about an axis through the central point with angular velocity

$$\sqrt{\left(\frac{5Gm}{4a^3}\right)}$$

7.8 Show that three particles which are acted on only by their mutual gravitational fields can be located at the vertices of an equilateral triangle with sides of constant length a, provided the triangle rotates about an axis through the centroid, perpendicular to its plane, with angular velocity

$$\sqrt{\left(\frac{G(m_1+m_2+m_3)}{a^3}\right)}$$

7.6. Motion of bodies with variable mass

In the theory of classical dynamics mass is regarded as an invariable characteristic of any particular body. It is not a variable quantity like kinetic energy or temperature. Thus it is only admissible to consider bodies with variable mass in situations in which additional particles are being added to a body or removed from it. For example, the mass of a raindrop may increase through condensation as it falls through a cloud, or the total mass of a vehicle may decrease as the fuel is burnt and its mass ejected through the exhaust.

It follows that, in order to develop a dynamical model to describe the motion of a body whose mass is varying, the mechanism by which mass is added or removed must be clearly specified. In particular, the dynamical properties of the mass that is added or removed must be known.

Consider first the situation where mass is being continuously added to a body, and that this mass, before it is joined to the body, is moving with velocity $\boldsymbol{u}$. It is appropriate here only to consider situations in which the body with varying mass can be represented by a particle. A model can now be developed by considering what happens in a short interval of time δt. At the beginning of this interval the mass of the body can be denoted by m, and its velocity by $\boldsymbol{v}$. A small mass δm moving with initial velocity $\boldsymbol{u}$ is now added to the body. It may be assumed that the total mass $m+\delta m$ subsequently moves together as a single particle whose velocity at the end of the interval is increased by an amount $\delta \boldsymbol{v}$. To achieve a general equation for the motion it is also appropriate to assume that the body is acted on during this interval by a number of external forces whose sum can be denoted by $\boldsymbol{F}$. It is now possible to apply proposition 7.5 to this situation. This states that the rate of change of the total linear momentum of the system is equal to the sum of the external forces that act on it. In the short interval being considered this can be interpreted by the equation

$$\frac{1}{\delta t}\{(m+\delta m)(\boldsymbol{v}+\delta \boldsymbol{v})-(m\boldsymbol{v}+\delta m\boldsymbol{u})\}=\boldsymbol{F}$$

in the limit as δt becomes arbitrarily small. This clearly leads to the equation of motion of the body in the form

$$m\dot{\boldsymbol{v}}=(\boldsymbol{u}-\boldsymbol{v})\dot{m}+\boldsymbol{F}$$

It can easily be shown that this equation also applies to the situation where mass is being discarded with velocity $\boldsymbol{u}$, but in this case it should be noticed that $\dot{m}$ is negative.

***Proposition* 7.8.** *The mass times the acceleration of a body whose mass is varying is equal to the rate of change of mass times the*

relative velocity of the mass that is added or discarded plus the external force that acts on the body.

It may also be observed of course that interactive forces will occur as the additional mass attaches itself to the body, or is ejected from it. These forces however are internal to the system and do not enter the equation of motion.

7.7. Rocket motion: examples

The result stated above can immediately be applied to the motion of a rocket, which obtains thrust by ejecting mass at a large relative speed. In practice this is achieved by burning fuel, the resulting gases being directed as a jet in a particular direction.

Assuming that the external forces are composed only of the gravitational force and a drag of magnitude D, the equation of rocket motion can be written as

$$m\ddot{\mathbf{r}} = \dot{m}\mathbf{c} + m\mathbf{g} - D\hat{\mathbf{v}}$$

where $\mathbf{c}$ is the velocity at which the exhaust fuel is ejected relative to the rocket.

Since $\dot{m}$ is negative, it can be seen that the rocket achieves a thrust in the opposite direction to that in which fuel is ejected. In addition, the thrust can be seen to be proportional both to the speed of the exhaust gases, and to the rate at which fuel is consumed.

Example 7.5: vertical ascent

Consider first a simple situation in which a rocket rises vertically. The total mass of the rocket can be denoted by M, and its payload by m_0. The initial mass of the fuel contained in the rocket can be denoted by εM where $0 < \varepsilon < 1$. It may now be assumed that, while the rocket is firing, fuel is expelled at a constant rate kM vertically downwards with constant speed c relative to the rocket. Since $\dot{m} = -kM$ the total mass of the rocket and payload while the fuel lasts is

$$m = M + m_0 - kMt$$

It can therefore be seen that the fuel will be exhausted after a time given by ε/k.

If the gravitational force is assumed to be constant, and the drag can be neglected, the equation of motion of the rocket can be written as

$$\frac{dv}{dt} = \frac{kc}{1 + \frac{m_0}{M} - kt} - g$$

which can immediately be integrated to give

$$v=-c\log\left(1-\frac{kMt}{M+m_0}\right)-gt+\text{const}$$

From this it can be seen that the rocket will only rise initially from a stationary state on the ground if

$$kcM>(M+m_0)g$$

It can also be seen that, while the fuel is being burnt at a constant rate, the acceleration of the rocket is increasing monotonically.

It may finally be pointed out that the final speed of the payload can be greater than c if

$$-\log\left(1-\frac{\varepsilon M}{M+m_0}\right)>1+\frac{g\varepsilon}{ck}$$

Example 7.6: multistage rockets

It is now possible to demonstrate the advantages of using multistage rockets. For this purpose it is convenient to neglect the gravitational force, and the rocket thrust may be considered to be the only force acting. It is also convenient only to consider motion in a straight line.

Consider a two-stage rocket in which the stages have mass M_1 and M_2. The mass of the payload is denoted by m_0. It may be assumed that both stages contain the same fraction ε of fuel, and that the exhaust gases in each case are ejected at the same relative speed c. However, the fuel may be burnt at a different rate in each stage, and these can be denoted by k_1M_1 and k_2M_2 respectively.

While the rockets in the first stage are burning, the total mass is

$$m=M_1+M_2+m_0-k_1M_1t$$

and the equation of motion implies that

$$\frac{dv}{dt}=\frac{k_1M_1c}{M_1+M_2+m_0-k_1M_1t}$$

The fuel in this stage will be exhausted after a time ε/k_1, and the integral of the equation of motion implies that the speed of the rocket at this time is

$$V_1=-c\log\left(1-\frac{\varepsilon M_1}{M_1+M_2+m_0}\right)$$

At this point the empty rocket case of the first stage can be discarded. It is convenient here to assume that it is simply unattached, so that it separates at zero relative speed and the procedure does not affect the speed of the remaining rocket.

The rocket motor of the second stage may now be ignited, and while fuel is being consumed the remaining total mass is

$$m=M_2+m_0-k_2M_2(t-\varepsilon/k_1)$$

The fuel in this stage is consumed after a further time ε/k_2. The empty rocket case of this stage may similarly be discarded, and the integral of the equation of motion implies that the payload has a final velocity given by

$$V_2 = V_1 - c \log\left(1 - \frac{\varepsilon M_2}{M_2 + m_0}\right)$$

It is now possible to attempt to optimise the rocket design by adjusting the size of each stage. It is appropriate to assume that the total mass $M_1 + M_2$ is a constant denoted by M. The final speed of the payload can then be expressed in terms of the single variable parameter M_2 in the form

$$V_2 = -c \log\left(1 - \frac{\varepsilon(M - M_2)}{M + m_0}\right) - c \log\left(1 - \frac{\varepsilon M_2}{M_2 + m_0}\right)$$

It can now be shown that the derivative of this with respect to M_2 is zero when

$$M_2 = \sqrt{(m_0 M + m_0^2)} - m_0$$

and with this value V_2 is a maximum. In situations, however, in which rockets are used to launch satellites, the mass of the payload is very small compared to the mass of the whole rocket. At this level of approximation the fractional size of the second stage to achieve maximum final speed is given approximately by

$$\frac{M_2}{M} = \sqrt{\left(\frac{m_0}{M}\right)}$$

To illustrate the advantage achieved in this way, substitute the values $M = M_1 + M_2 = 100\ m_0$ and $\varepsilon = \frac{5}{6}$. According to the above result, it would be best to choose $M_2 = 0.1M$ and $M_1 = 0.9M$. With these figures the final speed of the payload would be $2.77c$. By comparison, a similar single stage rocket with $M = 100m_0$ and $\varepsilon = \frac{5}{6}$, could only achieve a final speed of $1.74c$.

It has thus been demonstrated that the advantage achieved by using a two-stage rocket is very significant. In fact it is only by utilising this advantage that lunar and interplanetary space vehicles can be launched. It should also be pointed out, however, that, although a three-stage rocket would give a further improvement, the advantages achieved by subsequent stages are much less significant.

Exercises

7.9 A rocket whose total combined mass is M rises vertically by burning fuel at a constant rate kM, and ejects the exhaust gases vertically downwards at a constant speed c relative to the rocket. If the gravitational field is assumed constant and if resistance is neglected, show that, provided $kc > g$, the height reached when the fuel is all consumed is

$$\frac{c}{k}\varepsilon + \frac{c}{k}(1 - \varepsilon)\log(1 - \varepsilon) - \tfrac{1}{2}g\frac{\varepsilon^2}{k^2}$$

where ε is the fractional mass of fuel in the rocket initially.

7.10 Consider a three-stage rocket in which each stage contains the same proportion of fuel, the exhaust gases being ejected at the same relative speed. If the mass of each stage is denoted by M_1, M_2 and M_3 and the payload has mass m_0, show that, when the gravitational and resistance forces are ignored, the maximum possible speed that can be obtained for the payload for any given total mass of fuel is achieved when

$$M_2 = M_3\left(\frac{M_3}{m_0}+1\right)$$

$$M_1 = M_3\left(\frac{M_3}{m_0}+1\right)^2$$

Hence show that in this case the mass of the payload divided by the total initial mass is

$$\left(\frac{M_3}{m_0}-1\right)^{-3}$$

7.8. Constraints and degrees of freedom

When considering applications of the theory of systems of several particles, it is not usual for the individual particles to be permitted to move perfectly freely. For example, when analysing the motion of a body in terms of component particles, it is usually assumed that each particle must remain at a fixed point of the body. In such cases it is necessary to impose certain conditions on the motions of the particles. These conditions are referred to as *constraints.*

In the analysis of the motion of a system, the constraints have to be taken into account. It is clear that the imposition of constraints on the motion of particles reduces their freedom of motion and, since the constraints are necessarily satisfied, it is the remaining free motion that is required to be analysed. It is therefore very important to know exactly how much freedom each particle actually has. A convenient measure of this concept can now be introduced in terms of a number of degrees of freedom.

Basically, the number of degrees of freedom belonging to a given system can be taken to be equal to the number of independent parameters that are required to specify the system exactly. This approach gives rise to an initial definition that *a system of particles is said to possess n degrees of freedom if the position in space of every particle of the system is determined by n independent parameters.*

According to this definition a single particle that is free to move in space has three degrees of freedom. This can be seen since it requires three independent parameters to specify any point in a three-dimensional space. Similarly, it can be seen that for rectilinear motion, in which a particle moves along a fixed straight line, there is only one degree of freedom. In this latter example it is possible to introduce cartesian coordinates with the x axis being the line along which the motion occurs. The three parameters (x, y, z) then determine the position of the particle, but in this case two additional constraints, $y=0$ and $z=0$, have to be imposed. This illustrates the general property that the number of degrees of freedom of a system is equal to the number of parameters which specify it, less the number of independent constraints.

The initial definition of the number of degrees of freedom of a system is perfectly correct. But it only applies to systems which can be described completely in terms of n independent parameters. However, in practice many systems appear which cannot be described in terms of any number of independent parameters. Such systems have to be described in terms of a number of parameters, some of which are related according to certain constraints. Thus it is preferable to adopt a general definition of the number of degrees of freedom in the following way.

***Definition* 7.2.** *If the position in space of every particle of a system is determined by n parameters which are subject to p independent constraints, then the system is said to possess $n-p$ degrees of freedom.*

It is now appropriate to consider the subject of constraints in greater detail.

To start with we must consider what actually occurs when a constraint is imposed. A constraint is a condition which somehow limits the motion of a system. For example, a particle may be constrained to remain on a given surface or line. Or two particles may be constrained such that the distance between them remains constant. In such cases the constraints arise from the physical problem being analysed. They are introduced in an attempt to simplify the theoretical analysis by specifying an aspect of the motion that is already known. For example, the constraint that a particle remains on a surface may be introduced instead of a set of assumptions about the forces which act in order to keep the particle on the surface. Similarly, when a solid body is observed to roll over a surface, the rolling condition may be inserted as a constraint instead of introducing assumptions about the forces which act at the point of contact between the body and the surface. Thus it can be seen that constraints are often introduced in order to simplify the analysis by specifying the effect of forces which need not then be considered. However, it should be understood that there are always forces associated with any constraint. These are referred to as the *forces of constraint.*

In some of the above examples, the constraints that have been considered can be represented in terms of conditions on the parameters of the system. Such conditions can be expressed as functional relationships between the parameters and also, possibly, time. Each constraint of this type can be used to eliminate one of the parameters which describe the system. For example, in the simple case of rectilinear motion the constraints $y=0$ and $z=0$ can be used to specify the parameters y and z, and the only parameter which remains to describe the motion is x. Constraints of this type are described as being holonomic. This term may first be defined, and then clarified in the following discussion.

***Definition* 7.3.** *A constraint which can be expressed as a functional relationship between the parameters of a system including, possibly, time, is described as being holonomic. A constraint which cannot be expressed in this way is described as being non-holonomic.*

If the parameters of a system are denoted by $q_1, q_2, \ldots, q_n$, then a holonomic constraint is one which can be expressed in the form

$$f(q_1, q_2, \ldots, q_n, t)=0$$

It can immediately be seen that such a constraint can be used, at least implicitly, to determine one of the parameters in terms of the others. If a system satisfies p independent constraints of this form, then p of the parameters can be specified and the system can be described completely in terms of the remaining $n-p$ parameters which are then independent. Using this approach the system is finally described using the same number of parameters as it has degrees of freedom. Such a system is referred to as a *holonomic system.*

A nonholonomic system is one which has at least one constraint which cannot be written in the above form. It may be assumed that any holonomic constraints have been used to reduce the number of parameters of the system to n. The remaining number p of non-holonomic constraints implies that the system has $n-p$ degrees of freedom, but these constraints cannot be used to further reduce the number of parameters. Thus a nonholonomic system must necessarily be described in terms of more parameters than it has degrees of freedom.

Nonholonomic constraints have been defined above in a negative way. It is therefore appropriate to give some examples to illustrate some of the forms they may take. They have been defined to include every type of constraint which cannot be used to reduce the number of parameters of the system. Thus all types of inequality constraint are clearly nonholonomic. However, the most frequently quoted examples of nonholonomic constraints are rolling conditions. Constraints of this type are usually expressed in terms of relationships between the derivatives of the parameters which describe the system. Simple rolling conditions are represented as linear equations in the derivatives or velocities, but the coefficients are generally functions of the parameters. Normally, such equations cannot be integrated. Of course, if they could be integrated this would give a relation between the parameters and the constraints would be holonomic. Thus all nonintegrable constraints which relate the parameters and their derivatives are necessarily nonholonomic.

It should also be noticed that all possible configurations of a holonomic system with n degrees of freedom can be represented by points of an n-dimensional space. In contrast to this, a nonholonomic system also having n degrees of freedom, but subject to p nonholonomic constraints, requires a point in an $n+p$-dimensional space to represent its configuration at any time. In fact, for most nonholonomic systems no more configurations of the system would be accessible if the nonholonomic constraints were removed. Inequality constraints are an exception here, in that they sometimes restrict the boundary of the configuration space, but they do not reduce its dimensionality. This point may be summarised in the following result which also indicates the importance of carefully distinguishing between these two types of constraint.

***Proposition* 7.9.** *The dimension of the space describing the configurations of a system is reduced by the introduction of holonomic constraints, but not by nonholonomic constraints.*

In addition to the above classification, it is convenient to introduce another distinction between different types of constraint. This second classification is based on the question as to whether or not a constraint is time dependent. Again it is convenient to define first the terms used, and then to discuss their significance.

***Definition* 7.4.** *Constraints which do not explicitly depend on time are described as scleronomic. Time-dependent constraints are described as rheonomic.*

In the same way, a system may be classified either as rheonomic or scleronomic according to whether or not it contains a time-dependent constraint.

Scleronomic constraints are sometimes alternatively described as 'fixed' or 'workless', since the forces associated with these constraints do no net work in an arbitrary displacement of the system. For example, the constraint that two particles move in such a way that their distance apart is constant is clearly both holonomic and scleronomic. It can also be seen that, since the forces of constraint which act on the two particles are equal and opposite, the total work done by these forces is zero in any motion of the particles which satisfies the constraint.

Rheonomic constraints, on the other hand, always introduce an explicit time dependence into the motion of a system, and the forces associated with them necessarily do work. They are therefore sometimes referred to as 'moving' constraints. For example, the constraint that the distance between two particles is a given function of time is holonomic and rheonomic. In this case it can clearly be seen that the forces of constraint must do work, and since work is done the energy of the system is altered. It can therefore be seen that in rheonomic systems the total energy cannot be constant.

***Proposition* 7.10.** *An energy integral is not possible in any system which satisfies a rheonomic constraint.*

7.9. An approach to the dynamics of fluids

It is appropriate at this point to consider the application of the principles of classical dynamics to the motion of liquids and gases, which are together known as fluids. These particular applications have been considered ever since theories of dynamics were first proposed, and the classical theory of fluid dynamics has now become a fully developed theory in its own right. It is not an easy theory to apply in general, and it relies heavily on mathematical techniques.

It is therefore not appropriate here to do more than give a brief sketch of some of the basic equations.

It should, however, be pointed out that the contents of this section will not be referred to again in the remaining chapters of this book. They are not an essential part of the approach developed here but, rather, form an introduction to the more specialised theory of fluid dynamics. In addition, a deeper knowledge of vector field theory is assumed in this section than is required in the remainder of the book.

The basic approach to the study of the dynamics of fluids is that already suggested above, in which the fluid is represented by a continuum. This continuum is then broken down theoretically into component parts which may be taken to be arbitrarily small, so that each part may be represented by a classical particle. In this apparently contradictory approach it is possible to speak about particles of the fluid at the same time as regarding the fluid as a continuum. Both aspects are of course idealisations, and both can be justified provided the real discrete physical components of the fluid are sufficiently small in any particular situation.

According to the continuum approach to the study of fluids, mass is continuously distributed throughout the fluid. It is therefore convenient to consider the *density* of mass at any point. This may be defined as the ratio of the mass contained in a small closed surface containing the point to the volume enclosed by that surface, taken in the limit as the volume becomes infinitesimally small. Since mass can be assumed to be continuously distributed, such a limit exists and is unique at any instant of time. The density of a fluid thus defined is in general a function of position and time and is denoted by $\rho(\boldsymbol{r}, t)$.

It is now possible to analyse the motion of a fluid by considering its velocity and acceleration at any point. At every point within the fluid a unique velocity is determined at any time, and the velocity of the fluid thus defined may be represented by a vector field, denoted here by $\boldsymbol{v}(\boldsymbol{r}, t)$.

Having defined the velocity of the particles of the fluid, their acceleration is defined simply as their rate of change of velocity. However, this is not just the partial derivative of the velocity field with respect to time, since a steady flow is possible which does not vary with time but which varies with position in such a way that the particles of the fluid are all accelerating. Thus the acceleration of the fluid must include a term allowing for the way in which the velocity varies with position. This is given in the vector form

$$\frac{\mathrm{d}\boldsymbol{v}}{\mathrm{d}t}=\frac{\partial \boldsymbol{v}}{\partial t}+(\boldsymbol{v}\cdot\nabla)\boldsymbol{v}$$

which, mathematically, is the total time derivative of the velocity field. Now, using an identity which is well known in vector field theory, the acceleration may be rewritten as

$$\frac{\mathrm{d}\boldsymbol{v}}{\mathrm{d}t}=\frac{\partial \boldsymbol{v}}{\partial t}+\tfrac{1}{2}\operatorname{grad} v^2-\boldsymbol{v}\times\operatorname{curl}\boldsymbol{v}$$

This form is particularly useful in some types of fluid motion.

When the particles of a fluid are in motion, the density may be caused to vary. Some constraint must therefore exist which relates the velocity field to the rate of change of density. Such a constraint may be obtained by considering an imaginary fixed closed surface S containing a volume V in some region of the fluid. If it is assumed that no mass is 'created' inside the surface, then the rate of change of mass inside the volume must be equal to the rate at which mass flows across the surface. This can be expressed by the equation

$$\frac{\partial}{\partial t}\int_V \rho\,\mathrm{d}V=-\oint_S \rho\boldsymbol{v}\cdot\mathrm{d}\boldsymbol{S}$$

Now, since the surface is assumed not to vary with time, and using Gauss's divergence theorem, this equation may be rewritten in the form

$$\int_V\left\{\frac{\partial\rho}{\partial t}+\operatorname{div}(\rho\boldsymbol{v})\right\}\mathrm{d}V=0$$

Finally, since this must be satisfied for any arbitrary closed surface within the fluid, it is necessary that

$$\frac{\partial\rho}{\partial t}+\operatorname{div}(\rho\boldsymbol{v})=0$$

This result is known as the *continuity equation*. It can be seen by the above argument to be mathematically equivalent to the reasonable physical assumption that mass is not being created or destroyed at any point within the fluid.

In the general motion of fluids the continuity equation is usually assumed to hold universally. However, in some applications it may be convenient to introduce particular points to represent 'sources' or 'sinks' for the fluid. In such cases, it is presumed that some physical device is present which is capable of inserting or removing fluid at distinct points. The continuity equation for the fluid is then satisfied everywhere except at such points.

So far, the fluid motion has been discussed without any reference to the possible causes of the motion. Such causes must now be considered in terms of the various forces which act on the particles of the fluid. As well as external forces, internal pressures and stresses play an important rôle and may be introduced first.

Consider initially the case of a fluid which is at rest. If a small surface is inserted in the fluid, it is found that equal and opposite forces act on each side. The *pressure* at any point may be defined as the magnitude of the force acting on one side of a surface situated at that point, divided by the area of the surface, evaluated in the limit as the area becomes infinitesimally small. When a fluid is not in motion, the force acting on a surface is found to be always normal to the surface and also to be independent of the orientation of the surface. In such a case the pressure is said to be *isotropic*, and can be represented by a scalar field $p(\mathbf{r}, t)$ which depends at most on position and, if motion is included, on time as well.

Now it may intuitively be expected that pressure and density would be related quantities at any point within a fluid. Consider first the case of an *incompressible* fluid. In such a case the density remains constant whatever the pressure exerted on it. Such a situation occurs, at least approximately, with most liquids. In contrast to this, gases may usually be compressed very easily, and the density at any point is clearly related to the pressure. Unfortunately, the situation is further complicated since the local temperature of the gas must also be included in a general relationship between pressure and density. However, if it may be assumed that the gas expands without a transfer of heat, then it can be shown that the pressure and density are at least approximately related according to the equation

$$p = k\rho^{\gamma}$$

where k and γ are appropriate constants.

Having defined the concept of pressure in the static case, the dynamical situation may again be considered. In this case an additional complication appears in that, if an arbitrary surface is inserted in the fluid, that surface would interfere with the fluid flow. It is appropriate therefore only to consider the forces on small surfaces which are moving with the fluid.

It is also found in practice that the force acting on such a small surface is not necessarily normal to the surface. An additional tangential force component generally appears. Such a force may loosely be thought of as arising from frictional forces induced by particles of the fluid sliding past each other. Fluids, unlike solids, are set in motion when they are subjected to shearing forces, but they still provide some resistance. This ability to resist shearing motion is referred to as the *viscosity* of the fluid. It is this which gives rise to the tangential forces acting on surfaces which are moving with the fluid.

All fluids are viscous to some extent, and the equations that can be developed for the motion of viscous fluids are extremely complicated. However, in many physical situations the forces arising from the viscosity are negligibly small. In such situations it is appropriate to introduce the simplifying assumption that the viscosity of the fluid is zero. Fluids satisfying this assumption are referred to as being ideal or *inviscid*. In most applications, however, the viscosity only gives rise to significant effects when the velocity gradient is large. This occurs most commonly near the boundaries of the fluid flow. It is therefore often mathematically convenient to consider separately the boundary layers in which the forces arising from the viscosity are significant. The remainder of the fluid may then be represented by an inviscid fluid, which is much easier to analyse mathematically.

It is not appropriate here to proceed to derive general equations for viscous fluids. Instead, for the remainder of this section, only ideal or inviscid fluids are considered.

The equation of motion for an inviscid fluid may now be derived by considering a finite volume V of the fluid which is enclosed by a surface S. This volume may be divided into n small components $\delta V_i (i = 1, \ldots, n)$, where n is sufficiently large that the fluid in each component may be represented by a particle situated at a point with position vector $\boldsymbol{r}_i$ inside δV_i and having mass $\rho(\boldsymbol{r}_i, t)\delta V_i$. As before, the forces acting on each element may be distinguished as internal and external forces. In fluid dynamics it is convenient to assume that the external force can be represented by a vector field $\boldsymbol{f}(\boldsymbol{r}, t)$ which describes the force per unit mass acting on particles of the fluid having position vector $\boldsymbol{r}$. An external gravitational field, which is the usual source of the external forces that are considered, can easily be expressed in this form. With this notation the external force acting on each component is $\boldsymbol{f}(\boldsymbol{r}_i, t)\rho(\boldsymbol{r}_i, t)\, \delta V_i$. It is also convenient to assume that, if the viscosity is ignored, the internal forces can be represented purely in terms of the pressure at any point. Thus the sum of the internal forces acting on each element can be represented as the integral of the pressure over the surface δS_i of the volume δV_i. The equation of motion for each component is therefore given by

$$\rho(\boldsymbol{r}_i, t)\,\delta V_i \frac{\mathrm{d}\boldsymbol{v}_i}{\mathrm{d}t} = \boldsymbol{f}(\boldsymbol{r}_i, t)\rho(\boldsymbol{r}_i, t)\,\delta V_i - \oint_{\delta S_i} p\,\mathrm{d}\boldsymbol{S}$$

Summing this equation over all components of the volume V gives

$$\int_V \rho \frac{\mathrm{d}\boldsymbol{v}}{\mathrm{d}t}\,\mathrm{d}V = \int_V \rho \boldsymbol{f}\,\mathrm{d}V - \oint_S p\,\mathrm{d}\boldsymbol{S}$$

since the surface integrals over common surfaces cancel. Now, using an extension of Gauss' theorem, this equation may be written as

$$\int_V \left\{\rho \frac{\mathrm{d}\boldsymbol{v}}{\mathrm{d}t} - \rho \boldsymbol{f} + \mathrm{grad}\, p\right\} \mathrm{d}V = 0$$

And since this must be satisfied for any arbitrary volume at any time, the equation of motion for an inviscid fluid may be written in the form

$$\frac{\mathrm{d}\boldsymbol{v}}{\mathrm{d}t} = \boldsymbol{f} - \frac{1}{\rho}\,\mathrm{grad}\, p$$

where it must be remembered that $\boldsymbol{f}$ is the external force per unit mass that acts on the particles of the fluid. As may have been expected, the effect of the internal forces is represented in terms of the pressure gradient.

It has already been mentioned that the external forces which act on the particles of the fluid are most commonly due to a gravitational field. Such forces are conservative. It therefore often occurs that the field $\boldsymbol{f}(\boldsymbol{r}, t)$ is irrotational and it is therefore possible to introduce a scalar field $\Omega(\boldsymbol{r}, t)$ such that

$$\boldsymbol{f} = -\mathrm{grad}\,\Omega$$

The field $\Omega(\boldsymbol{r}, t)$ may be interpreted as the potential energy per unit mass of the particles of the fluid.

Another quantity that is considered in the analysis of fluids is the *vorticity*. Denoting this by $\boldsymbol{\zeta}(\boldsymbol{r}, t)$, it is a vector field defined by

$$\boldsymbol{\zeta} = \mathrm{curl}\,\boldsymbol{v}$$

It can easily be shown that this may be interpreted as twice the local angular velocity of a small element of the fluid at any point about that same point.

It follows from the definition that a fluid has zero vorticity at all points if, and only if, the velocity field is irrotational. In such a case it is possible to write the velocity as the gradient of some potential function $-\phi(\boldsymbol{r}, t)$.

$$\boldsymbol{\zeta}=0 \quad \Leftrightarrow \quad \boldsymbol{v}=-\text{grad}\,\phi$$

Fluid flow which satisfies these conditions is known as potential flow. It may be thought that these conditions are so restrictive that they would only very rarely occur in applications. However, this is not so. In situations where the viscosity may be neglected, the flow is often irrotational. Thus, outside boundary regions, fluid motion may often be treated as potential flow.

Finally, it is appropriate to state the basic equations for incompressible fluids. Liquids are, at least approximately, incompressible, and therefore this condition is frequently used in applications. If the fluid is also homogeneous the density of the fluid can be regarded as a constant, and the continuity equation simply reduces to the condition

$$\text{div}\,\boldsymbol{v}=0$$

Also, if the external force field is conservative and viscosity is neglected, the equation of motion can be written in the form

$$\frac{\partial \boldsymbol{v}}{\partial t}-\boldsymbol{v}\times\boldsymbol{\zeta}=-\text{grad}\left(\frac{P}{\rho}+\tfrac{1}{2}v^2+\Omega\right)$$

It can easily be shown that this equation is integrable provided

$$\frac{\partial \boldsymbol{\zeta}}{\partial t}-\text{curl}\,(\boldsymbol{v}\times\boldsymbol{\zeta})=0$$

This equation incidentally indicates that if the vorticity is zero at any time, then it remains zero for all time. Irrotational flow therefore remains irrotational in this case.

8

Rigid body dynamics

8.1. The concept of a rigid body

Most simple elements and compounds exist in three possible states, gaseous, liquid and solid. According to the methods described in the previous chapter, the motion of bodies composed of matter in any of these states can be described in terms of a large number of small particles. The motions of gases and liquids have been briefly discussed at the end of the last chapter, and we turn now to consider the motion of solid bodies.

In general, solid bodies have a very complicated structure, and in their motion some parts move relative to other parts. However, in many applications, particularly in mechanical problems, the component parts of a body move in an approximately rigid way. Thus it is convenient to introduce the concept of a rigid body as a theoretical idealisation. This may be defined as follows.

***Definition* 8.1.** *A body is said to be rigid if the distance between each of its constituent points remains constant, irrespective of the motion of the body as a whole or the forces that act upon it.*

Of course no physical body is exactly rigid. Even objects made of solid steel bend or deform slightly when acted on by external forces. But when considering the motion of solid objects, it is often convenient to regard them as being perfectly rigid, at least as a first approximation. Thus, in all applications of the study of rigid body dynamics, there must always be an initial simplifying assumption that the body under consideration should be regarded as being perfectly rigid.

According to the classical approach, physical bodies or their theoretical representation can be divided into a large number of small parts which are sufficiently small for each part to be represented by a particle. Sometimes it is convenient to let each part become arbitrarily small so that the mathematical techniques of differentiation and integration can be applied. For this it is necessary to introduce the simplifying assumption that mass is continuously distributed throughout the body. However, whether this additional assumption is included or not, rigid bodies are regarded as being composed of particles.

In a rigid body the distribution of mass does not alter relative to the body. Thus, once a body is regarded as a collection of particles, the condition of rigidity can be stated more clearly. A system of particles is said to be rigid if, and only if, the distance between any pair of particles remains constant, even though the system as a whole may move. This can be expressed in terms of the set of constraints

$$(\boldsymbol{r}_i-\boldsymbol{r}_j)^2-c_{ij}=0$$

where $\boldsymbol{r}_i$ and $\boldsymbol{r}_j$ are the position vectors of any two particles of the body and c_{ij} is the square of the distance between them. These constraints are both holonomic, since they each express a functional relationship between the coordinates, and scleronomic since they do not involve time explicitly. However, they are not all independent. This can be shown by first considering the number of degrees of freedom that a body has in general.

The position of a rigid body in space is uniquely determined if the position of three noncolinear points of it are known. Each of these points can be defined in terms of three coordinates. However, these nine parameters are not independent. They are subject to the three constraints that the distance between each pair of points is constant. Thus, a rigid body which is otherwise free to move in space has six degrees of freedom.

It can now be seen that six parameters are required to specify the position and orientation of a rigid body in space. Once this is determined the above constraints can be used to determine the position of every particle of the body. In particular, it should be noticed that for a rigid body the centroid of the particles composing the body is always at a fixed point relative to the body.

In the general analysis of systems of several particles, it has been found convenient to distinguish between the internal and external forces which act on the individual particles. For a solid body, however, it is the internal forces which keep the body approximately rigid. Thus, in the idealisation of a perfectly rigid body, it is convenient to introduce the constraint of rigidity as a way of expressing the effects of the internal forces. In the analysis of the motion of a rigid body, it is therefore only the external forces that need to be considered, together with the constraint of rigidity. The internal forces appear only as the forces of constraint, and in practice these rarely need to be known.

8.2. The possible motions of a rigid body

It is now convenient to consider the equations of motion for a rigid body. These can be quoted directly from the previous chapter since a rigid body is regarded as a collection of particles. It is also convenient at this point to assume that an inertial frame of reference is being used.

The first equation, then, according to proposition 7.5, is that the total mass times the acceleration of the centroid of the body is equal to the sum of the external forces

$$m\ddot{\bar{\boldsymbol{r}}} = \boldsymbol{F}$$

According to proposition 7.1, the total linear momentum of the body is equal to its total mass times the velocity of its centroid. The above result can therefore be stated in terms of the rate of change of the linear momentum of the body.

The second equation, according to proposition 7.7, is that the rate of change of the angular momentum of the body about its centroid is equal to the sum of the moments about the centroid of the external forces which act on it

$$\frac{\mathrm{d}}{\mathrm{d}t}\boldsymbol{L}_{\mathrm{c}} = \boldsymbol{M}_{\mathrm{c}}$$

The general motion of a rigid body in space has six degrees of freedom. In this case the two equations above are independent and, since they are both vector equations, they provide six equations for the six parameters required. Thus all the possible motions of a rigid body in space are determined completely by these equations. It is the combination of the simplicity of this pair of equations with their completeness, which gives the subject of rigid body dynamics its importance. It has already been pointed out that no physical body is exactly rigid. However, it can now be seen that the assumption of rigidity not only simplifies the physical analysis, but also its mathematical formulation.

The main difference between the analysis of the motion of a rigid body and that of a particle now appears in that the orientation of the rigid body is important. The motion of a particle is determined entirely in terms of its position at any time. In its unconstrained motion in space it has three degrees of freedom, and its motion is determined by one vector equation $m\ddot{\boldsymbol{r}} = \boldsymbol{F}$. In contrast to this, in order to determine the motion of a rigid body it is necessary to specify at all times both the position of some point of the body and also its orientation. It is now appropriate to divide the six degrees of freedom for its general motion into two groups of three, associated respectively with its translational and rotational motion.

It is well known that a general displacement of a rigid body can be represented as the combination of a translation and a rotation. Moreover, these two aspects are independent. So that it does not matter whether the rotation follows or precedes the translation, or whether they occur simultaneously. In the continuous motion of a rigid body, therefore, these two aspects may be treated separately.

In a general displacement, the translation can be regarded as a motion in which every particle is displaced by an equal amount parallel to that of some assumed point of the body. A general translation is thus a vector quantity having three degrees of freedom. The rotation can then be regarded as a rotation of the whole body about an axis through this assumed point. This also has three degrees of freedom since two parameters are necessary to define the axis and one more is required to determine the magnitude of the rotation about this axis.

When analysing the motion of a rigid body, it can now be seen to be generally convenient to take the assumed point mentioned above to be the centroid of the body. The translational motion of the body is then just the motion of its centroid, and this is determined by the first of the above equations of motion. This determines the motion of the particle representation of the body.

The rotational motion of the body is now analysed in terms of the rotational motion about its centroid, and this is determined by the second of the above equations of motion. This also is a vector equation whose solution specifies the three parameters which describe the orientation of the body at any time.

The otherwise unconstrained motion of a rigid body can thus be analysed in two parts. There is one equation for the motion of the centroid, and another for the rotational motion about the centroid. However, if the motion is constrained in some way, then it is not always convenient to make this particular distinction. For example, if the motion of the body is such that one point in it is constrained to remain at a fixed point, the motion has only three degrees of freedom. These may be associated with the bodies of freedom of orientation about the fixed point. Since there are only three degrees of freedom, only one vector equation of motion is required. According to proposition 7.6, this may be taken in the form that the rate of change of angular momentum about the fixed point is equal to the sum of the moments of the external forces about that point.

$$\frac{\mathrm{d}}{\mathrm{d}t}\boldsymbol{L}=\boldsymbol{M}$$

In this equation it has been assumed that the fixed point is chosen as the origin. In most applications of this equation, it may be assumed that the frame of reference being used is inertial. However, if this is not so, then the additional fictitious forces that arise have to be included with the external forces.

8.3. Moments of inertia

Before proceeding with a general analysis it is appropriate first to consider the motion of a rigid body which is rotating about a fixed axis through it. In this case the body has only one degree of freedom, which is the freedom of rotation about the axis.

It is convenient to choose the origin for a frame of reference to be some point on the axis. Also, if the angular velocity of the body is represented by the vector $\boldsymbol{\omega}$, then the direction of this vector is the direction of the axis. Thus, representing a unit vector along the axis by $\boldsymbol{u}$, the angular velocity is

$$\boldsymbol{\omega}=\omega\boldsymbol{u}$$

where ω is the rate of rotation about the axis.

Now regarding the body as a system of n particles, each with position vector $\boldsymbol{r}_i$, the velocity of each particle is given by the expression

$$\boldsymbol{v}_i=\boldsymbol{\omega}\times\boldsymbol{r}_i \qquad (i=1,2,\ldots,n)$$

The total angular momentum vector of the body can now be obtained from section 7.2 in the form

$$\boldsymbol{L}=\sum_{i=1}^{n}\boldsymbol{r}_i\times m_i(\boldsymbol{\omega}\times\boldsymbol{r}_i)$$

However this expression may be a little misleading at this point, since it is not necessarily in the direction of the axis about which the body rotates. It is therefore appropriate here to consider the total angular momentum of the body about the axis of rotation. This is just the component of the vector $\boldsymbol{L}$ in the direction of the axis. Using the above notation, this is

$$\begin{aligned} L = \boldsymbol{L}\cdot\boldsymbol{u} &= \sum_{i=1}^{n} \boldsymbol{u}\cdot\{\boldsymbol{r}_i\times m_i(\boldsymbol{\omega}\times\boldsymbol{r}_i)\} \\ &= \sum_{i=1}^{n} m_i(\boldsymbol{u}\times\boldsymbol{r}_i)\cdot(\boldsymbol{u}\times\boldsymbol{r}_i)\omega \\ &= \sum_{i=1}^{n} m_i d_i^2\omega \end{aligned}$$

where d_i is the perpendicular distance of the ith particle from the axis of rotation (see figure 8.1). It should be noticed that at this point L is not the magnitude of the vector $\boldsymbol{L}$, but the component of $\boldsymbol{L}$ in the direction of the fixed axis of rotation.

This equation can now be conveniently written in the form

$$L = I\omega$$

where I is called the moment of inertia of the body about the axis. This is given by the expression

$$I = \sum_{i=1}^{n} m_i d_i^2$$

Fig. 8.1. A body rotates with angular velocity $\boldsymbol{\omega}$ about a fixed axis in direction $\boldsymbol{u}$. The origin O is a point on the axis and P is a particle in the body with position vector $\boldsymbol{r}$; $d = |\boldsymbol{u}\times\boldsymbol{r}|$ is the perpendicular distance of P from the axis.

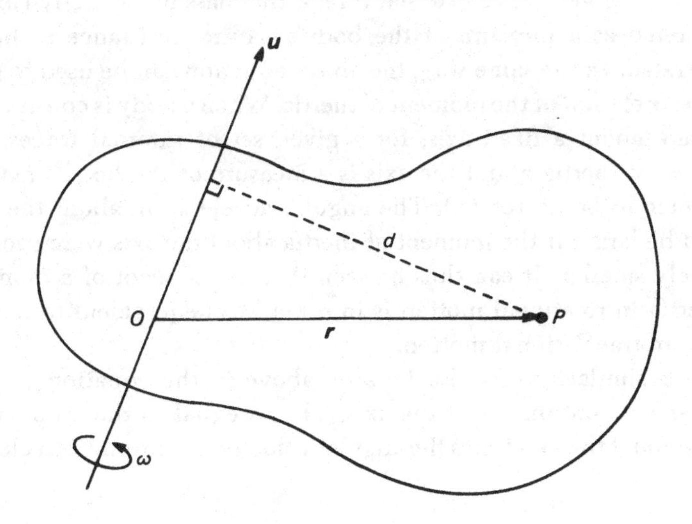

It can immediately be seen that, if the body is rigid, this quantity is a constant for any particular body and any fixed axis within it. It does not depend in any way on the motion of the body. Rather, it is a characteristic of the body's mass distribution about the axis. It may now be defined formally.

***Definition* 8.2.** *The moment of inertia of a body composed of particles about a fixed axis is the sum of the products of the mass of each particle and the square of its perpendicular distance from the axis.*

In order to show the significance of the concept of a moment of inertia, it is necessary to consider the equation of motion. Since the body is assumed to be rotating about a fixed axis through the origin, this may now be written in the form

$$\frac{\mathrm{d}}{\mathrm{d}t} L = \boldsymbol{u} \cdot \boldsymbol{M}$$

This states that the rate of change of the angular momentum about the axis of rotation is equal to the sum of the moments of the external forces about the axis. However, it can now be seen that the rate of change of the angular momentum about the axis is equal to the moment of inertia times the angular acceleration about the axis

$$I\dot{\omega} = \boldsymbol{u} \cdot \boldsymbol{M}$$

This equation may now be compared with the equation for the linear motion of a body. This states that the mass times the acceleration of the centroid of the body is equal to the sum of the external forces. Thus, for a given set of external forces, the mass of the body can be interpreted as a measure of the body's natural resistance to being accelerated. In the same way, the above equation can be used to give an interpretation of the moment of inertia. When a body is constrained to rotate about a fixed axis, for a given set of external forces, the moment of inertia about the axis is a measure of the body's natural resistance to being rotated. The angular acceleration about the axis would be larger if the moment of inertia about the axis were proportionately smaller. It can thus be seen that the concept of a moment of inertia in rotational motion is in some ways equivalent to that of a mass in translational motion.

This similarity can also be seen above in the equation for the angular momentum about the axis. This is equal to the moment of inertia about the axis times the angular velocity. This result can clearly be seen to parallel the definition of linear momentum as the mass times the translational velocity.

While considering the rotational motion of a rigid body about a fixed axis within it, it is also convenient to obtain an expression for its kinetic energy. This is given by

$$\begin{aligned} T &= \sum_{i=1}^{n} \tfrac{1}{2} m_i v_i^2 \\ &= \sum_{i=1}^{n} \tfrac{1}{2} m_i (\boldsymbol{\omega} \times \boldsymbol{r}_i) \cdot (\boldsymbol{\omega} \times \boldsymbol{r}_i) \\ &= \tfrac{1}{2} \sum_{i=1}^{n} m_i d_i^2 \omega^2 \\ &= \tfrac{1}{2} I \omega^2 \end{aligned}$$

This expression again illustrates the apparent equivalence between moments of inertia and angular velocity in rotational motion on the one hand, and mass and linear velocity in translation motion on the other.

It is obvious that a knowledge of the moment of inertia of a body about an axis does not lead uniquely to a knowledge of its mass distribution. Distinct bodies with different mass distributions may well have the same moment of inertia about an agreed axis. And, if the moments of inertia are the same, so also is the equation of motion. Thus, when subjected to the same external forces, their motion would be identical. Thus, bodies having the same moments of inertia are indistinguishable dynamically in their rotational motion. They are then said to be *equimomental*, a concept that will be clarified later. The analogue of this in translational motion is simply that it is not possible to distinguish dynamically between different bodies of equal total mass by considering the motion of their centroids.

Another term that is associated with the moment of inertia is the radius of gyration. This may be defined as follows.

***Definition* 8.3.** *If a rigid body has a total mass m and its moment of inertia about a fixed axis is I, then the radius of gyration k about that axis is defined such that*

$$I = mk^2$$

It can immediately be seen that the radius of gyration of a body about an axis is a measure of the distribution of mass about the axis.

From the definition of the moment of inertia, it can be expressed in the form

$$k^2 = \frac{\sum_{i=1}^{n} m_i d_i^2}{\sum_{i=1}^{n} m_i}$$

This is just the weighted second moment of the perpendicular distances of the particles from the axis. This provides a physical interpretation of the radius of gyration. However, the moment of inertia is a measure of essentially the same property, and the radius of gyration is rarely referred to in practice.

It now remains in this section simply to make a few remarks on the way in which moments of inertia are evaluated in practice. Of course, if a body can be represented as a small number of distinct particles, then the appropriate terms can be summed according to the definition. However, it is more likely that the component parts of the body can be represented in terms of continuous mass distributions, and in this case techniques of integration have to be employed. In such cases, the definition of the moment of inertia can be written in the form

$$I = \int d^2 \, \mathrm{d}m$$

where d is the perpendicular distance from the axis of the infinitesimal mass $\mathrm{d}m$, and it is understood that the integration is taken over the whole body.

It should be obvious that the moment of inertia of a body about one particular axis, in general differs from that about another. It is therefore essential when stating a moment of inertia to state also the particular axis to which it refers. There are however a number of cases in which moments of inertia about different axes are simply related. These cases are described in the following two propositions.

***Proposition* 8.1.** *The moment of inertia of a body about a given axis is equal to the moment of inertia of the same body about another axis which is parallel to the given one and passes through the centroid, together with the product of the mass of the body and the square of the perpendicular distance between the two axes.*

The result is known as the *parallel axis theorem.* It can be proved by first writing the moment of inertia in the form

$$I = \sum_{i=1}^{n} m_i(\boldsymbol{u} \times \boldsymbol{r}_i)^2$$

where an origin has been chosen at a point on the required axis whose direction is that of the unit vector $\boldsymbol{u}$. The position of each particle relative to the centroid can now be introduced by putting

$$\boldsymbol{r}_i = \boldsymbol{\rho}_i + \bar{\boldsymbol{r}}$$

and the moment of inertia can be expanded as

$$\begin{aligned} I &= \sum_{i=1}^{n} m_i(\boldsymbol{u} \times \boldsymbol{\rho}_i + \boldsymbol{u} \times \bar{\boldsymbol{r}})^2 \\ &= \sum_{i=1}^{n} m_i(\boldsymbol{u} \times \boldsymbol{\rho}_i)^2 + 2(\boldsymbol{u} \times \bar{\boldsymbol{r}}) \cdot \sum_{i=1}^{n} m_i(\boldsymbol{u} \times \boldsymbol{\rho}_i) + m(\boldsymbol{u} \times \bar{\boldsymbol{r}})^2 \end{aligned}$$

It is clear that the middle term must be zero from the definition of the centroid. The remaining terms then give the required equation

$$I = I_c + md^2$$

where I_c is the moment of inertia of the body about an axis through the centroid parallel to $\boldsymbol{u}$, and $d = |\boldsymbol{u} \times \bar{\boldsymbol{r}}|$ is the perpendicular distance of the centroid from the original axis.

The parallel axis theorem is equivalent to the statement that the moment of inertia of a body about an axis is equal to its moment of inertia about a parallel axis through the centroid, together with the moment of inertia about the original axis of a particle of equal mass situated at the centroid of the body.

Proposition 8.1 clearly only applies with reference to a parallel axis through the centroid. However, if the moment of inertia of a body about one axis is known, then that about any axis parallel to it can be determined from the theorem by first using it to find the moment of inertia about the parallel axis through the centroid.

***Proposition* 8.2.** *The moment of inertia of a lamina about an axis perpendicular to its plane is equal to the sum of the moments of inertia about any two mutually perpendicular axes in the plane passing through the common point.*

This proposition is less useful than the previous one since it applies only to a lamina, or a thin plane body. It is sometimes known as the perpendicular axis theorem, but it is safer and more common to refer to it as the *lamina theorem*, since this emphasises that it does not apply to a general body.

Proposition 8.2 can be simply proved by choosing a set of cartesian axes, such that the plane of the body is given by $z=0$. The x and y axes are then an arbitrary pair of perpendicular axes in the plane of the body passing through the common point which is now the origin. The moment of inertia about the z axis is given by

$$\begin{aligned} I_3 &= \sum_{i=1}^{n} m_i d_i^2 \\ &= \sum_{i=1}^{n} m_i x_i^2 + \sum_{i=1}^{n} m_i y_i^2 \\ &= I_2 + I_1 \end{aligned}$$

where I_1 and I_2 are the moments of inertia about the x and y axes respectively, and take this form since $z=0$ for all particles of the body.

These two propositions can be used to obtain the moments of inertia of a body about a number of axes if those about one or two axes are known or can easily be calculated. It is therefore not necessary to evaluate all moments of inertia from first principles. In addition, the moment of inertia of a body is equal to the sum of the moments of inertia of its component parts taken about the same axis. A body may therefore sometimes be divided into standard component parts whose moments of inertia are known, and the total moment of inertia obtained by summation.

8.4. Evaluating moments of inertia: examples

The theoretical techniques that may be used to estimate the moment of inertia of a body about a particular axis are illustrated in this section, first by determining a number of standard results, and then by illustrating the use of the parallel axis and the lamina theorems.

Example 8.1: *uniform rod*

Consider here a thin rigid rod. Its mass may be denoted by m, and its length by $2a$. It is required to determine the moment of inertia of the rod about a line through its centre perpendicular to its length.

To achieve this it is convenient to theoretically divide the rod into n small pieces. Each piece may have length δx_i, and its centre located at a distance x_i from the line through the centre of the rod, as illustrated in figure 8.2. If it is assumed that mass is uniformly and continuously distributed along

Fig. 8.2. A uniform rod divided into n small pieces of length δx_i a distance x_i from the centre.

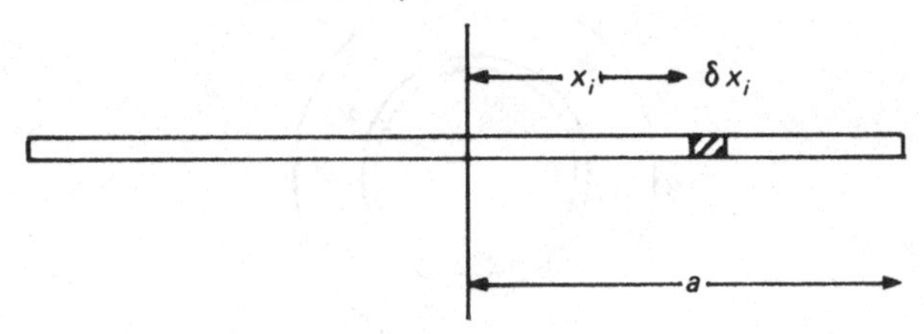

the length of the rod, then the mass of each piece can be expressed as

$$m_i = \frac{m}{2a}\,\delta x_i$$

The moment of inertia of the rod may now be obtained directly from the definition, by evaluating the sum

$$\sum_{i=1}^{n} m_i d_i^2 = \sum_{i=1}^{n} \frac{m}{2a}\,\delta x_i x_i^2$$

In the limit as each piece is considered to become arbitrarily small, this can be expressed as the integral, giving

$$I = \frac{m}{2a}\int_{-a}^{a} x^2\,\mathrm{d}x = \tfrac{1}{3}ma^2$$

Example 8.2: uniform disc

Determine the moment of inertia of a thin flat disc of radius a about an axis through its centre perpendicular to its plane. It may be assumed that mass is uniformly distributed across the disc and its total mass may be denoted by m.

As in the previous example, it is possible to theoretically divide the disc into a large number of small pieces. The required moment of inertia may then be evaluated from the sum by integrating over the area of the disc using the technique of double integration. This technique, however, may be avoided by considering together all the pieces that are the same distance from the axis, thus forming a set of concentric rings. Now, since the moment of inertia is a simple sum, the required result is the sum of the moments of inertia of each ring, and these are simply the mass of each ring times the square of its radius, since this is the perpendicular distance of each component particle of the ring from the axis.

The disc may thus be theoretically divided into a set of n concentric rings, each of radius r_i and thickness δr_i, as illustrated in figure 8.3. The mass of each ring is thus

$$m_i = \frac{m}{\pi a^2}\,2\pi r_i \delta r_i = \frac{2m}{a^2}\,r_i \delta r_i$$

Fig. 8.3. A uniform thin disc considered as a set of thin rings.

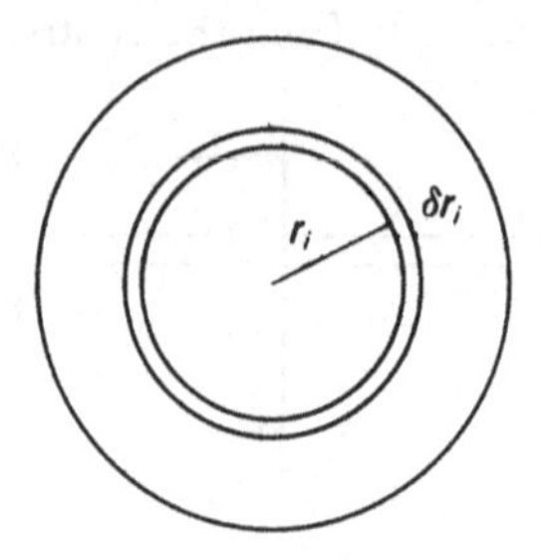

and

$$I = \sum_{i=1}^{n} \frac{2m}{a^2} r_i \delta r_i r_i^2$$

which in the limit becomes

$$I = \frac{2m}{a^2} \int_0^a r^3 \, \mathrm{d}r = \tfrac{1}{2} m a^2$$

Example 8.3: uniform spherical shell

Determine the moment of inertia of a thin spherical shell of radius a about an axis through its centre. It may be assumed that mass is uniformly distributed over the surface, and the total mass may be denoted by m.

The technique described in the previous example may again be used here. By joining together all parts of the shell that are the same distance from the axis, the shell may also be divided into a set of rings. In this case, however, it is convenient to use a parameter θ, which is the angle that the position vector of any point on the ring relative to the centre makes with the axis.

Fig. 8.4. A thin spherical shell considered as a set of thin rings of radius $a \sin \theta_i$.

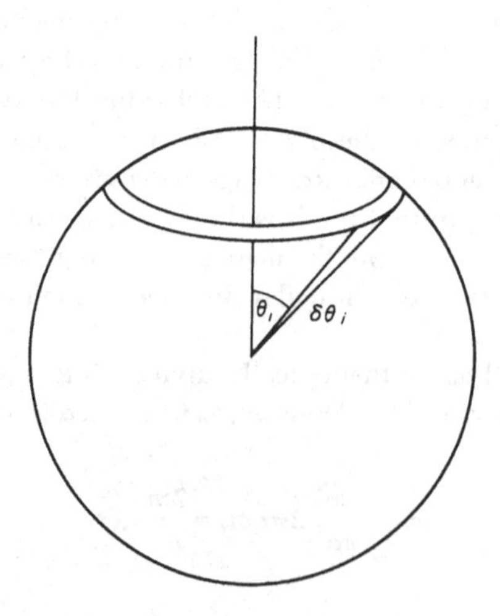

Using figure 8.4 it can be seen that the mass of each ring is given by

$$m_i = \frac{m}{4\pi a^2} 2\pi a \sin\theta_i a\delta\theta_i = \frac{m}{2}\sin\theta_i\delta\theta_i$$

and

$$I = \sum_{i=1}^{n} \frac{m}{2}\sin\theta_i\delta\theta_i a^2 \sin^2\theta_i$$

which in the limit becomes

$$I = \frac{ma^2}{2}\int_0^{\pi} \sin^3\theta\, d\theta = \tfrac{2}{3}ma^2$$

Example 8.4: uniform solid sphere

Determine the moment of inertia of a solid sphere of radius a and mass m about an axis through its centre. It may be assumed that mass is uniformly and continuously distributed throughout its volume.

It is appropriate to present two different methods of obtaining the required result.

Method 1

The technique of the previous examples may be used, to join together the component pieces of the sphere that are the same distance from the axis. In this approach the sphere can be considered to be divided into n thin concentric cylindrical shells. It is convenient to denote the radius of each cylinder by x_i and its thickness by δx_i, as illustrated in figure 8.5. Its height is then given by $2\sqrt{(a^2 - x_i^2)}$.

Fig. 8.5. A solid sphere considered as a set of n cylindrical shells of radius x_i and thickness δx_i.

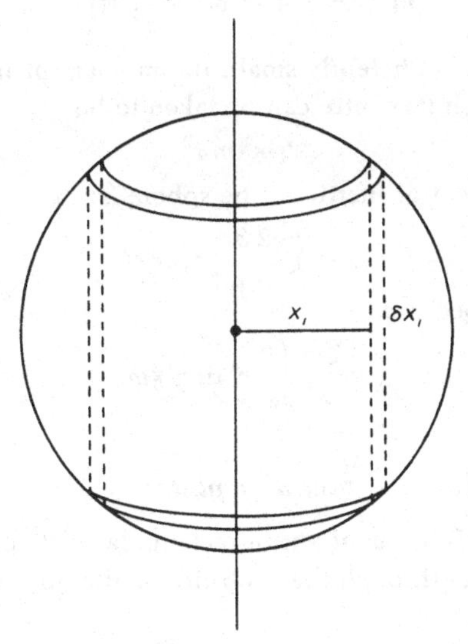

The mass of each shell is

$$m_i = \frac{m}{\frac{4}{3}\pi a^3} 2\pi x_i 2\sqrt{(a^2 - x_i^2)}\, \delta x_i$$

$$= \frac{3m}{a^3} x_i\sqrt{(a^2 - x_i^2)}\, \delta x_i$$

and

$$I = \sum_{i=1}^{n} \frac{3m}{a^3} x_i\sqrt{(a^2 - x_i^2)}\, \delta x_i x_i^2$$

which in the limit becomes

$$I = \frac{3m}{a^3} \int_0^a x^3\sqrt{(a^2 - x^2)}\, \mathrm{d}x = \tfrac{2}{5}ma^2$$

Method 2

It is now appropriate to illustrate yet another alternative to using multiple integration. This also depends on the fact that the total moment of inertia of a body about a particular axis is the sum of the moments of inertia of the component parts of the body about that axis. The technique used previously has been to divide a body into components in which all the mass is at the same distance from the axis. This condition, however, is not necessary. The body can be divided into any component parts for which expressions for their moments of inertia can easily be obtained. For example, a uniform solid sphere may be considered as a set of thin discs, or as a set of thin spherical shells, and the results of examples 8.2 or 8.3 may be used.

To illustrate this approach, consider the uniform solid sphere to be composed of n concentric spherical shells, each of radius r_i and thickness δr_i. The mass of each shell is

$$m_i = \frac{m}{\frac{4}{3}\pi a^3} 4\pi r_i^2 \delta r_i = \frac{3m}{a^3} r_i^2 \delta r_i$$

Now, provided δr_i is sufficiently small, the moment of inertia of each shell about an axis through its centre can be taken to be

$$I_i = \tfrac{2}{3} m_i r_i^2$$

Thus the total moment of inertia of the sphere is

$$I = \sum_{i=1}^{n} \frac{2}{3}\frac{3m}{a^3} r_i^2 \delta r_i r_i^2$$

which in the limit becomes

$$I = \frac{2m}{a^3} \int_0^a r^4\, \mathrm{d}r = \tfrac{2}{5}ma^2$$

as before.

Example* 8.5: *uniform rectangular plate

Find the moments of inertia of a thin rectangular plate of mass m and sides $2a$ and $2b$, about axes through the centroid parallel to the sides and perpendicular to the plane.

Fig. 8.6. A uniform rectangular plate. Cartesian axes are chosen for convenience.

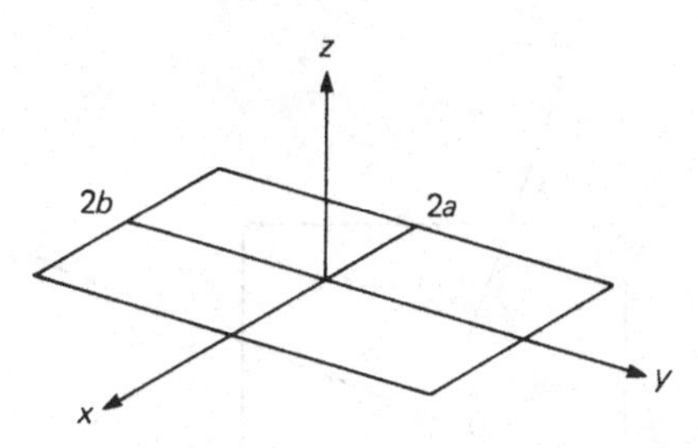

It is convenient to introduce cartesian axes such that the plate is in the plane $z=0$, and the x and y axes are parallel to the sides of length $2b$ and $2a$ respectively (see figure 8.6).

To determine the moment of inertia of the plate about the x axis, it is convenient to divide the plate into thin strips parallel to the axis. The calculation which then follows is identical to that of example 8.1 where a uniform rod is considered. In fact it can immediately be seen that the distribution of the mass of the plate about the x axis is equivalent to that of a uniform rod of mass m and length $2a$, since the distribution of mass along the axis is irrelevant. It follows that

$$I_x = \tfrac{1}{3}ma^2$$

and similarly that

$$I_y = \tfrac{1}{3}mb^2$$

The lamina theorem can then be used to show that

$$I_z = \tfrac{1}{3}m(a^2+b^2)$$

Example 8.6: *uniform square plate*

Find the moment of inertia of a thin square plate about any axis in its plane which passes through its centre.

Denote the length of the sides of the square by $2a$, and choose an arbitrary pair of cartesian axes in the plane (see figure 8.7). It follows from the previous example that the moment of inertia of the plate about an axis through its centre perpendicular to its plane is

$$I_z = \tfrac{2}{3}ma^2$$

It may now be argued from the obvious symmetry properties of the square that the moments of inertia about the x and y axes must be the same for an arbitrary pair of orthogonal axes in the plane. It therefore follows from the lamina theorem that

$$I_x = I_y = \tfrac{1}{3}ma^2$$

It may therefore be concluded that the moment of inertia of a square plate is the same for any axis in its plane which passes through the centroid.

Fig. 8.7. A uniform square plate, and an arbitrary pair of axes in its plane and passing through its centre.

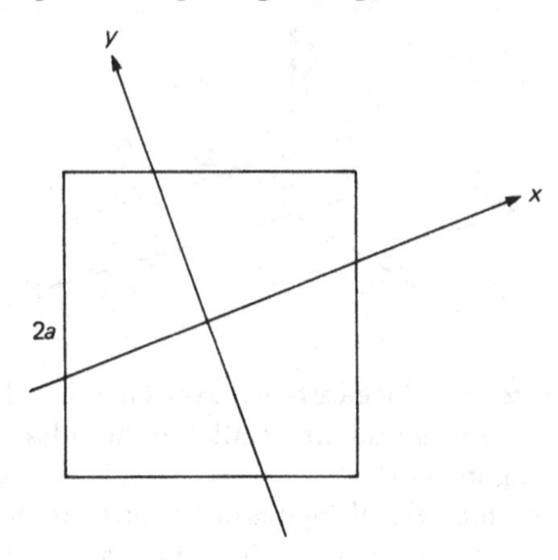

***Example* 8.7**

Find the moment of inertia of a uniform disc of radius a and mass m about a tangent on the rim.

It is convenient to introduce cartesian axes with origin at the centre of the disc, the z axis perpendicular to the plane, and the y axis parallel to the required tangent. This is illustrated in figure 8.8.

According to exercise 8.2

$$I_z = \tfrac{1}{2}ma^2$$

Then, since by symmetry $I_x = I_y$, the lamina theorem implies that

$$I_y = \tfrac{1}{4}ma^2$$

The parallel axis theorem can finally be used to obtain the moment of inertia about the tangent on the rim as

$$I_T = \tfrac{5}{4}ma^2$$

Fig. 8.8. A uniform circular disc with axes that are chosen for convenience.

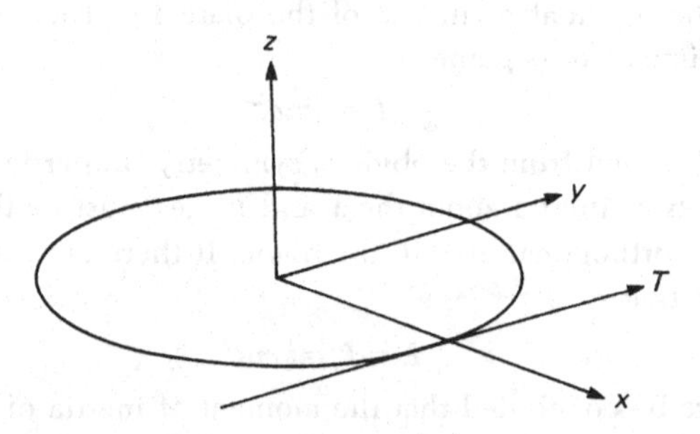

Exercises

8.1 Particles of mass m, $2m$, $3m$ and $4m$ are held in a rigid light framework at points $(0, 1, 1)$, $(1, 1, -1)$, $(1, -1, 0)$ and $(-1, 0, 1)$ respectively. Show that the moments of inertia of the system about the x, y and z axes are $13m$, $16m$ and $15m$ respectively.

8.2 Show that the moment of inertia of a uniform rigid rod of mass m and length $2a$ about an axis through an end, inclined at an angle α to the rod is $\frac{4}{3}ma^2\sin^2\alpha$.

8.3 If a three-legged stool can be represented as a thin circular disc of mass m and radius a with legs equally spaced round the rim of the disc, and if the legs are of length $2a$ and mass $m/4$ and are mounted parallel to each other, show that the moment of inertia about any diameter of the disc is $\frac{39}{24}ma^2$.

8.4 Show that the moment of inertia of a uniform solid cone of height h and circular base of radius a, about a diameter of the base is $m(3a^2+2h^2)/20$.

8.5 Show that the moments of inertia of a uniform solid cuboid with sides of length $2a$, $2b$ and $2c$ and mass m about axes through the centroid parallel to the sides are

$\frac{1}{3}m(b^2+c^2)$ $\frac{1}{3}m(c^2+a^2)$ and $\frac{1}{3}m(a^2+b^2)$

8.6 Show that the moment of inertia of a uniform thin triangular plate about a side is $mh^2/6$, where m is its mass and h is the perpendicular distance of the opposite vertex from the given side.

8.7 Show that the moment of inertia of a uniform solid cube of mass m and with sides of length $2a$ about a diagonal of a face is $\frac{5}{3}ma^2$.

8.8 If the density of a sphere of mass m and radius a at a distance r from its centre is proportional to $2a^2-r^2$, show that the moment of inertia about any tangent line is $\frac{67}{49}ma^2$.

8.5. Equations for motion in a plane

The equations for the general motion of a rigid body have already been stated in section 8.2, quoting propositions 7.5, 7.6 and 7.7. However, when learning how to apply them it is sensible first to develop skills in the analysis of two-dimensional problems before moving on to general three-dimensional motion. For this reason a brief description of the equations for the motion of a rigid body in a plane is now given.

In the two-dimensional motion considered here, it is assumed that the centroid of the body is constrained to remain in a fixed plane. In many applications, this plane is vertical. It is also assumed that the body is only free to rotate about an axis perpendicular to the plane.

Thus, there are in general three degrees of freedom, two of which may be associated with the position of the centroid, and the other may be associated with the freedom of rotation in the plane.

It is convenient to introduce a frame of reference with origin in the plane containing the centroid of the body, and to denote a unit vector perpendicular to this plane by $\boldsymbol{u}$. The position of the centroid at any time can now be determined by the vector $\bar{\boldsymbol{r}}(t)$, which is subject to the constraint that $\bar{\boldsymbol{r}} \cdot \boldsymbol{u} = 0$. Thus $\bar{\boldsymbol{r}}$ has two independent components, and two equations for the motion can be obtained from proposition 7.5 in the form

$$m\ddot{\bar{\boldsymbol{r}}} = \boldsymbol{F}$$

In order for the motion to remain in a plane, it is necessary that the sum of the external forces is a vector in the plane. This is expressed by the condition that

$$\boldsymbol{u} \cdot \boldsymbol{F} = 0$$

The position of the centroid at any time is determined by the above equation, and it is now necessary to consider the freedom of rotation about the centroid. This can be described by the angular velocity vector $\boldsymbol{\omega} = \omega \boldsymbol{u}$, where ω is the rate of rotation of the body in the plane of motion. The equation of motion regulating this freedom is that stated in proposition 7.7, which involves the angular momentum of the body $\boldsymbol{L}_c$ about the centroid. However, there is here only a freedom of rotation about an axis through the centroid parallel to $\boldsymbol{u}$. The moment of inertia about this axis, which may be denoted by I_c, is therefore a constant, and the angular momentum about this axis is given by

$$\boldsymbol{L}_c \cdot \boldsymbol{u} = I_c \omega$$

The remaining equation of motion therefore takes the form

$$I_c \dot{\omega} = \boldsymbol{u} \cdot \boldsymbol{M}_c$$

It is of course assumed that the forces acting do not induce a rotation which is inconsistent with the constraints. They must therefore satisfy the condition that

$$\boldsymbol{u} \times \boldsymbol{M}_c = 0$$

It may be thought that a possible alternative equation may be obtained by considering the rate of change of the angular momentum of the body about a fixed point, according to proposition 7.6. However, this is not convenient if the fixed point does not also happen to be a stationary point of the body. The moment of inertia of the body about an axis through a fixed point perpendicular to the plane is not constant in general through the motion, and so the equation does not simplify. It is more convenient in this case to consider the motion of the centroid and the angular motion about the centroid, as described above.

However the case of a rigid body which is constrained to rotate about a fixed axis is a special case of the motion considered above. In this case, an origin may be chosen on the axis whose direction may be described by the unit vector $\boldsymbol{u}$. The moment of inertia I about this axis is a constant, and since there is only one degree of freedom the only equation of motion required is that obtained in the previous section in the form

$$I\dot{\omega} = \boldsymbol{u} \cdot \boldsymbol{M}$$

This has been obtained via proposition 7.6 and gives an equation for the rate of change of the angular velocity ω in terms of the moments of the external forces about the axis.

The equations of motion described above are sufficient to determine the two-dimensional motion of a rigid body. For any given set of forces they are a set of second-order differential equations, whose solution would describe the position of the body at any time if it were subject only to the forces assumed. In practice, however, it is not always neccessary to start with these equations of motion. It often happens that a first integral of the equations can immediately be obtained. This may be an integral associated with the constancy of linear momentum, angular momentum or energy. These may easily be evaluated in any particular situation. At this point it is only appropriate to give a general expression for the kinetic energy.

It can be seen from proposition 7.2 that the kinetic energy of a rigid body can be evaluated as the sum of that relative to the centroid and that of a hypothetical particle of equal mass situated always at the centroid. For the two-dimensional motion considered here the kinetic energy relative to the centroid is that which is associated with the rotational motion of the body about an axis which is perpendicular to the plane in which motion takes place, and which moves such that at any time it passes through the centroid of the body. The moment of inertia of the body about this axis is a constant which may be denoted by I_c. The total kinetic energy can then be expressed in the form

$$T = \tfrac{1}{2}I_c\omega^2 + \tfrac{1}{2}m\dot{\bar{r}}^2$$

If the body does not have a fixed centre of rotation, its kinetic energy is best evaluated in this way.

8.6. 2D rigid body motion: examples

Example 8.8: the compound pendulum

The motion of a simple pendulum has been considered previously in example 6.1. This was the highly idealised model of a single particle attached to a fixed point by a light inelastic string. It is now appropriate to consider the slightly more realistic situation in which a body of finite size is free to rotate rigidly about a fixed point which can be labelled O.

With the new simplifying assumption of a rigid rotation about O, the actual size of the object is irrelevant. For example, it does not matter whether the body extends to include O or not. All that is required is a knowledge of its mass and its moment of inertia either about the centroid or about O.

It may be assumed that the mechanism by which the body rotates about O is reasonably smooth, so that any frictional couple about O may be ignored. The only forces which are considered to act on the body are the gravitational force which acts at the centroid C, and either the reactional forces at O or the forces which constrain the body to rotate rigidly about O. The former case may be assumed here, where the body extends to include the point O whose distance from the centroid is denoted by h.

Here the body is considered only to rotate in a fixed vertical plane through O. It is convenient to introduce the parameter θ to denote the angle that OC makes with the vertical. The angular speed of the body is then $\omega = \dot{\theta}$. It is also convenient initially to resolve the reactional force at O into components

Fig. 8.9. The compound pendulum. The reactional force at the pivot is divided into components parallel and normal to OC.

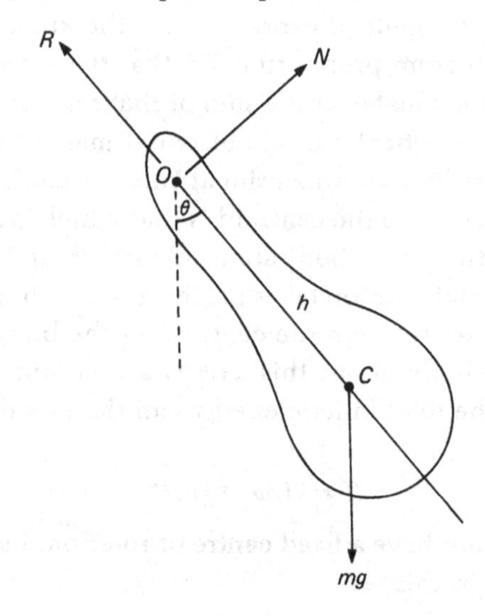

R and N which are respectively aligned with OC and perpendicular to it, as shown in figure 8.9.

The equation for the rotational motion of the body about O is

$$I_0\ddot{\theta} = -mgh \sin \theta$$

Now, the moment of inertia of the body about O can be expressed as

$$I_0 = I_c + mh^2$$

and, writing $I_c = mk^2$, where k is the radius of gyration of the body about C, the equation for rotational motion becomes

$$\ddot{\theta} = -\frac{gh}{k^2+h^2} \sin \theta$$

Since the body has only one degree of freedom, this equation is sufficient to determine the position of the body at any time given any particular boundary conditions.

It can immediately be seen from this equation that, for small oscillations in which $\sin \theta$ can be approximated by θ, the oscillations are periodic with frequency $\sqrt{(gh/(k^2+h^2))}$. This result clearly reduces to that for a simple pendulum in the case when $k = 0$.

In the general case, the above equation can be integrated, after multiplying by $2\dot{\theta}$, to yield

$$\dot{\theta}^2 = \frac{2gh}{k^2+h^2}(\cos \theta - \cos \theta_0)$$

where the constant of integration has been evaluated using the boundary condition that the body oscillates with θ varying between $-\theta_0$ and θ_0. It may be observed that this equation is equivalent to the energy integral $T + V =$ const., since the system is clearly conservative and the kinetic energy and potential energy relative to the horizontal through O are given by

$$T = \tfrac{1}{2}m(k^2+h^2)\dot{\theta}^2$$

$$V = -mgh \cos \theta$$

Thus the above equations could alternatively be obtained by initially writing down the energy integral, and then differentiating it to obtain an equation for $\ddot{\theta}$.

The above equations, together with appropriate boundary conditions, are sufficient to determine θ and hence the position of the body at any given time. In this sense the problem is already solved, and it only remains to integrate the final equation numerically to obtain θ explicitly at any required time. However, the reactional forces at O are still unknown. These can be interpreted as the forces of constraint that are required to permit a rigid rotation about O. To determine them it is necessary to consider the equations for translational motion, that the mass times the acceleration of the centroid is equal to the sum of the external forces. Two alternative methods may be used.

One method is to use polar coordinates in which the position of the centroid relative to the fixed point O is determined in terms of the polar angle θ. The centroid, however, remains at a fixed distance h from O, so that $\bar{r} = he_r$. In this approach it is convenient to represent the reactional force by the components R and N, as illustrated in the above diagram. The equation for translational motion then reads

$$m(-h\dot{\theta}^2 e_r + h\ddot{\theta} e_\theta) = mg(\cos\theta e_r - \sin\theta e_\theta) - Re_r + Ne_\theta$$

whose components give

$$R = mg\cos\theta + mh\dot{\theta}^2$$
$$N = mg\sin\theta + mh\ddot{\theta}$$

The above equations for $\ddot{\theta}$ and $\dot{\theta}^2$ may now be used to obtain expressions for these components of the reactional force in terms of angle θ only as

$$R = \frac{mg}{k^2 + h^2}\{(k^2 + 3h^2)\cos\theta - 2h^2\cos\theta_0\}$$

$$T = \frac{mgk^2}{k^2 + h^2}\sin\theta$$

An alternative approach is to use cartesian coordinates with origin at O. In this case it is convenient to represent the reactional force in terms of its horizontal and vertical components H and V, as illustrated in figure 8.10.

The horizontal and vertical components of the equation of translational motion now read

$$m\ddot{\bar{x}} = -H$$
$$m\ddot{\bar{y}} = V - mg$$

Fig. 8.10. The compound pendulum. The reactional force at the pivot is divided into its horizontal and vertical components.

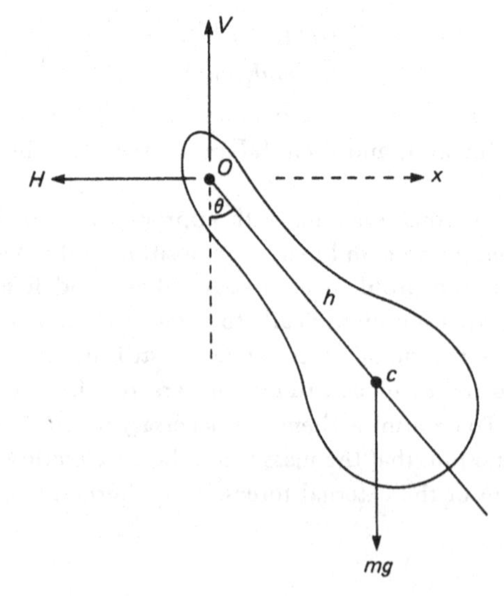

Although these equations appear simpler than those of the previous method, it is clear that $\bar{x} = h \sin\theta$ and $\bar{y} = -h \cos \theta$, so that these equations give

$$H = -mh \cos \theta\ddot{\theta} + mh \sin \theta\dot{\theta}^2$$

$$V = mg + mh \sin \theta\ddot{\theta} + mh \cos \theta\dot{\theta}^2$$

Again substituting the above expressions for $\ddot{\theta}$ and $\dot{\theta}^2$ determines these components of the reactional force at O in the form

$$H = \frac{mgh^2}{k^2 + h^2} \sin \theta \,(3 \cos \theta - 2 \cos \theta_0)$$

$$V = \frac{mg}{k^2 + h^2} \{k^2 + h^2 \cos \theta(3 \cos \theta - 2 \cos \theta_0)\}$$

Example 8.9: *the sliding sphere*

Consider a situation in which a sphere is projected along a horizontal plane with no initial angular velocity. Such a situation occurs, for example, in the game of bowls and is sometimes possible in snooker. In such cases, although the initial motion is one of sliding over a rough surface, the sphere starts to roll after a short time. The situation is illustrated in figure 8.11.

Assuming a simple two-dimensional problem, the position of the body at any time can be determined by the two parameters x and θ which measure the horizontal displacement of the centroid and the rotation of the sphere respectively. The main forces which act on the body are the gravitational force, the vertical reaction R from the plane acting on the body, and the frictional force F which is induced while the sphere slides over the surface and acts in a horizontal direction on that point of the sphere which is in contact with the plane.

Assuming that the horizontal plane is flat, the centroid will not move in the vertical direction, and there can therefore be no net component of the external forces in this direction. Thus it is necessary that

$$R = mg$$

It is also found in practice that the frictional force is approximately constant. For a highly simplified problem such as this, it is appropriate to assume that it is related to the reactional force according to the equation

$$F = \mu R$$

Fig. 8.11. A sphere sliding on a horizontal surface.

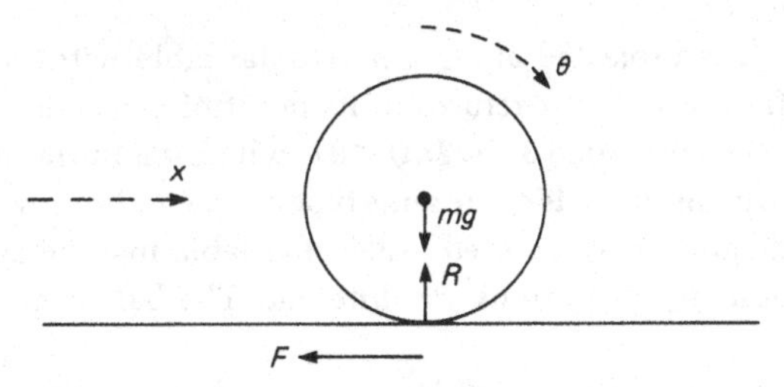

where μ is the coefficient of friction. For a more sophisticated model the theory behind this equation may be more fully developed with different values of the coefficient being chosen in different circumstances.

To specify a particular problem, it may be assumed that the body is a uniform solid sphere of mass m and radius a, and therefore its moment of inertia about an axis through its centre is $\frac{2}{5}ma^2$. The equations of motion for translational motion in the horizontal direction and rotational motion about the centroid are now

$$m\ddot{x} = -F$$
$$\tfrac{2}{5}ma^2\ddot{\theta} = aF$$

These immediately yield the differential equations

$$\ddot{x} = -\mu g$$
$$a\ddot{\theta} = \tfrac{5}{2}\mu g$$

which can be integrated to give

$$\dot{x} = V - \mu g t$$
$$a\dot{\theta} = \tfrac{5}{2}\mu g t$$

where the boundary conditions have been applied that the sphere is projected initially with speed $\dot{x} = V$ and no angular velocity $\dot{\theta} = 0$.

It can be seen that, while the frictional force is acting, the speed of the centroid is decreasing and the angular speed is increasing. This will continue until $\dot{x} = a\dot{\theta}$, at which point the motion is one of pure rolling. At this point the sphere ceases to slip, the frictional force vanishes, and the sphere then continues to roll. By substituting the above expressions into the condition $\dot{x} = a\dot{\theta}$, it can be predicted that the sphere starts to roll after a time

$$t = \frac{2V}{7\mu g}$$

and that its speed is then $\frac{5}{7}V$. It may also be observed that the kinetic energy given by

$$T = \tfrac{1}{2}m\dot{x}^2 + \tfrac{12}{25}ma^2\dot{\theta}^2$$

is reduced to

$$T = \tfrac{5}{14}mV^2$$

when the sphere starts to roll. Thus the kinetic energy is also reduced by a factor of $\frac{5}{7}$ while the frictional force acts.

Example 8.10

A ping-pong ball is projected along a horizontal table with initial speed V and backspin Ω. Show that it returns to its point of projection rolling with speed $(2a\Omega - 3V)/5$, provided $3V < 2a\Omega < 8V$, where a is the radius of the ball.

This particular problem is of the same type as that of the previous example, and the same diagram may be used. Again the table may be assumed to be flat, but the boundary conditions are different. The ball is given an initial horizontal velocity of magnitude V, and an initial angular velocity about a horizontal axis through the centroid perpendicular to the direction of its motion in a vertical plane and initially $\dot{x} = V$ and $\dot{\theta} = -\Omega$. There is no vertical motion so $R = mg$ and it may be assumed that $F = \mu R$. Omitting the vertical forces, a simplified diagram describing the situation while slipping occurs is given in figure 8.12.

The mass of the ping pong ball may be denoted by m, and it is reasonable to represent it by the idealised 'thin spherical shell' so that its moment of inertia about an axis through the centre can be taken to be $\frac{2}{3}ma^2$. The equations for translational and rotational motion are thus

$$m\ddot{x} = -\mu mg$$
$$\tfrac{2}{3}ma^2\ddot{\theta} = \mu mga$$

which integrate to give

$$\dot{x} = V - \mu gt$$
$$\dot{\theta} = -\Omega + \frac{3}{2}\frac{\mu g}{a}t$$

after applying the boundary conditions. A second integration also gives

$$x = Vt - \tfrac{1}{2}\mu gt^2$$

It can thus be seen that, provided the ball is still slipping, it would start to return after a time $t = V/\mu g$ and would return to its point of projection after a time $t = 2V/\mu g$.

Clearly, the speed $\dot{x}$ continuously decreases while slipping occurs, and $\dot{\theta}$ increases from its initial negative value. The ball starts to roll at the point when $\dot{x} = a\dot{\theta}$ whether these values are positive or negative. This occurs when

$$t = \frac{2}{5}\frac{(V + a\Omega)}{\mu g}$$

and then

$$\dot{x} = a\dot{\theta} = -\tfrac{1}{5}(2a\Omega - 3V)$$

Now, for the ball to return to its point of projection rolling, it must start to roll at some point after it has started to return but before it reaches its

Fig. 8.12. A ball sliding on a horizontal surface. Vertical forces are not shown.

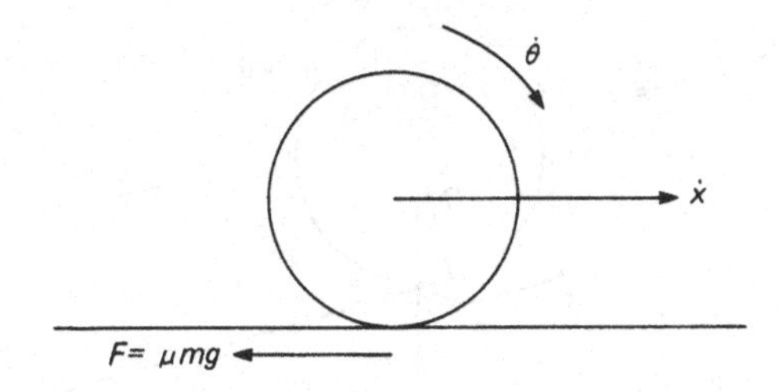

initial position. This can be expressed by the condition

$$\frac{V}{\mu g} < \frac{2}{5}\frac{(V+a\Omega)}{\mu g} < \frac{2V}{\mu g}$$

which yields the required condition that $3V < 2a\Omega < 8V$.

Example* 8.11: *sphere rolling down a plane

Consider the motion of a uniform solid sphere rolling down a rough plane that is inclined at an angle α to the vertical.

It is convenient to consider the sphere to start from rest, so that the subsequent motion is two-dimensional in a fixed vertical plane. It may be assumed that the only forces which act on the sphere are the gravitational force and the reaction from the plane which comprises a normal component denoted by R and a frictional component F. The motion can be described by two parameters x, the distance moved by the centroid, and θ, the angle through which the body rotates. These features are illustrated in figure 8.13.

Since the sphere remains in contact with the plane, there can be no net component of force normal to the plane. It is therefore necessary that

$$R = mg \cos \alpha$$

The equation of motion for translational motion down the plane and rotational motion about the centroid are in this case

$$m\ddot{x} = mg \sin \alpha - F$$

$$\tfrac{2}{5}ma^2\ddot{\theta} = aF$$

It is, however, also necessary to impose the constraint that the sphere is rolling down the plane. This is expressed by the condition

$$\dot{x} = a\dot{\theta}$$

which clearly implies that $m\ddot{x} = ma\ddot{\theta}$. It can be seen that the above equations can only be consistent with this constraint if

$$F = \tfrac{2}{7}mg \sin \alpha$$

This is the frictional force that is required to prevent the sphere sliding down the plane.

Fig. 8.13. A sphere rolling down an inclined plane.

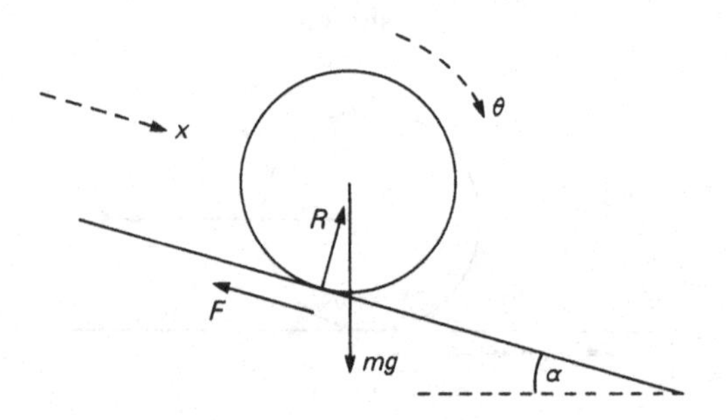

It may immediately be observed that

$$\frac{F}{R} = \tfrac{2}{7} \tan \alpha$$

This however must be less than the coefficient of friction μ. It may therefore be concluded that the sphere will slip down the plane if

$$\mu < \tfrac{2}{7} \tan \alpha$$

or if

$$\alpha < \tan^{-1} \frac{7\mu}{2}$$

It may also be observed that while the sphere is rolling down the plane

$$\ddot{x} = \tfrac{5}{7} g \sin \alpha$$

and it may be concluded that the acceleration down the plane does not depend on the mass of the sphere. This forms the theoretical basis for a set of traditional experiments that can be performed to demonstrate that the acceleration due to gravitation does not depend on the mass of the body being acted on. However, it should be emphasised that, for rolling motion, the acceleration down the plane is *not* $g \sin \alpha$.

It may finally be pointed out that, although the frictional force is nonzero in this situation, the system is still conservative. In a pure rolling motion the point instantaneously in contact with the plane is always stationary. The reactional force from the plane, which includes its frictional component, therefore does no work. The energy integral is in fact

$$\tfrac{7}{10} m\dot{x}^2 - mgx \sin \alpha = \text{const}$$

which can also be seen to be an integral of the above equations of motion when the rolling condition is applied.

Exercises

8.9 A thin uniform rod of mass m which is free to turn about one end is held horizontally and released. Show that in the subsequent motion, when the rod makes an angle θ with the vertical, the reactional force required at the fixed end has magnitude $\frac{1}{4} mg\sqrt{(1 + 99 \cos^2 \theta)}$.

8.10 A uniform circular disc is free to rotate in its plane about a point on its circumference. If it is slightly disturbed from rest in its position of unstable equilibrium, show that its supports must be able to bear at least $\frac{11}{3}$ times the gravitational force which acts on the disc.

8.11 A uniform rod which rests vertically on a rough horizontal table is slightly displaced and falls to the right. Show that when the rod makes an angle θ with the vertical, if it does not slip, the reaction from the table is inclined to the right at an angle α to the vertical where

$$\tan \alpha = \frac{3 \sin \theta (3 \cos \theta - 2)}{(3 \cos \theta - 1)^2}$$

Show also that if $\mu < \frac{15}{64} \sqrt{(\frac{5}{2})}$, where μ is the coefficient of friction, the rod will slip to the left when $0<\theta<\cos^{-1}\frac{9}{11}$, and that if $\mu\geqslant\frac{15}{64}\sqrt{(\frac{5}{2})}$ the rod will slip to the right when $\cos^{-1}\frac{2}{3}<\theta<\cos^{-1}\frac{1}{3}$.

8.12 A uniform thin hollow cylinder of radius a and mass m has a particle also of mass m attached to a point on its surface. The cylinder is placed on a rough horizontal plane in a position of unstable equilibrium and slightly displaced. Show that when the cylinder has turned through an angle θ

$$a\dot{\theta}^2=g\frac{(1-\cos\theta)}{(2+\cos\theta)}$$

provided it does not slip. Show that when $\theta=\pi/2$, the normal reaction on the cylinder is $13\,mg/8$ and the frictional reaction is $mg/4$.

8.13 A uniform solid cylinder is rolling down a rough plane which is inclined at an angle α to the horizontal. Show that its acceleration is $\frac{2}{3}g\sin\alpha$, and that the coefficient of friction is greater than $\frac{1}{3}\tan\alpha$.

8.14 A uniform rod of mass m and length $2a$ is suspended at one end from a fixed point by a light string of length l. When it oscillates in a vertical plane show that its kinetic energy is

$$T=\tfrac{1}{2}m\{l^2\dot{\theta}^2+\tfrac{4}{3}a^2\dot{\phi}^2+2al\dot{\theta}\dot{\phi}\cos(\phi-\theta)\}$$

where θ and ϕ are, respectively, the inclinations of the string and the rod to the vertical.

8.15 Four equal uniform rods each of length $2a$ are freely jointed at their ends to form a rhombus $ABCD$ which is freely suspended from the corner A which is fixed. The corner C is raised until it almost reaches A, with B and D on opposite sides of AC in the same vertical plane, and the system is then released from rest. If during the subsequent motion θ denotes the acute angle between any rod and the vertical, show that

$$a\dot{\theta}^2=3g\cos\theta/(1+3\sin^2\theta)$$

8.7. Moments and products of inertia

It has already been stated that the general motion of a rigid body can be analysed in terms of the motion of its centroid and its rotational motion about the centroid. In the two-dimensional motion considered above, the direction of the axis of rotation through the centroid is constant and so also is the moment of inertia of the body about it. This enables the equations of motion to be written in the simple form given in the previous sections. In general, however, the direction of the instantaneous axis of rotation varies with the motion and, before considering the equations for such a motion, it is necessary first to extend the concept of a moment of inertia.

The concept of a moment of inertia was first introduced by considering the rotational motion of a rigid body about a fixed axis.

It is now convenient to consider the more general rotational motion of a body about a fixed point. The point, which is not necessarily the centroid of the body, may be taken to be the origin of a frame of reference, and the rotational motion of the body about it has in general three degrees of freedom.

In this case the angular velocity vector $\boldsymbol{\omega}(t)$ is a vector through the origin whose direction is that of the instantaneous axis of rotation. Now, regarding the body as a system of n particles, each with position vector $\boldsymbol{r}_i$ relative to the origin, the total angular momentum of the the body is given by

$$\boldsymbol{L} = \sum_{i=1}^{n} \boldsymbol{r}_i \times m_i(\boldsymbol{\omega} \times \boldsymbol{r}_i)$$

$$= \sum_{i=1}^{n} m_i r_i^2 \boldsymbol{\omega} - \sum_{i=1}^{n} m_i(\boldsymbol{r}_i \cdot \boldsymbol{\omega})\boldsymbol{r}_i$$

It can immediately be seen that this vector is not necessarily in the direction of the instantaneous axis of rotation.

It is convenient to introduce a cartesian frame of reference centred at the origin. It is then possible to represent the position vectors of the particles and the angular velocity and angular momentum vectors of the body, respectively, in terms of their cartesian components in the form

$$\boldsymbol{r}_i = \begin{pmatrix} x_i \\ y_i \\ z_i \end{pmatrix} \qquad \boldsymbol{\omega} = \begin{pmatrix} \omega_1 \\ \omega_2 \\ \omega_3 \end{pmatrix} \qquad \boldsymbol{L} = \begin{pmatrix} L_1 \\ L_2 \\ L_3 \end{pmatrix}$$

The above expression for the angular momentum vector can now be evaluated explicitly to obtain the components

$$L_1 = \sum_{i=1}^{n} m_i\{\omega_1(y_i^2 + z_i^2) - \omega_2 x_i y_i - \omega_3 z_i x_i\}$$

$$L_2 = \sum_{i=1}^{n} m_i\{\omega_2(z_i^2 + x_i^2) - \omega_3 y_i z_i - \omega_1 x_i y_i\}$$

$$L_3 = \sum_{i=1}^{n} m_i\{\omega_3(x_i^2 + y_i^2) - \omega_1 z_i x_i - \omega_2 y_i z_i\}$$

These expressions may be simplified by making the following definitions.

Definition 8.4. *Taking cartesian axes at a fixed point, the moments of inertia of a set of n particles with respect to the x, y and z axes are given respectively by*

$$I_{xx} = \sum_{i=1}^{n} m_i(y_i^2 + z_i^2)$$

$$I_{yy} = \sum_{i=1}^{n} m_i(z_i^2 + x_i^2)$$

$$I_{zz} = \sum_{i=1}^{n} m_i(x_i^2 + y_i^2)$$

and the products of inertia with respect to the same axes are defined by

$$I_{xy} = I_{yx} = -\sum_{i=1}^{n} m_i x_i y_i$$

$$I_{xz} = I_{zx} = -\sum_{i=1}^{n} m_i x_i z_i$$

$$I_{yz} = I_{zy} = -\sum_{i=1}^{n} m_i y_i z_i$$

It can be seen from definition 8.2 that the terms I_{xx}, I_{yy} and I_{zz} are the moments of inertia about the x, y and z axes respectively. Apart from introducing a new notation, the purpose of definition 8.4 is the introduction of the products of inertia. It should, however, be noticed that, whereas the moments of inertia are necessarily positive, the products of inertia may be positive, negative or zero.

With these definitions, the components of the angular momentum vector can now be conveniently expressed in matrix notation in the form

$$\begin{pmatrix} L_1 \\ L_2 \\ L_3 \end{pmatrix} = \begin{pmatrix} I_{xx} & I_{xy} & I_{xz} \\ I_{yx} & I_{yy} & I_{yz} \\ I_{zx} & I_{zy} & I_{zz} \end{pmatrix} \begin{pmatrix} \omega_1 \\ \omega_2 \\ \omega_3 \end{pmatrix}$$

This matrix equation can be seen to generalise the similar scalar equation $L = I\omega$ for the angular momentum of a rigid body rotating about a fixed axis.

The matrix I_{ij}, where i and j denote cartesian components, is referred to as the *inertia matrix*. Strictly speaking, it is a tensor quantity and so may more accurately be termed the *inertia tensor*. It has been defined relative to a set of cartesian coordinates and, since these have been chosen arbitrarily, the above expression is independent of the particular coordinates used. It can be shown that, under a change of coordinates, I_{ij} transforms as a tensor quantity, while L_i and ω_i transform as vector quantities. It is not necessary, however, to introduce tensor analysis here and I_{ij} may loosely be thought of as a matrix for any given set of cartesian coordinate axes.

The above expressions have been obtained relative to an arbitrary set of cartesian axes whose origin is a fixed point about which the body may rotate. It is clear, however, that if the body moves relative to the coordinate system, then the moments and products of inertia would probably vary with the motion. It is therefore usually convenient to adopt a cartesian frame of reference which is fixed relative to the body. Thus, as the body rotates, the coordinate frame also rotates with the same angular velocity. Although in this case some complications arise because the frame of reference is noninertial, this is compensated for by the considerable simplifications that appear because the components of the inertia matrix are all constants.

Having decided to adopt a frame of reference which is fixed relative to the body, there is still the choice of which particular frame to adopt. In practice a particular set of axes is chosen for convenience, making use perhaps of some symmetry property of the body. However, any one frame can be obtained from any other by a simple rotation of the coordinate axes. Under such a coordinate transformation, the inertia tensor I_{ij} transforms as a tensor quantity. However, a rotation of coordinates has a simple matrix representation, and the new inertia matrix can be obtained from the old by a well-known matrix transformation.

In the evaluation of the moment of inertia of a body about a particular axis, the results stated as propositions 8.1 and 8.2 are often useful. The parallel axis theorem can in fact easily be generalised to include products of inertia. This then enables the inertia matrix with respect to one point of the body to be obtained from the inertia matrix evaluated at the centroid, and hence that at any other point; it being understood that the cartesian axes at each point are parallel to each other. This generalised parallel axis theorem can be stated as follows.

***Proposition* 8.3.** *The moments and products of inertia of a body with respect to a set of cartesian axes are equal to those of the same body with respect to a set of parallel axes through the centroid, together with those of a particle with mass equal to that of the body situated at its centroid with respect to the original set of axes.*

This can be proved simply by substituting $r_i = \boldsymbol{\rho}_i + \bar{r}$ into the definitions 8.4, and using the property of the centroid that $\sum_{i=1}^{n} m_i \boldsymbol{\rho}_i = 0$. It is therefore clear that this result applies only with respect to the centroid of the body. Denoting the cartesian coordinates of the centroid by $\bar{x}$, $\bar{y}$, $\bar{z}$, proposition 8.3 can be expressed in the form

$$I_{xx} = I_{cxx} + m(\bar{y}^2 + \bar{z}^2)$$
$$I_{yy} = I_{cyy} + m(\bar{z}^2 + \bar{x}^2)$$

$$I_{zz} = I_{czz} + m(\bar{x}^2 + \bar{y}^2)$$
$$I_{xy} = I_{cxy} - m\bar{x}\bar{y}$$
$$I_{xz} = I_{cxz} - m\bar{x}\bar{z}$$
$$I_{yz} = I_{cyz} - m\bar{y}\bar{z}$$

where I_{cij} denotes the inertia matrix of the body with respect to a set of parallel axes at the centroid.

This result can be used to describe the change in the inertia tensor of a body under a translation of coordinates, since the translation can always be considered in two stages via the centroid. The change due to a rotation of the coordinates can also easily be described. Thus, if the inertia tensor of a body is known with respect to one frame of reference, then that with respect to all other frames can be obtained.

8.8 Principal axes of inertia

It has already been emphasised that, in the general motion of a rigid body, the angular momentum vector is not always parallel to the instantaneous angular velocity vector. It has also been indicated that the difference between these two vectors depends on the mass distribution of the body. There are, however, a number of particular situations in which the two vectors are aligned. Relative to the point of the body about which it is assumed to rotate, there are always a distinct number of directions which are fixed relative to the body in which the angular momentum vector can be aligned with the angular velocity vector. Such directions or axes are referred to as principal axes of inertia, or sometimes as axes of spontaneous rotation.

***Definition* 8.5.** *If the rotation of a rigid body about an axis through a point O is such that the angular momentum vector of the body relative to the point O is aligned with the axis of rotation, then that axis is called a principal axis of inertia of the body at O.*

The principal axes of inertia of a body at any point are of both physical and mathematical interest.

Physically, the principal axes of the body are those of axes about which it may rotate in the absence of any external couples. It is understood that there is a point of reference which may be taken as an origin, which is that point of the body about which it is assumed to rotate. In this case the sum of the moments of the external forces about this point is zero, and so the angular momentum vector of the body is constant. If this vector is aligned with a principal axis, then the body also rotates about this axis. On the other hand, if the constant angular momentum vector is not aligned with a principal axis, then the instantaneous angular velocity vector is not in this constant direction. Such a motion can be observed, for example, in the wobble of the axis of rotation of a buckled or unbalanced wheel. In a balanced wheel, the axis is a principal axis at the centroid, and in the absence of external couples the wheel may freely rotate about it with constant angular velocity.

From this interpretation it can immediately be seen that *if an axis through a point O of a rigid body is a principal axis of inertia, then it is also a principal axis of the body with respect to all points along that axis.* Since the axis is one about which the body is free to rotate with angular velocity and angular momentum vectors parallel, no unique point along the axis is singled out as the one about which the body is rotating.

Apart from such a physical interpretation, the concept of a principal axis is also of very deep theoretical interest because it leads to a considerable simplification in the mathematical formulation of the equations of rotational motion. Consider the motion of a body about a fixed point. The condition for a principal axis through this point is the condition that the angular momentum and the angular velocity vectors are aligned. This can be expressed in the form

$$\mathbf{L} = \lambda \boldsymbol{\omega}$$

where λ is an appropriate constant. Taking a set of cartesian axes which are fixed in the body with origin at the fixed point, this equation can be expressed in terms of the moments and products of inertia of the body in the form

$$\begin{pmatrix} I_{xx} & I_{xy} & I_{xz} \\ I_{yx} & I_{yy} & I_{yz} \\ I_{zx} & I_{zy} & I_{zz} \end{pmatrix} \begin{pmatrix} \omega \\ \omega_2 \\ \omega_3 \end{pmatrix} = \lambda \begin{pmatrix} \omega_1 \\ \omega_2 \\ \omega_3 \end{pmatrix}$$

This can now be rewritten as the matrix form of three linear equations

$$\begin{pmatrix} I_{xx} - \lambda & I_{xy} & I_{xz} \\ I_{yx} & I_{yy} - \lambda & I_{yz} \\ I_{zx} & I_{zy} & I_{zz} - \lambda \end{pmatrix} \begin{pmatrix} \omega_1 \\ \omega_2 \\ \omega_3 \end{pmatrix} = 0$$

It is well known that these equations can only be solved if the determinant of the matrix is zero. This condition gives an equation for λ which is a cubic equation having three real roots which are not necessarily distinct. These roots are referred to as the *eigenvalues* of the inertia matrix I_{ij}, and for each eigenvalue a vector $\boldsymbol{\omega}$ can be determined up to a scale factor. Such vectors are referred to as the *eigenvectors* of the inertia matrix corresponding to the appropriate eigenvalue.

It is also clear that the constant λ introduced above can be interpreted as the moment of inertia of the body about the axis of rotation. Since the axis, however, is a principal axis, it is convenient to refer to λ as a principal moment of inertia.

***Definition* 8.6.** *The moment of inertia of a body about a principal axis of inertia is called a principal moment of inertia.*

The argument presented above may now be stated more rigorously in the form of a proposition.

***Proposition* 8.4.** *An axis through a point O of a rigid body is a principal axis of inertia if, and only if, it corresponds to an eigenvector of the inertia matrix of the body at O. The corresponding eigenvalue is the principal moment of inertia of the body about that axis.*

It can also be seen from the definition of the moments and products of inertia, given in definition 8.4, that the inertia matrix is a positive definite symmetric matrix. Matrices of this type are well known in mathematics and have a number of properties which are of direct importance here. For example, it is known that the eigenvectors corresponding to distinct eigenvalues are mutually orthogonal. Also, if there is a repeated eigenvalue, then there exists a plane through O which is such that all vectors in the plane are eigenvectors corresponding to the same eigenvalue. Finally, if all three eigenvalues are the same, then all vectors through O are eigenvectors. From these properties the following statement can immediately be inferred.

***Proposition* 8.5.** *Relative to any point of a rigid body, there exist three mutually perpendicular principal axes of inertia, corresponding to three principal moments of inertia which are not necessarily distinct.*

It is possible to use such a set of principal axes to make considerable simplifications in the mathematical formulation of the equations for the rotational motion of a rigid body. This is considered in the following section, but it can be used here to obtain a simplified expression for the angular momentum vector.

A further property which is well known in mathematics is that a rotation of coordinate axes can be used to transform a symmetric matrix into diagonal form in which the diagonal elements are the eigenvalues of the matrix. This can now be applied to the inertia matrix. According to proposition 8.5, relative to any point of a body there exist three mutually orthogonal principal axes. If the coordinate axes are rotated to coincide with such a set of principal axes, then it can be shown that the inertia matrix must take diagonal form. The moments of inertia are then the principal moments, and the products of inertia all vanish. This is a clear corollary to proposition 8.5.

Corollary. *Relative to any point of a rigid body a set of principal axes can be chosen with respect to which the moments of inertia about the axes are the principal moments and the products of inertia are all zero.*

It is convenient to denote the principal moments of inertia of a body at some point by the symbols I_1, I_2 and I_3. These are of course not necessarily distinct. With this notation the general expression for the angular momentum vector, relative to a set of principal axes defined in the body at the stated point, can be written in the form

$$\begin{pmatrix} L_1 \\ L_2 \\ L_3 \end{pmatrix} = \begin{pmatrix} I_1 & 0 & 0 \\ 0 & I_2 & 0 \\ 0 & 0 & I_3 \end{pmatrix} \begin{pmatrix} \omega_1 \\ \omega_2 \\ \omega_3 \end{pmatrix} = \begin{pmatrix} I_1\omega_1 \\ I_2\omega_2 \\ I_3\omega_3 \end{pmatrix}$$

It can be seen that, in this simplified form, the angular momentum of the body can be regarded as the sum of the angular momentum components about each of the principal axes.

In practice it is not always necessary to obtain the principal axes of a body at a point by first evaluating all the moments and products of inertia relative to an initially assumed base, and then finding the eigenvectors of the inertia matrix. Sometimes a set of principal axes at a point can be spotted from the known dynamical behaviour of the body in its rotational motion about that point. More reliably it is sometimes possible to spot a set of principal axes from their known property that, with respect to such a set, the products of inertia all vanish. For example, it can easily be seen that the principal axes at the centroid of a uniform rectangular cuboid are those which are parallel to the edges, since with respect to such axes the products of inertia are all zero.

In this way it can also be seen that *if a rigid body posesses an axis of symmetry, then that axis is a principal axis of inertia with respect to any point along it.* At any such point, coordinate axes may be chosen with one as the axis of symmetry and the other two perpendicular to it. It is then a simple matter to demonstrate that with respect to such axes the products of inertia are all zero. In this particular case, it should be noticed that the principal moments about the axes perpendicular to the axis of symmetry are equal. In fact, because of the symmetry, *all axes in the body which pass through a point on the axis of symmetry and are perpendicular to it are also principal axes of inertia having the same principal moment.* In fact, such axes need not even be fixed in the body but may rotate about the axis of symmetry.

One further mathematical device must be presented at this stage. This can be introduced by considering the moment of inertia of a rigid body about a given axis through a point, relative to a set of axes at that point. If the body rotates about this axis with angular velocity $\boldsymbol{\omega}$, then its angular momentum is given by

$$\begin{pmatrix} L_1 \\ L_2 \\ L_3 \end{pmatrix} = \begin{pmatrix} I_{xx} & I_{xy} & I_{xz} \\ I_{yx} & I_{yy} & I_{yz} \\ I_{zx} & I_{zy} & I_{zz} \end{pmatrix} \begin{pmatrix} \omega_1 \\ \omega_2 \\ \omega_3 \end{pmatrix}$$

Now let $\boldsymbol{u}$ be a unit vector in the direction of the axis, then $\boldsymbol{\omega} = \omega \boldsymbol{u}$ and the angular momentum about this axis is the scalar product $L = \boldsymbol{u} \cdot \boldsymbol{L}$. The moment of inertia I about the axis is defined by the scalar equation $L = I\omega$, and this can now be obtained in the form

$$I = (u_1 \ \ u_2 \ \ u_3) \begin{pmatrix} I_{xx} & I_{xy} & I_{xz} \\ I_{yx} & I_{yy} & I_{yz} \\ I_{zx} & I_{zy} & I_{zz} \end{pmatrix} \begin{pmatrix} u_1 \\ u_2 \\ u_3 \end{pmatrix}$$

This expression represents the moment of inertia as a quadratic form involving the components of the inertia matrix evaluated relative to a point and a set of cartesian axes at that point.

In the above expression a moment of inertia is obtained uniquely for any given direction with unit vector $\boldsymbol{u}$. Now, consider the vector $\boldsymbol{r}$ in the direction of $\boldsymbol{u}$ having magnitude $I^{-1/2}$

$$\boldsymbol{r} = 1/\sqrt{I}\boldsymbol{u}$$

Letting the vector $\boldsymbol{r}$ have components (x, y, z), the above equation may now be written in the form

$$(x \ \ y \ \ z) \begin{pmatrix} I_{xx} & I_{xy} & I_{xz} \\ I_{yx} & I_{yy} & I_{yz} \\ I_{zx} & I_{zy} & I_{zz} \end{pmatrix} \begin{pmatrix} x \\ y \\ z \end{pmatrix} = 1$$

This quadratic form can be written out explicitly as the equation

$$I_{xx}x^2 + I_{yy}y^2 + I_{zz}z^2 + 2I_{xy}xy + 2I_{xz}xz + 2I_{yz}yz = 1$$

This is clearly the equation of an ellipsoid centred at the origin of coordinates. It is known as the *ellipsoid of inertia*, or the *momental ellipsoid* of the body relative to the point of the body being considered. Each point on the ellipsoid determines a vector $\boldsymbol{r}$ whose square is the reciprocal of the moment of inertia of the body about an axis in the direction of the vector $\boldsymbol{r}$.

If the coordinate axes used are principal axes of inertia, then the above equation takes the form

$$I_1x^2 + I_2y^2 + I_3z^2 = 1$$

Thus the principal axes of the ellipsoid can be seen to coincide with the principal axes of inertia. These results may now be summarised as follows.

***Proposition* 8.6.** *The quadratic form, whose coefficients are the elements of the inertia tensor of a rigid body at a point O,* $I_{xx}x^2 + I_{yy}y^2 + I_{zz}z^2 + 2I_{xy}xy + 2I_{xz}xz + 2I_{yz}yz = 1$ *describes the locus of a point P on an ellipsoid centred at O, whose principal axes coincide with the principal axes of inertia of the body at O, and is such that the square of the length OP is inversely proportional to the moment of inertia of the body about the line OP.*

It should be noticed that a knowledge of the ellipsoid of inertia at a point is sufficient to determine the moments of inertia of the body about all axes passing through that point, as well as the principal axes of inertia. Thus instead of considering the inertia matrix, the dynamical properties of the body can alternatively be obtained from a consideration of the geometrical quantity, the ellipsoid of inertia.

Apart from the neat mathematical interpretation of this approach, it can also sometimes be used to give a geometrical representation of the rotational motion of a rigid body under certain conditions.

8.9. The inertia tensor: examples

Example 8.12

A triangular plate is made of a uniform material and has sides of length a, $2a$ and $\sqrt{3}a$. Determine the principal axes and the principal moments of inertia about the 30° corner.

The mass of the plate may be denoted by m and its corners by A, B and C, where A is the 30° corner and C is 60°. A set of cartesian axes may conveniently be chosen with origin at A, the triangle in the plane $z=0$ and AB along the x axis. The line AC is then given by $y=1/\sqrt{3}x$, $z=0$. This is illustrated in figure 8.14.

It is now necessary to determine the inertia tensor at A relative to these coordinates. The moment of inertia about the x axis may be obtained by direct calculation, but here may simply be quoted as the answer to exercise 8.6 as

$$I_{xx}=\tfrac{1}{6}ma^2$$

Similarly the moment of inertia about BC is $\tfrac{1}{6}m3a^2$ and, therefore, using the parallel axis theorem, proposition 8.1, the moment of inertia about the y axis is

$$\begin{aligned} I_{yy}&=\tfrac{1}{6}m3a^2-m\frac{a^2}{3}+m\frac{4a^2}{3} \\ &=\tfrac{3}{2}ma^2 \end{aligned}$$

Then, using the lamina theorem, proposition 8.2, the moment of inertia about the z axis is

$$I_{zz}=\tfrac{1}{6}ma^2+\tfrac{3}{2}ma^2=\tfrac{5}{3}ma^2$$

The problem now is to find the products of inertia. Clearly, I_{xz} and I_{yz} are both zero since $z=0$ at all points on the plate, and I_{xy} may be obtained from first principles by dividing the plate into n small rectangular pieces of size $\delta x_i\times\delta y_i$. Since the total area of the plate is $\tfrac{1}{2}\sqrt{3}a^2$, the mass of each piece is

$$m_i=\frac{2m}{\sqrt{3}a^2}\,\delta x_i\,\delta y_i$$

and therefore the product of inertia

$$I_{xy}=-\sum_{i=1}^{n} m_i x_i y_i$$

Fig. 8.14. The thin triangular plate considered in example 8.12.

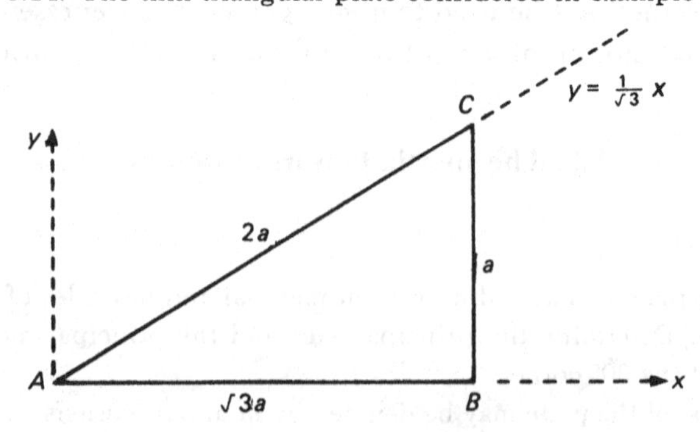

in the limit becomes the double integral

$$I_{xy} = -\int_0^{\sqrt{3}a} dx \int_{y=0}^{y=(1/\sqrt{3}x)} \frac{2m}{\sqrt{3}a^2} xy\, dy$$
$$= -\frac{\sqrt{3}}{4} ma^2$$

The components of the inertia tensor at A relative to these axes are now all known and can be written in the form

$$I_{ij} = \frac{ma^2}{12}\begin{pmatrix} 2 & -3\sqrt{3} & 0 \\ -3\sqrt{3} & 18 & 0 \\ 0 & 0 & 20 \end{pmatrix}$$

The required principal moments and principal axes can now be obtained from the eignevalues and eigenvectors of this matrix according to proposition 8.4. It is convenient to write the eigenvalues as $ma^2\kappa/12$, where κ is a root of the equation

$$\begin{vmatrix} 2-\kappa & -3\sqrt{3} & 0 \\ -3\sqrt{3} & 18-\kappa & 0 \\ 0 & 0 & 20-\kappa \end{vmatrix} = 0$$

which becomes

$$(20-\kappa)(9-20\kappa+\kappa^2)=0$$

Thus $\kappa = 20$ or $10 \pm \sqrt{(91)}$, and so the principal moments of inertia are

$$\lambda_1 = (10+\sqrt{(91)})\frac{ma^2}{12} \qquad \lambda_2 = (10-\sqrt{(91)})\frac{ma^2}{12} \qquad \lambda_3 = \tfrac{5}{3}ma^2$$

By substituting these values back, the corresponding eigenvectors are found to be

$$\boldsymbol{e}_1 = 3\sqrt{3}\boldsymbol{i} - (8+\sqrt{(91)})\boldsymbol{j} \qquad \boldsymbol{e}_2 = (8+\sqrt{(91)})\boldsymbol{i} + 3\sqrt{3}\boldsymbol{j} \qquad \boldsymbol{e}_3 = \boldsymbol{k}$$

These vectors determine the principal axes of the triangular plate at A.

Example 8.13

Find the principal moments of inertia and the principal axes of a uniform solid hemisphere about a point on its rim.

Denote the mass of the hemisphere by m, and its radius by a. It can easily be determined that its centroid is located on the axis of symmetry a distance $3a/8$ from the base. A point on the rim of the base at which principal moments and axes are required may be labelled O.

Now, it is not immediately clear how to determine the components of the inertia tensor at O. It is therefore appropriate to consider whether or not it would be easier first to determine the inertia tensor at the centroid C and then to obtain that at O using the extended parallel axis theorem stated here as proposition 8.3. In this case, this approach would clearly be advantageous, since an axis of symmetry passes through C. This may be chosen as a coordinate axis, and the products of inertia must then all be zero. Also, the moment of inertia about this axis is immediately known, but the other principal moments are not so clear.

In this particular case, however, it may be observed that the inertia tensor at the point A, which is at the centre of the base, can immediately be obtained. It is therefore appropriate here to start with this and then to find the inertia tensors at C and O using proposition 8.3.

Accordingly, it is convenient to choose cartesian axes $Oxyz$ at O with Ox as a diameter of the base and Oy a tangent to it. Sets of parallel axes $Ax'y'z'$ and $Cx''y''z''$ may then be chosen at A and C, as illustrated in the figure 8.15.

Clearly the axes $Ax'y'z'$ are principal axes at A, and the principal moments are all equal to $\frac{2}{5}ma^2$. This can be seen by imagining the complete sphere. By dividing this sphere along the plane $z=0$, the moment of inertia is halved. However, the mass is also halved, so the expressions for these principal moments remains the same as the moment of inertia of a sphere about a line through its centre. Thus, relative to these axes,

$$I_{Aij}=\frac{2}{5}ma^2\begin{pmatrix}1&0&0\\0&1&0\\0&0&1\end{pmatrix}$$

Now the position of C from A is $\frac{3}{8}a\mathbf{k}'$ and so, using proposition 8.3,

$$I_{Cij}=\frac{2}{5}ma^2\begin{pmatrix}1&0&0\\0&1&0\\0&0&1\end{pmatrix}-\frac{9}{64}ma^2\begin{pmatrix}1&0&0\\0&1&0\\0&0&0\end{pmatrix}$$

Finally, the position of C from O is $\bar{\mathbf{r}}=a\mathbf{i}+\frac{3}{8}a\mathbf{k}$ so, again using proposition

Fig. 8.15. A uniform solid hemisphere. Sets of parallel cartesian axes are chosen for convenience.

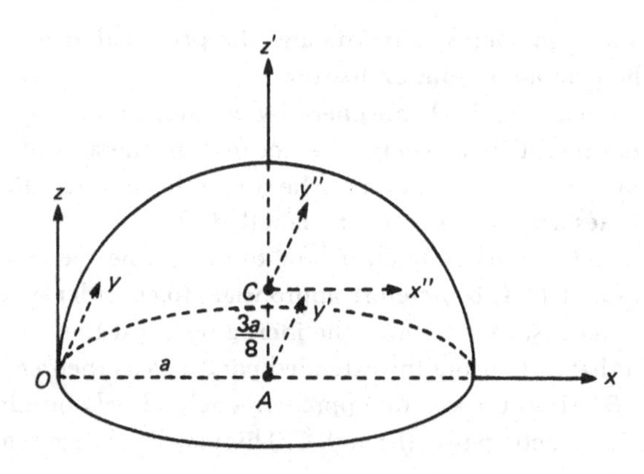

8.3, the inertia matrix at O relative to the axes $Oxyz$ is given by

$$I_{Oij}=\tfrac{2}{5}ma^2\begin{pmatrix}1&0&0\\0&1&0\\0&0&1\end{pmatrix}-ma^2\begin{pmatrix}\frac{9}{64}&0&0\\0&\frac{9}{64}&0\\0&0&0\end{pmatrix}+ma^2\begin{pmatrix}\frac{9}{64}&0&-\frac{3}{8}\\0&\frac{73}{64}&0\\-\frac{3}{8}&0&1\end{pmatrix}$$

Dropping the suffix 0 this gives

$$I_{ij}=\frac{ma^2}{40}\begin{pmatrix}16&0&-15\\0&56&0\\-15&0&56\end{pmatrix}$$

According to proposition 8.4 the required principal moments and principal axes at O can now be obtained from the eigenvalues and eigenvectors of this matrix. Writing the eigenvalues as $ma^2\kappa/40$, these can be obtained from the roots of the determinant

$$\begin{vmatrix}16-\kappa&0&-15\\0&56-\kappa&0\\-15&0&56-\kappa\end{vmatrix}=0$$

which becomes a cubic which in this case factorises to

$$(56-\kappa)(61-\kappa)(11-\kappa)=0$$

The principal moments of inertia are thus

$$\tfrac{7}{5}ma^2 \qquad \tfrac{61}{40}ma^2 \qquad \tfrac{11}{40}ma^2$$

By substituting these values back, it can be seen that the principal axis associated with the principal moment $\frac{7}{5}ma^2$ is $\boldsymbol{j}$, which is the y axis or tangent to the rim at O. The principal axes associated with the principal moments $\frac{61}{40}ma^2$ and $\frac{11}{40}ma^2$ are given respectively by $10^{-1/2}(\boldsymbol{i}-3\boldsymbol{k})$ and $10^{-1/2}(3\boldsymbol{i}+\boldsymbol{k})$. It may immediately be observed that these three principal axes are mutually orthogonal as is required.

Exercises

8.16 Four particles of masses m, $2m$, $3m$ and $4m$ are located at the points (a, a, a), $(a, -a, -a)$, $(-a, a, -a)$ and $(-a, -a, a)$, respectively, and are rigidly connected to one another by a light framework. Using the basic definitions of moments and products of inertia, show that the principal moments of inertia of the system at the origin are
$20\ ma^2$; $2(10+\sqrt{5})ma^2$, $2(10-\sqrt{5})ma^2$

8.17 Three uniform rods OA, OB and OC are each of unit length and unit mass. Choosing a cartesian frame of reference with origin at O, the coordinates of the points A, B and C are respectively $(1, 0, 0)$, $(0, 0, 1)$ and $(\sqrt{3}/2, \frac{1}{2}, 0)$. Show that the principal moments of inertia of the system at O are $\frac{2}{3}$, $\frac{2}{3}+1/(2\sqrt{3})$ and $\frac{2}{3}-1/(2\sqrt{3})$.

8.18 A uniform solid rectagular block of mass m has a square base with edges of length $2a$, and its height is a. Using cartesian axes parallel to the height and to the diagonals of the base, show that the principal moments of inertia of the block relative to a corner of the base are

$\frac{8}{3}ma^2$, $(\frac{5}{3}+\sqrt{(\frac{3}{2})})ma^2$, $(\frac{5}{3}-\sqrt{(\frac{3}{2})})ma^2$

Find also the principal axes.

8.19 A uniform square plate $OABC$ which has sides of length $2a$, is cut in half along the diagonal OB. Show that the principal moments of inertia of the triangular plate OAB relative to the corner O are

$\frac{8}{3}ma^2$, $\frac{1}{3}(4+\sqrt{(13)})ma^2$, $\frac{1}{3}(4-\sqrt{(13)})ma^2$

where m is its mass. Determine also the principal axes.

8.20 A uniform solid cylinder has a circular base of radius a and a height $\sqrt{3}a$. Show that its principal moments of inertia relative to a point on the rim of the base are $\frac{1}{2}ma^2$, $\frac{9}{4}ma^2$ and $\frac{9}{4}ma^2$. Show also that, relative to this point, one of the principal axes passes through the centroid.

8.21 Show that the principal axes of a uniform circular cone with semi vertical angle $\pi/4$ relative to a point on the rim of the base are the tangent to the base at that point and two lines perpendicular to it, one being inclined to the diameter of the base passing through the point at an angle $\frac{1}{2}\tan^{-1}\frac{10}{21}$.

8.10. Equations of motion

The equations for the motion of a rigid body have already been stated in section 8.2. The first equation, obtained initially as proposition 7.5, states that the mass of the body times the acceleration of its centroid is equal to the sum of the external forces acting on the body.

$$m\ddot{\bar{\boldsymbol{r}}} = \boldsymbol{F}$$

This is an equation for the translational motion of the body. It is essentially the equation for its particle representation, although it clarifies the point that the body can be represented in this case as a particle of equal mass located at the centroid. The techniques developed initially for particle dynamics can all be applied to the analysis of this equation, and so no further discussion of it is required here.

The other equation which is required for the analysis of the dynamics of a rigid body is that which relates to its rotational motion. Such an equation was obtained in proposition 7.7 which states that the rate of change of its angular momentum about the centroid is equal to the sum of the moments about the centroid of the external forces which act on the body.

$$\frac{\mathrm{d}}{\mathrm{d}t}\boldsymbol{L}_c = \boldsymbol{M}_c$$

It is this equation which requires further discussion.

The problem which arises in this equation is in the evaluation of the angular momentum vector. It is usually comparatively easy to evaluate the moments of the external forces. However, it has been shown that the angular momentum vector must be evaluated in terms of the moments and products of inertia of the body and, if an inertial frame of reference is used, then the moments and products of inertia will normally vary as the body rotates. It is possible to describe this mathematically by the introduction of matrix operators which describe the rotation of a frame of reference which is fixed in the body. This is then applied to the inertia matrix of the body relative to that frame, and thus an expression is obtained for the angular momentum vector of the body. The resulting equation of motion then contains the parameters which describe the rotation and their first derivatives. However, in general, the integration of these equations is very difficult, but, when it can be done, this approach has the advantage that the orientation of the body can immediately be deduced from the rotation parameters.

There is, however, an alternative approach which is frequently more convenient. The essential feature in this case is to work in terms of a frame of reference which is rotating with the body. With respect to such a frame the moments and products of inertia are constants, but a complication is introduced because the frame of reference is rotating. However, applying lemma 4.1, and using a dot to denote the apparent time derivative of a quantity with respect to the frame of reference being used, which in this case is rotating with the same angular velocity as the body, the equation for its rotational motion can now be written in the form

$$\boldsymbol{L}_{\mathrm{c}} + \boldsymbol{\omega} \times \boldsymbol{L}_{\mathrm{c}} = \boldsymbol{M}_{\mathrm{c}}$$

The equations stated above are sufficient to describe the motion of a rigid body which has, in general, six degrees of freedom. They determine the motion of the centroid and the rotational motion of the body about the centroid. Even when the motion of the body is subject to a number of constraints, these equations may still be used, although alternative equations may then sometimes be more convenient. In particular, when the body is constrained to rotate about a point which is fixed relative to an inertial frame of reference, it is convenient to take that point as an origin. According to proposition 7.6, the rate of change of angular momentum about that point is then equal to the sum of the moments of the external forces about it. Choosing axes fixed in the body, and using the above notation, the equation for the rotational motion of a rigid body about a fixed point may be written in the form

$$\dot{\boldsymbol{L}} + \boldsymbol{\omega} \times \boldsymbol{L} = \boldsymbol{M}$$

This clearly has the same mathematical form as the above equation for the rotational motion of a rigid body which does not have a fixed point, but may be considered to rotate about its centroid.

In the latter case it has been assumed that the origin is a fixed point in an inertial frame of reference. This assumption, however, is not necessary provided that the additional fictitious forces which arise are included as external forces acting on all the particles of the body. Although this is theoretically possible, it is usually inconvenient. Thus the latter equation above is normally applied only to the motion of a body which rotates about a point which is fixed in an inertial frame of reference. In cases in which the point about which the body is considered to rotate accelerates relative to an inertial frame, it is usually more convenient to use the previous equations which describe the motion of its centroid and its rotational motion about the centroid.

In the approach just described it was assumed to be convenient to use a set of coordinate axes which are fixed relative to the body. Thus the angular velocity of the coordinate frame is the same as the angular velocity of the body. However, this choice of coordinate axes is not the most convenient in all cases. It was adopted above simply because the inertia matrix is constant with respect to such axes. Now it is most convenient to have a constant inertia matrix, but this does not always have to be achieved by fixing the axes relative to the body.

Consider, for example, a body which posesses an axis of symmetry and is free to rotate about a fixed point on that axis. Principal axes at that point are the axis of symmetry and an arbitrary pair of axes perpendicular to it having the same principal moment of inertia. Thus, for such a body, the inertia matrix is constant relative to any frame of reference which has one axis aligned with the axis of symmetry, even though the body may rotate about that axis relative to the coordinate frame. This approach is used, for example, in the analysis of the motion of a symmetrical top about a point on its axis. One coordinate axis is chosen to be the axis of symmetry and it is convenient to choose a second coordinate axis which is perpendicular to the first and lies in the horizontal plane. In such cases, it is necessary to distinguish between the angular velocity $\boldsymbol{\Omega}$ of the coordinate system and the angular velocity $\boldsymbol{\omega}$ of the body.

Bearing this point in mind, the above discussion may now be summarised in the form of a proposition.

Proposition 8.7. *The equation for the rotational motion of a rigid body about a point O, which is either the centroid or a stationary point in an inertial frame of reference, relative to a set of axes which have origin at O and are rotating with angular velocity* $\mathbf{\Omega}$ *relative to an inertial frame, is given by*

$$\dot{\boldsymbol{L}}+\mathbf{\Omega}\times\boldsymbol{L}=\boldsymbol{M}$$

where $\mathbf{L}$ *is the angular momentum of the body which may be expressed as the product of the inertia matrix and the angular velocity vector of the body* $\boldsymbol{\omega}$, *and* $\boldsymbol{M}$ *is the sum of the moments of the external forces about O.*

This equation of motion takes a particularly simple form if the axes chosen are a set of principal axes which are fixed in the body. In this case the angular velocity of the axes is the same as that of the body so the inertia matrix is constant, the products of inertia are all zero, and the components of the angular momentum vector are the products of the principal moments and the components of the angular velocity vector. The explicit components of the equation of motion, in this case, take the particularly simple form

$$I_1\dot{\omega}_1-(I_2-I_3)\omega_2\omega_3=M_1$$
$$I_2\dot{\omega}_2-(I_3-I_1)\omega_3\omega_1=M_2$$
$$I_3\dot{\omega}_3-(I_1-I_2)\omega_1\omega_2=M_3$$

These are known as *Euler's equations of motion.* It must, however, be emphasised that they apply only with respect to a set of principal axes which are fixed relative to the body.

Euler's equations of motion are widely used in the analysis of the rotational notion of a rigid body. The basic approach used is to attempt to integrate these equations to obtain an expression for instantaneous angular velocity vector $\boldsymbol{\omega}$ as a function of time. However, even having obtained this vector, it is still not an easy matter to determine the orientation of the body at any time. The particular techniques that may be used to determine the orientation depend heavily on the nature of the particular problem.

Having discussed the equations of rigid body motion, the following observation may now be made. The two vector equations which involve the mass and the moments and products of inertia are sufficient to determine uniquely the motion of a rigid body under the action of a given set of external forces. Thus the characteristics of a rigid body which determine its dynamical behaviour are its total mass and either the components of its inertia matrix or its ellipsoid of inertia evaluated at its centroid. If these are known, then the components of its inertia matrix at any other point can be determined from the parallel axis theorem, proposition 8.3. These quantities, however, do not uniquely determine the mass distribution within the body. In fact it is quite clear that a number of different distributions of mass may well have the same total mass and the same inertia matrix or ellipsoid of inertia relative to its centroid. Such systems are referred to as *Equimomental systems.* Although their mass distributions may differ, the dynamical properties of equimomental systems must be identical.

8.11. The energy integral

The equations for the motion of a rigid body have now been determined, and the remaining problem is how to integrate them. In practice it turns out that only very rarely is it possible to obtain complete integrals which are given explicitly in terms of simple functions. However, a number of first integrals may frequently be obtained. These usually have some simple physical significance and therefore may be stated without necessarily integrating the equations of motion. They may be integrals of linear momentum, angular momentum or energy. Any component of the linear or angular momentum vectors may easily be obtained using expressions already given, but the energy integral requires some further clarification.

Let us first consider the kinetic energy of a rigid body which is constrained to rotate about a fixed point. Assuming that the angular velocity of the body at any instant is $\boldsymbol{\omega}$, then relative to a frame of reference at that point its kinetic energy is given by

$$\begin{aligned} T &= \sum_{i=1}^{n} \tfrac{1}{2} m_i v_i^2 \\ &= \sum_{i=1}^{n} \tfrac{1}{2} m_i (\boldsymbol{\omega} \times \boldsymbol{r}_i) \cdot (\boldsymbol{\omega} \times \boldsymbol{r}_i) \\ &= \tfrac{1}{2} \boldsymbol{\omega} \cdot \left\{ \sum_{i=1}^{n} \boldsymbol{r}_i \times m_i (\boldsymbol{\omega} \times \boldsymbol{r}_i) \right\} \\ &= \tfrac{1}{2} \boldsymbol{\omega} \cdot \boldsymbol{L} \end{aligned}$$

Thus the kinetic energy of the body can be seen to be one-half of the scalar product of the angular velocity and the angular momentum vectors.

The angular momentum vector, however, may be expressed in terms of the components of the inertia matrix. The kinetic energy can therefore be seen to be one-half of the quadratic form formed by the angular velocity vector and the inertia matrix.

$$T=\tfrac{1}{2}I_{xx}\omega_1^2+\tfrac{1}{2}I_{yy}\omega_2^2+\tfrac{1}{2}I_{zz}\omega_3^2+I_{xy}\omega_1\omega_2+I_{xz}\omega_1\omega_3+I_{yz}\omega_2\omega_3$$

This expression is in fact independent of a rotation of the coordinate axes. It is therefore always possible to choose a frame of reference which is fixed in the body so that, with respect to such a frame, the components of the inertia matrix are all constants. Further simplification can of couse be made if the axes chosen are the set of principal axes at the point. In this case the kinetic energy of the body can be expressed as

$$T=\tfrac{1}{2}I_1\omega_1^2+\tfrac{1}{2}I_2\omega_2^2+\tfrac{1}{2}I_3\omega_3^2$$

In this form it can be seen that the kinetic energy of a rigid body is equal to the sum of the kinetic energies of rotation about the three principal axes.

The above expressions are for the kinetic energy of a rigid body which is free to rotate about a fixed point. In the general motion of a body, however, according to proposition 7.2, the kinetic energy is the sum of that associated both with the linear motion of the centroid and the rotational motion about the centroid. This can be simply given by the expression

$$T=\tfrac{1}{2}\boldsymbol{\omega}\cdot\mathbf{L}_c+\tfrac{1}{2}m\dot{\mathbf{r}}^2$$

The first term, which describes the rotational kinetic energy about the centroid, may also be expanded in terms of the components of the inertia matrix in exactly the same way as that given above, but it must be remembered in this case that the inertia matrix must be evaluated relative to the centroid.

With the above formulae, expressions for the kinetic energy of a rigid body can always be obtained, and its potential energy must now be considered.

Of course, an energy integral can always be obtained whenever all the external forces acting on the body which do work are conservative. In any such general case the potential energy may be evaluated appropriately for each application. However, most of the applications of rigid body dynamics concern the motion of solid objects in the environment in which we live. Since this is confined to the region near the earth's surface, it is usual to consider bodies moving in a gravitational field which is approximately constant. At this level of approximation the results of section 7.2 as stated in proposition 7.4 may be used. The potential energy can then be expressed as

$$V = mg\bar{z} + c$$

General expressions for the kinetic energy and the potential energy of a rigid body in a constant gravitational field have now been given. These will usually be included in any energy integral which may be obtained. However, it must be remembered that an energy integral will not exist if any nonconservative forces do work, and that, if any other conservative forces are applied, then potential energies associated with these forces must also be included.

8.12. 3D rigid body motion: examples

This section contains a set of miscellaneous examples in which the orientation of a rigid body varies with time. In each case the direction of the instantaneous axis of rotation does not remain constant.

Example 8.14: *the falling door*

Consider here the motion of a door if the upper hinge were suddenly to break or be removed. Clearly, in practice there would be a number of complicating factors, so only a highly idealised model is considered here.

The door may initially be represented as a uniform rectangular plate, its mass may be denoted by m, its width by $2a$ and its height by $2b$. Its hinges may be assumed to be located at an upper and a lower corner, so that when the upper hinge is removed the door will rotate about a lower corner. The frictional forces as the lower edge slides along the floor may initially be ignored. It may therefore be assumed that the only forces acting on the door are the gravitational force, the reaction at the fixed corner O, and the normal reaction on the lower edge of the door which ensures that it remains in a fixed horizontal plane through O.

A set of fixed cartesian axes may be chosen with the door initially in the plane $y = 0$, and the vertical z axis aligned initially with the hinged edge of the door. As the door falls, however, it is convenient to use a new set of axes $\hat{a}$, $\hat{b}$, $\hat{c}$ which are fixed relative to the door with $\hat{a}$ and $\hat{b}$ aligned with the initially hinged edge and the lower edge respectively. These axes are illustrated in figure 8.16.

Since the lower edge is constrained to lie in the horizontal plane through O, the body has two degrees of freedom. Thus, as it falls, the motion of the door can be described by two parameters θ and ϕ, where θ is the angle between the initially hinged edge of the door $\hat{a}$ and the vertical, and ϕ is the angle through which the lower edge has rotated.

In this highly idealised situation it is found to be unnecessary to consider the full equations of motion, as two first integrals can easily be obtained. Since all frictional forces have been ignored, the reactional forces at O and along the lower edge do no work, and thus the energy integral can be obtained. Also, since the gravitational force and the reactional forces along the lower edge act vertically, the angular momentum of the door about the vertical axis through O is constant. Since there are only two degrees of freedom, the two integrals $T+V=\text{const.}$ and $\boldsymbol{L}_O \cdot \boldsymbol{k}=\text{const.}$ are sufficient to determine the motion.

To use these equations it is first necessary to determine the inertia matrix of the door relative to the axes $\hat{a}$, $\hat{b}$, $\hat{c}$. Clearly,

$$I_{aa}=\tfrac{4}{3}ma^2 \qquad I_{bb}=\tfrac{4}{3}mb^2 \qquad I_{cc}=\tfrac{4}{3}m(a^2+b^2)$$

Fig. 8.16. A falling rectangular plate with corner O fixed. Axes $\hat{a}$, $\hat{b}$ and $\hat{c}$ are fixed relative to the plate.

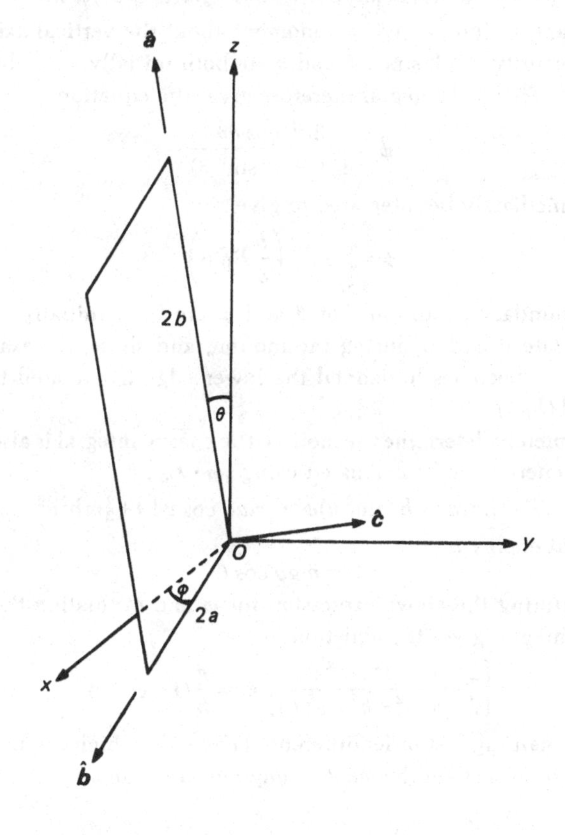

It can also be quickly shown that

$$I_{ab} = -mab \qquad I_{ac} = 0 \qquad I_{bc} = 0$$

so that the inertia matrix relative to these axes can be written as

$$I_{ij} = \frac{m}{3}\begin{pmatrix} 4a^2 & -3ab & 0 \\ -3ab & 4b^2 & 0 \\ 0 & 0 & 4(a^2+b^2) \end{pmatrix}$$

Now the door, and hence the coordinate frame, clearly rotates with angular velocity $\dot{\phi}$ about the vertical $\boldsymbol{k}$ and simultaneously with angular velocity $\dot{\theta}$ about the lower edge $\hat{\boldsymbol{b}}$. But

$$\boldsymbol{k} = \cos\theta\hat{\boldsymbol{a}} + \sin\theta\hat{\boldsymbol{c}}$$

Thus the angular velocity of the door is

$$\boldsymbol{\omega} = \dot{\phi}\cos\theta\hat{\boldsymbol{a}} + \dot{\theta}\hat{\boldsymbol{b}} + \dot{\phi}\sin\theta\hat{\boldsymbol{c}}$$

Using the inertia matrix, the angular momentum of the door about O is therefore

$$\boldsymbol{L}_0 = ma(\tfrac{4}{3}a\dot{\phi}\cos\theta - b\dot{\theta})\hat{\boldsymbol{a}} + mb(\tfrac{4}{3}b\dot{\theta} - a\dot{\phi}\cos\theta)\hat{\boldsymbol{b}} + \tfrac{4}{3}m(a^2+b^2)\dot{\phi}\sin\theta\hat{\boldsymbol{c}}$$

Thus

$$\boldsymbol{L}_0 \cdot \boldsymbol{k} = ma\cos\theta(\tfrac{4}{3}a\dot{\phi}\cos\theta - b\dot{\theta}) + \tfrac{4}{3}m(a^2+b^2)\dot{\phi}\sin^2\theta$$

Since there are no forces having a moment about the vertical axis, this is a conserved quantity. And, since $\dot{\phi}$ and $\dot{\theta}$ are both initially zero, this constant must be zero. This first integral therefore gives the equation

$$\dot{\phi} = \frac{3ab\cos\theta\dot{\theta}}{4(a^2+b^2\sin^2\theta)}$$

This can immediately be integrated to give

$$\phi = \tfrac{3}{4}\tan^{-1}\left(\frac{b}{a}\sin\theta\right)$$

using the boundary condition that θ and ϕ are both initially zero. These equations relate θ and ϕ during the motion, and show, for example, that when the door becomes horizontal the lower edge has rotated through an angle $\frac{3}{4}\tan^{-1}(b/a)$.

To completely determine the motion, the energy integral is also required. The kinetic energy can be evaluated using $\frac{1}{2}\boldsymbol{\omega}\cdot\boldsymbol{L}_0$ as

$$T = \tfrac{2}{3}m(a^2+b^2\sin^2\theta)\dot{\phi}^2 - mab\cos\theta\dot{\phi}\dot{\theta} + \tfrac{2}{3}mb^2\dot{\theta}^2$$

The potential energy is

$$V = mgb\cos\theta$$

Then, substituting the above expression for $\dot{\phi}$ and evaluating the constant, the energy integral gives the equation

$$\left\{\frac{2}{3} - \frac{3}{8}\frac{a^2\cos^2\theta}{(a^2+b^2\sin^2\theta)}\right\}\dot{\theta}^2 = \frac{g}{b}(1-\cos\theta)$$

This is the remaining first-order differential equation which can be integrated numerically to give θ and hence ϕ at any required time.

Example 8.15. Free motion of a body with an axis of symmetry

Consider the rotational motion of a body having an axis of symmetry under the action of a set of forces which have no net moment about the centroid. Such a situation sometimes occurs for example in the free motion of an artificial satellite. Although a satellite does not strictly have an axis of symmetry, it could possibly be equimomental with such a body, and so its motion is dynamically equivalent. In free satellite motion, the only force acting is the gravitational force which is equivalent to a single force acting at the centroid. There is no net moment about the centroid and, consequently, the angular momentum of the body is constant.

To analyse the rotational motion of such a body it is convenient to choose a set of axes that are fixed in the body, having origin at the centroid and z axis as the axis of symmetry. Such axes are clearly principal axes with $I_1 = I_2$. In this situation Euler's equations reduce to

$$I_1\dot{\omega}_1 = (I_1 - I_3)\omega_2\omega_3$$
$$I_1\dot{\omega}_2 = (I_3 - I_1)\omega_3\omega_1$$
$$I_3\dot{\omega}_3 = 0$$

Clearly, the last of these equations implies that

$$\omega_2 = \text{const}$$

Also, by multiplying the first equation by ω_1, the second by ω_2 and adding, it can be seen that

$$\omega_1\dot{\omega}_1 + \omega_2\dot{\omega}_2 = 0$$

which implies that

$$\omega_1^2 + \omega_2^2 = \text{const}$$

Thus it can be seen that the components of the angular velocity vector both in the direction of the axis of symmetry, and in the plane perpendicular to it, are constants. The angular velocity vector must therefore be inclined at a constant angle to the axis of symmetry. This angle may be denoted by θ. It is then convenient to denote by ϕ the angle between the component in the plane perpendicular to the axis of symmetry and the axis; θ and ϕ are then the normal spherical polar coordinates defining the direction of the axis of rotation relative to the chosen coordinates which are fixed in the body (see figure 8.17).

The cartesian components of the angular velocity vector may thus be expressed as

$$\omega_1 = \omega \sin\theta \cos\phi$$
$$\omega_2 = \omega \sin\theta \sin\phi$$
$$\omega_3 = \omega \cos\theta$$

where θ is a constant. The angle ϕ, however, is not constant, and an equation for it can be obtained by substituting these expressions back into one of the first two of Euler's equations. The resulting equation is

$$\dot{\phi} = \frac{(I_3 - I_1)}{I_1} \omega \cos \theta$$

It can thus be seen that in the general free motion of a body with an axis of symmetry, the angular velocity vector has constant magnitude, is inclined at a constant angle to the axis of symmetry and rotates about it with constant angular speed. However, this result only defines the position of the angular velocity vector relative to axes that are fixed in the body, and how these axes rotate with the body has not yet been determined.

In order to fully describe the rotation of the body, it is necessary to use the fact that the angular momentum vector is constant, since the external forces have no net moment about the centroid. Since the axes adopted are principal axes, the angular momentum vector can be expressed as

$$\boldsymbol{L} = I_1 \omega_1 \boldsymbol{i} + I_1 \omega_2 \boldsymbol{j} + I_3 \omega_3 \boldsymbol{k}$$

or

$$\boldsymbol{L} = \omega\{I_1 \sin \theta(\cos \phi \boldsymbol{i} + \sin \phi \boldsymbol{j}) + I_3 \cos \theta \boldsymbol{k}\}$$

From this, or by observing that $\boldsymbol{L} = I_1 \boldsymbol{\omega} + (I_3 - I_1)\boldsymbol{k}$, It can be seen that the angular momentum vector is in the same plane as the axis of symmetry and the instantaneous axis of rotation, and is inclined at an angle β to the axis of symmetry where

$$\tan \beta = \frac{I_1}{I_3} \tan \theta$$

It is, however, the angular momentum vector which is constant. Both its magnitude and its direction in space are fixed. The angular velocity vector and the axis of symmetry of the body therefore rotate about this direction with constant angular speed $\dot{\phi}$, as expressed above. The body itself, however, always rotates about the instantaneous angular velocity vector whose direction rotates about that of the angular momentum vector.

Fig. 8.17. The angular velocity relative to axes fixed in the body.

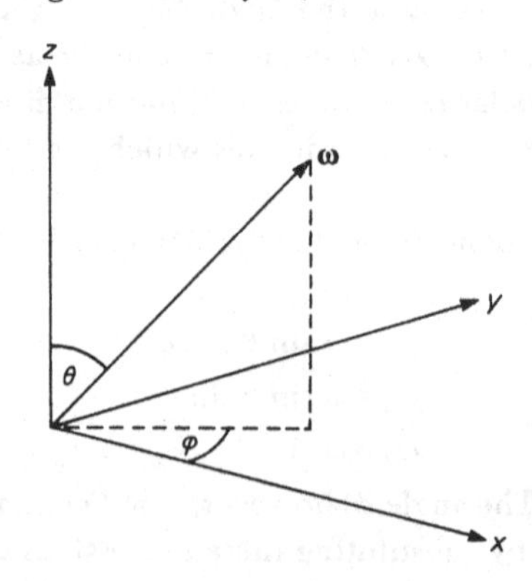

This type of motion can be pictured as the rolling motion of one cone over another. The rotation of the instantaneous angular velocity vector around the angular momentum vector generates a cone of semiangle $|\theta - \beta|$ which is fixed in space and is known as the *space cone*. The rotation of the body about a line on this cone is, then, like the rolling of a cone of semiangle θ around the space cone. This second cone is known as the *body cone* since it is fixed relative to the body. Its motion therefore illustrates the motion of the body. There are two cases to be considered according as to whether I_1 is less than or greater than I_3. These correspond to the cases when β is less than or greater than θ, and the space cone may therefore be inside or outside the body cone. These situations are illustrated in figures 8.18 and 8.19.

Fig. 8.18. When $I_1 < I_3$ the body cone rolls around the space cone.

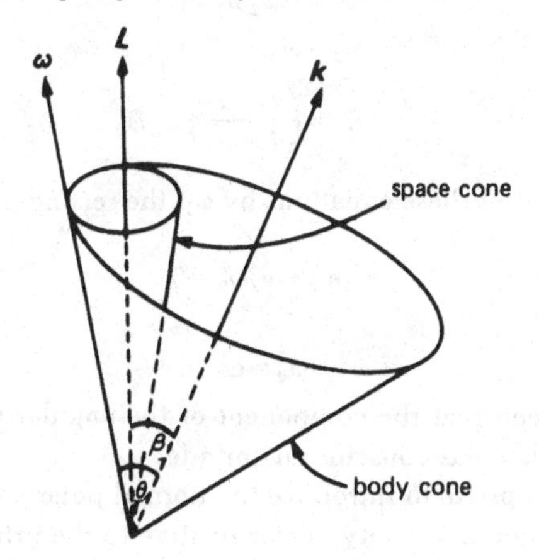

Fig. 8.19. When $I_1 > I_3$ the body cone rolls on the outside of the space cone.

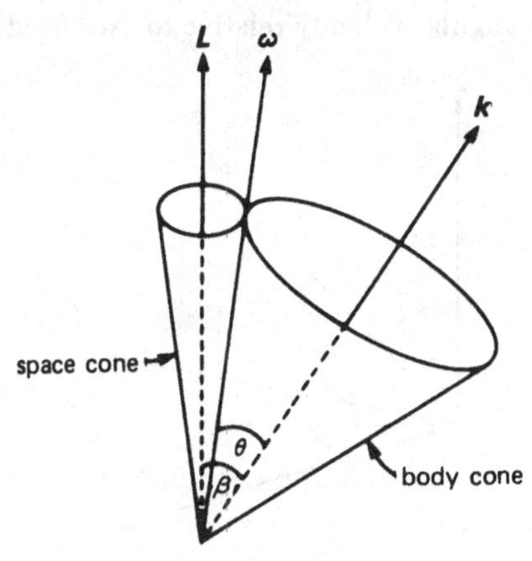

Example* 8.16: *free motion of a plane body

Consider the motion of a plane body under the action of a set of forces which have no net moment about the centroid. As in the previous example, the sum of the external forces determines the motion of the centroid of the body, but it is only the rotational motion of the body about the centroid which is to be considered here. Again, the external forces have no net moment about the centroid, so the angular momentum of the body is constant.

For any plane body, two principal axes can be found in the plane of the body. The third principal axis is then perpendicular to plane and, by the lamina theorem,

$$I_3 = I_1 + I_2$$

In this case, Eulers equations reduce to the following

$$\dot{\omega}_1 = -\omega_2\omega_3$$
$$\dot{\omega}_2 = \omega_3\omega_1$$
$$\dot{\omega}_3 = \left(\frac{I_1 - I_2}{I_1 + I_2}\right)\omega_1\omega_2$$

Multiplying the first of these equations by ω_1, the second by ω_2, and adding, gives

$$\omega_1\dot{\omega}_1 + \omega_2\dot{\omega}_2 = 0$$

which implies that

$$\omega_1^2 + \omega_2^2 = \text{const}$$

Hence it can be seen that the component of the angular velocity vector in the plane of the plate has constant magnitude.

It is now appropriate to introduce the normal polar angles to define the direction of the angular velocity vector relative to the principal axes. Let θ denote the angle between the angular velocity vector and the normal to the

Fig. 8.20. The angular velocity relative to axes fixed in the body.

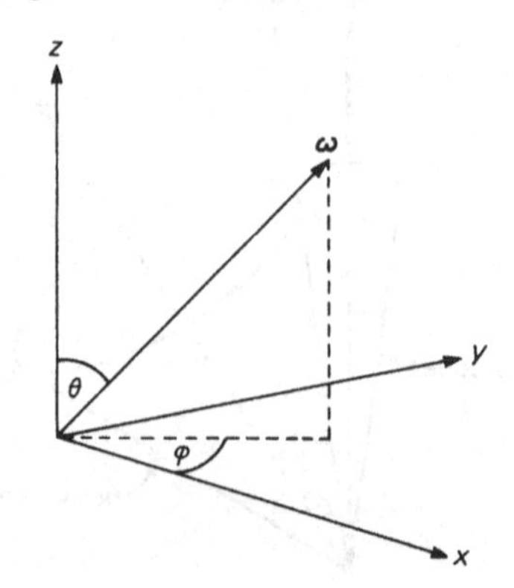

plane, and ϕ the angle between the component of the angular velocity vector in the plane and the x axis, as illustrated in figure 8.20.

Denoting the initial values of θ, ϕ and ω by θ_0, ϕ_0 and ω_0, the components of the angular velocity vector in the directions of the principal axes in the plane can be expressed as

$$\omega_1 = \omega_0 \sin\theta_0 \cos\phi$$
$$\omega_2 = \omega_0 \sin\theta_0 \sin\phi$$

Substituting these back into the first of Euler's equations implies that

$$\omega_3 = \dot{\phi}$$

And substitution into the last of Euler's equations then gives

$$\ddot{\phi} = \left(\frac{I_1 - I_2}{I_1 + I_2}\right) \omega_0^2 \sin^2\theta_0 \cos\phi \sin\phi$$

This is a second-order differential equation which determines the rotational motion of the body. It can immediately be integrated to give

$$\dot{\phi}^2 = \left(\frac{I_1 - I_2}{I_1 + I_2}\right) \omega_0^2 \sin^2\theta_0 \sin^2\phi + \text{const}$$

and hence

$$\omega_3^2 = \omega_0^2 \left\{\cos^2\theta_0 + \left(\frac{I_1 - I_2}{I_1 + I_2}\right) \sin^2\theta_0 \,(\sin^2\phi - \sin^2\phi_0)\right\}$$

This equation may now be further integrated in any particular application to determine ϕ at any time. This would then determine the three components of the angular velocity vector. If required, θ could be obtained from the result that

$$\omega_3 \tan\theta = \omega_0 \sin\theta_0$$

Finally, the directions of the principal axes at any time can be discussed using the fact that the angular momentum vector is a constant of the motion.

Example 8.17

As a final example of the application of Euler's equations to the rotational motion of a rigid body, the following particular situation may be considered.

A uniform solid cuboid of mass m whose sides are of lengths $2a$, $4a$ and $6a$ is acted on by a set of forces which have no net moment about the centroid. If it is initially spinning about an axis which is perpendicular to the sides of length $4a$ and inclined to the sides of length $2a$ at an angle α where $\cos\alpha = \frac{5}{8}$, determine the subsequent rotational motion.

The principal axes of the body at the centroid are clearly those which are parallel to the edges. Choosing the x, y and z axes parallel to the sides of length $2a$, $4a$ and $6a$ respectively (see figure 8.21), the principal moments of inertia are (see exercise 8.5)

$$I_1 = \tfrac{13}{3}ma^2 \qquad I_2 = \tfrac{10}{3}ma^2 \qquad I_3 = \tfrac{5}{3}ma^2$$

In this case Euler's equations reduce to

$$13\dot{\omega}_1 = 5\omega_2\omega_3$$

$$5\dot{\omega}_2 = -4\omega_3\omega_1$$

$$5\dot{\omega}_3 = 3\omega_1\omega_2$$

Any pair of these equations can be used to obtain an immediate first integral, but for the convenience of avoiding large integers and hyperbolic functions the last pair are the most convenient here. They give

$$3\omega_2\dot{\omega}_2 + 4\omega_3\dot{\omega}_3 = 0$$

which implies that

$$3\omega_2^2 + 4\omega_3^2 = \text{const}$$

The constant here can be evaluated using the initial condition implicit in the question that, when $t = 0$,

$$\omega = \frac{\omega_0}{8}(5\boldsymbol{i}_0 + \sqrt{(39)}\boldsymbol{k}_0)$$

relative to the adopted coordinate frame. The above constant is therefore $\frac{39}{16}\omega_0^2$.

The above integral implies that the component of the angular velocity vector in the coordinate plane $x = 0$ traces out an ellipse. It is therefore convenient to introduce a parameter ϕ such that

$$\omega_2 = \frac{\sqrt{(13)}}{4}\omega_0 \sin\phi$$

$$\omega_3 = \frac{\sqrt{(39)}}{8}\omega_0 \cos\phi$$

From the given boundary conditions it can be seen that the parameter ϕ is initially zero.

Fig. 8.21. A uniform solid cuboid showing principal axes at the centroid.

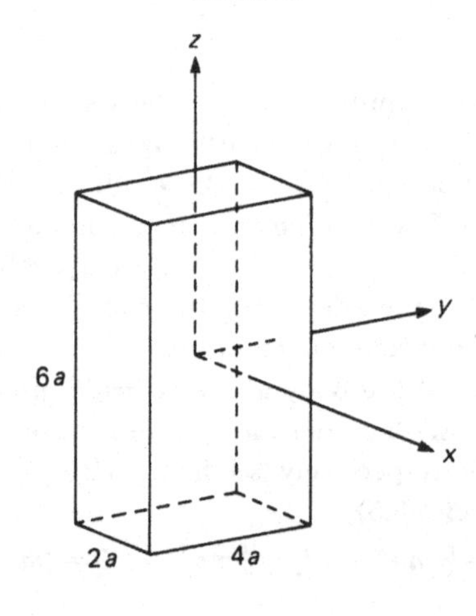

These expressions may now be substituted back into either of the components of Euler's equations that have already been used, to obtain an expression for ω_1 in terms of the same parameter. This gives

$$\omega_1 = -\frac{5}{2\sqrt{3}}\dot{\phi}$$

By again looking at the boundary conditions, it can be seen that initially

$$\phi = 0 \qquad \dot{\phi} = -\frac{\sqrt{3}}{4}\omega_0$$

The various expressions for the components of the angular velocity vector may now be substituted back into the remaining component of Euler's equation, which here is the first. The result is a single differential equation involving only the parameter ϕ. This reduces to

$$\ddot{\phi} = -\tfrac{3}{16}\omega_0^2 \sin\phi \cos\phi$$

which can be integrated to give

$$\dot{\phi}^2 = \tfrac{3}{16}\,\omega_0^2 \cos^2\phi + \text{const}$$

From the initial conditions it can be seen that the constant is zero and therefore

$$\dot{\phi} = -\frac{\sqrt{3}}{4}\,\omega_0 \cos\phi$$

where the negative sign of the root is also required by the initial conditions.

Although in most problems of this type it is not possible to proceed analytically beyond this point, in this case the above equation can again be integrated to give

$$\log(\sec\phi - \tan\phi) = \frac{\sqrt{3}}{4}\,\omega_0 t$$

which then enables each component of the angular velocity vector to be written explicitly as functions of time

$$\omega_1 = \tfrac{5}{8}\omega_0 \operatorname{sech}\left(\frac{\sqrt{3}}{4}\,\omega_0 t\right)$$

$$\omega_2 = -\frac{\sqrt{13}}{4}\,\omega_0 \tanh\left(\frac{\sqrt{3}}{4}\,\omega_0 t\right)$$

$$\omega_3 = \frac{\sqrt{39}}{8}\,\omega_0 \operatorname{sech}\left(\frac{\sqrt{3}}{4}\,\omega_0 t\right)$$

It may immediately be seen from this that, although the body initially rotates about an axis in the xz plane, the direction of the angular velocity vector asymptotically approaches the negative y axis. The coordinate axes, however, rotate with the body, and to determine the orientation of the body in space it is necessary to consider the angular momentum vector which here is a constant of the motion

$$\boldsymbol{L} = I_1\omega_1\boldsymbol{i} + I_2\omega_2\boldsymbol{j} + I_3\omega_3\boldsymbol{k}$$

For example, by substituting $t=0$ and $t=\infty$, the direction of the asymptotic angular velocity can be determined in terms of the initial orientation.

$$\tfrac{13}{3}ma^2\tfrac{25}{8}\omega_0\boldsymbol{i}_0+\tfrac{5}{8}ma^2\frac{\sqrt{(39)}}{8}\omega_0\boldsymbol{k}_0=\frac{-10}{3}ma^2\frac{\sqrt{(13)}}{4}\omega_0\boldsymbol{j}_\infty$$

which implies that

$$\boldsymbol{j}_\infty=-\tfrac{1}{4}(\sqrt{(13)}\boldsymbol{i}_0+\sqrt{3}\boldsymbol{k}_0)$$

This may now be compared with the initial direction of the axis of rotation $\frac{1}{8}(5\boldsymbol{i}_0+\sqrt{(39)}\boldsymbol{k}_0)$ and, by taking scalar products, it can be seen that the final direction of the angular velocity vector is inclined to its initial direction at an angle $\cos^{-1}(\frac{1}{4}\sqrt{(13)})$.

Example 8.18: *the spinning top*

A top can be represented simply as a rigid body with an axis of symmetry which is free to move about a point on that axis, subject only to the force of a constant gravitational field. In this idealisation all other forces including resistance are ignored.

Clearly, the principal axes of the top at the fixed point, which will be labelled O, are the axis of symmetry and any other axis through O which is perpendicular to the axis of symmetry. The axis of symmetry may be defined at any time by the unit vector $\hat{\boldsymbol{c}}$ and the moment of inertia about it by I_3. It is now convenient to choose two other axes $\hat{\boldsymbol{a}}$ and $\hat{\boldsymbol{b}}$ orthogonal to $\hat{\boldsymbol{c}}$ such that $\hat{\boldsymbol{a}}$ is always horizontal. These are illustrated in figure 8.22. The moments of inertia about these axes are always the same constant denoted by I_1, even though these axes are not fixed in the body.

Fig. 8.22. A spinning top. Principal axes $\hat{\boldsymbol{a}}$, $\hat{\boldsymbol{b}}$ and $\hat{\boldsymbol{c}}$ are chosen for convenience, and the orientation of the top at any time is defined in terms of Euler's angles θ, ϕ and ψ.

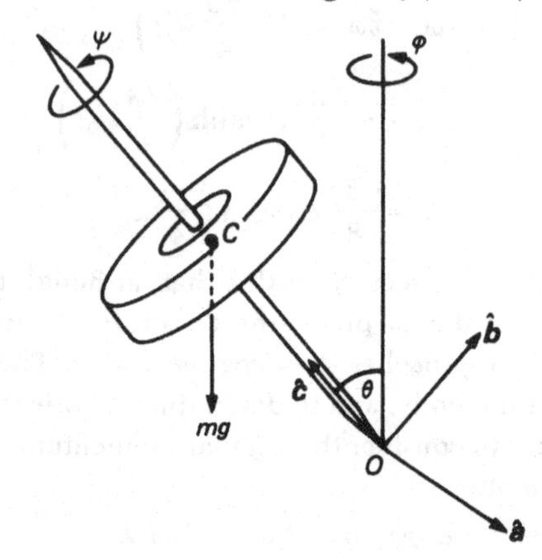

The orientation of the top at any time can be defined in terms of three angles θ, ϕ and ψ which are generally known as *Euler's angular coordinates.* The angle θ is the inclination of the axis of symmetry to the vertical. Here, axes have been chosen such that an increase in θ corresponds to a positive rotation about the $\hat{\boldsymbol{a}}$ axis. The angle ϕ is the angular displacement of the axis of symmetry about the vertical axis. Thus θ and ϕ can be seen to be identical to the familiar components of spherical polar coordinates. They simply define the direction of the axis of symmetry. Finally, ψ is the angular displacement of the body about the axis of symmetry. These angles are also illustrated in figure 8.22.

The equation for the motion of the top is that stated in proposition 8.7. The motion consists of a change in the direction of the axis of the top while it is also spinning about that axis. The angular velocity $\boldsymbol{\Omega}$ of the coordinate axes consists of two components $\dot{\theta}\hat{\boldsymbol{a}}$ and $\dot{\phi}\boldsymbol{k}$. In terms of the axes $\hat{\boldsymbol{a}}$, $\hat{\boldsymbol{b}}$ and $\hat{\boldsymbol{c}}$ this is expressed as

$$\boldsymbol{\Omega} = \dot{\theta}\hat{\boldsymbol{a}} + \dot{\phi} \sin\theta\hat{\boldsymbol{b}} + \dot{\phi}\cos\theta\hat{\boldsymbol{c}}$$

The angular velocity of the top itself however is composed of three components, $\dot{\theta}\hat{\boldsymbol{a}}$ and $\dot{\phi}\boldsymbol{k}$ which describe the changing direction of the axis, and also $\dot{\psi}\hat{\boldsymbol{c}}$ which describes the additional rotation of the top about the axis.

$$\boldsymbol{\omega} = \dot{\theta}\hat{\boldsymbol{a}} + \dot{\phi}\sin\theta\hat{\boldsymbol{b}} + (\dot{\psi} + \dot{\phi}\cos\theta)\hat{\boldsymbol{c}}$$

The angular momentum of the top is therefore

$$\boldsymbol{L} = I_1\dot{\theta}\hat{\boldsymbol{a}} + I_1\dot{\phi}\sin\theta\hat{\boldsymbol{b}} + I_3(\dot{\psi} + \dot{\phi}\cos\theta)\hat{\boldsymbol{c}}$$

Finally, the only external force which has a nonzero moment about O is the gravitational force which effectively acts at the centroid C. Denoting the distance between C and O by h, and the mass of the top by m, the moment of this force about O is

$$\boldsymbol{M} = mgh\sin\theta\hat{\boldsymbol{a}}$$

The above components may now be substituted into the equation of motion (proposition 8.7)

$$\dot{\boldsymbol{L}} + \boldsymbol{\Omega}\times\boldsymbol{L} = \boldsymbol{M}$$

The $\hat{\boldsymbol{c}}$ component of this equation is

$$I_3\frac{\mathrm{d}}{\mathrm{d}t}(\dot{\psi} + \dot{\phi}\cos\theta) = 0$$

which immediately implies that

$$\dot{\psi} + \dot{\phi}\cos\theta = s$$

where s is a constant known as the *spin* of the top. Clearly, the spin is the angular velocity of the top about its axis. This also infers that the angular momentum of the top about its axis is a constant of the motion: a result which is not immediately obvious. This result may however be confirmed by considering

$$\frac{\mathrm{d}}{\mathrm{d}t}(\boldsymbol{L}\cdot\hat{\boldsymbol{c}})=\hat{\boldsymbol{c}}\cdot\frac{\mathrm{d}}{\mathrm{d}t}\boldsymbol{L}+\boldsymbol{L}\cdot\frac{\mathrm{d}}{\mathrm{d}t}\hat{\boldsymbol{c}}$$
$$=\hat{\boldsymbol{c}}\cdot\boldsymbol{M}+\boldsymbol{L}\cdot(\boldsymbol{\Omega}\times\hat{\boldsymbol{c}})$$
$$=0$$

The complication arises here because $\hat{\boldsymbol{c}}$ is not a constant vector.

Using this first result, the $\hat{\boldsymbol{b}}$ component of the equation of motion can be integrated to give another first integral

$$I_1\dot{\phi}\sin^2\theta+I_3 s\cos\theta=L_z$$

where L_z is a constant. This constant is the angular momentum of the top about the vertical axis $L_z=\boldsymbol{L}\cdot\boldsymbol{k}$. This result could therefore have been obtained without considering the equation of motion, simply by observing that $\boldsymbol{M}\cdot\boldsymbol{k}=0$, where $\boldsymbol{k}$ is a constant vector.

The remaining $\hat{\boldsymbol{a}}$ component of the equation of motion may now be written in the form

$$I_1\ddot{\theta}=(I_1\dot{\phi}^2\cos\theta-I_3 s\dot{\phi}+mgh)\sin\theta$$

Using the above results, this component may also be integrated. The resulting expression, however, is in fact the energy integral

$$\tfrac{1}{2}I_1\dot{\theta}^2+\tfrac{1}{2}I_1\dot{\phi}^2\sin^2\theta+\tfrac{1}{2}I_3 s^2+mgh\cos\theta=E$$

where E is a constant. This also could have been deduced initially just from the observation that the system is conservative.

The first of these integrals has effectively been used to eliminate reference to $\dot{\psi}$ in the remaining equations. The second result may similarly be used to eliminate $\dot{\phi}$, leaving a single first-order equation involving θ only,

$$I_1\dot{\theta}^2=2E-I_3 s^2-\frac{(L_z-I_3 s\cos\theta)^2}{I_1\sin^2\theta}-2mgh\cos\theta$$

No analytic solution to this equation can be obtained in general, but it can be simplified by multiplying by $\sin^2\theta/I_1$ and changing the variable to $x=\cos\theta$. It is also convenient to relabel the constant parameters by putting

$$a=\frac{2mgh}{I_1}\qquad b=\frac{I_3}{I_1}s\qquad d=\frac{L_z}{I_1}\qquad e=\frac{2E-I_3 s^2}{I_1}$$

With these simplifications the above equation becomes

$$\dot{x}^2=ax^3-(e+b^2)x^2-(a-2bd)x+(e-d^2)$$

It may now be observed that the right-hand side of this equation is a cubic in x having, at most, two roots in the range $-1\leqslant x<1$, which is the required range for a real, nonzero, value of θ. Also, since the left-hand side is a perfect square, the right-hand side must be nongegative, and so the solution is restricted to the range of x between its two appropriate roots. It may thus be concluded that, in the general motion of a top θ, the inclination of the axis to the vertical is restricted to a certain range whose limits are determined by the above cubic. At these limits, which may be denoted by $\theta=\alpha$ and $\theta=\beta$ with $\beta>\alpha$, the above equation also implies that $\dot{\theta}=0$.

The motion of the axis of the top may now be classified into three distinct cases. It is also convenient to illustrate each case by considering the motion of a point on the axis on a spherical surface centred at O.

Case 1: $L_z > I_3 s \cos\alpha$
In this case $\dot{\phi}>0$ for all $\alpha \leqslant \theta \leqslant \beta$, and a point on the axis traces out a sinusoidal path on the surface (see figure 8.23).

Case 2: $L_z = I_3 s \cos\alpha$
In this case $\dot{\phi}>0$ when $\alpha<\theta\leqslant\beta$, but $\dot{\phi}=0$ when $\theta=\alpha$. In this case a point on the axis traces out a path with cusps (see figure 8.24). At each cusp the axis of the top is instantaneously stationary. This type of motion occurs whenever a top is given a spin about its axis and the axis is initially held and released from rest.

Case 3: $I_3 s \cos\beta < L_z < I_3 s \cos\alpha$
In this case $\dot{\phi}$ is negative when $\theta=\alpha$ and positive when $\theta=\beta$. In motion of this type a point on the axis traces out a trocoidal path on the spherical surface (see figure 8.25).

These types of motion correspond exactly with the observed motion of real tops. They clearly include the steady precessional motion in which θ is a constant, as this is included in case 1 with the two roots being equal ($\alpha=\beta$).

Fig. 8.23. Path of the axis of a top when $L_z > I_3 s \cos\alpha$.

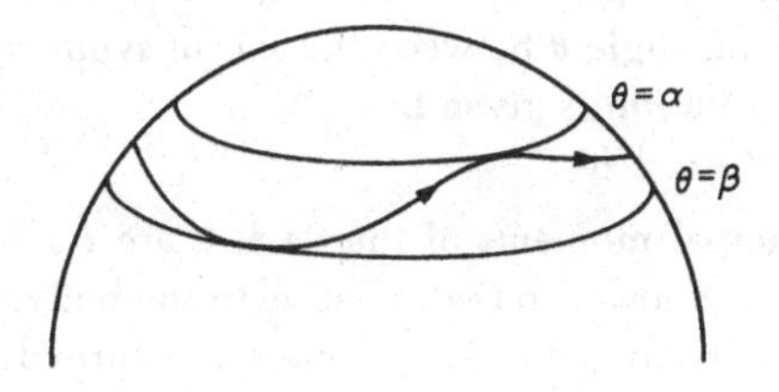

Fig. 8.24. Path of the axis of a top when $L_z = I_3 s \cos\alpha$.

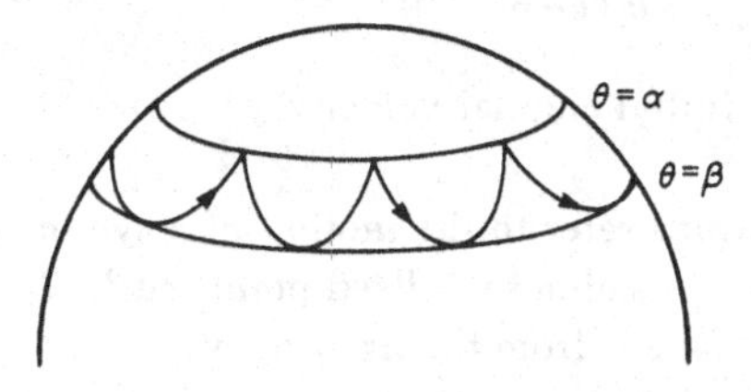

Fig. 8.25. Path of the axis of a top when $I_3 s \cos\beta < L_z < I_3 s \cos\alpha$.

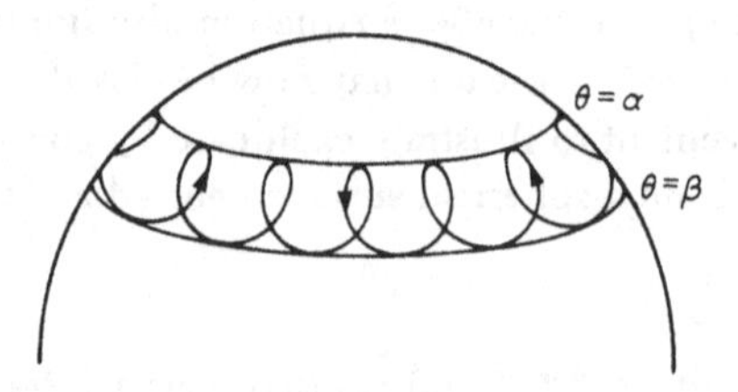

Exercises

8.22 A uniform thin rectangular plate whose length is twice its width can freely rotate about its centroid. Initially, the angular velocity vector has magnitude ω_0 and is equally inclined to the three principal axes at the centroid. Show that in the subsequent motion the magnitude of the angular velocity vector varies between $2\omega_0/\sqrt{5}$ and $\sqrt{(\frac{6}{5})}\omega_0$, and that the angle between the angular velocity vector and the plane of the plate varies between $\cot^{-1}\sqrt{5}$ and $\cot^{-1}\sqrt{5}/2$.

8.23 A uniform rectangular plate whose sides are of lengths $2a$ and $\sqrt{2}a$ freely rotates about its centroid. It is set rotating with angular velocity ω_0 about an axis which is perpendicular to the shorter sides, and which is inclined at 30° to the plane of the plate. Show that at a subsequent time t the components of the angular velocity vector of the plate about principal axes at the centroid are

$$\frac{\sqrt{3}}{2}\omega_0 \operatorname{sech}\frac{\omega_0 t}{2} \quad \frac{\sqrt{3}}{2}\omega_0 \tanh\frac{\omega_0 t}{2} \quad \frac{1}{2}\omega_0 \operatorname{sech}\frac{\omega_0 t}{2}.$$

8.24 Consider the rotational motion of a rigid body with an axis of symmetry about a point O on its axis, in the case when there is a retarding couple $\boldsymbol{M} = -\lambda\boldsymbol{\omega}$ which is proportional to the angular velocity of the body. Show in general that the angle θ between the axis of symmetry and the instantaneous axis of rotation is given by

$$\tan\theta = \tan\theta_0 \, e^{\lambda t(1/I_3 - 1/I_1)}$$

where the principal moments of inertia at 0 are I_1, I_1, I_3 and θ_0 is the initial value of θ. Show also that, relative to the body, the instantaneous axis of rotation at any subsequent time t has turned about the axis of symmetry through an angle

$$\phi = \frac{I_3(I_3 - I_1)}{\lambda I_1}\omega_0 \cos\theta_0 (1 - e^{-\lambda t/I_3})$$

where ω_0 is the initial angular velocity.

The following questions refer to the motion of a symmetric top of mass m whose principal moments about the fixed point are I_1, I_1, I_3, and for which the centroid is a distance h from the fixed point.

8.25 A top is given a spin s about its axis where $s^2 = 6mghI_1I_3^{-2}$, and its axis is initially held at an angle $\cos^{-1}(3/4)$ to the vertical and then released. Show that in the subsequent motion $\frac{1}{2} \leqslant \cos\theta \leqslant \frac{3}{4}$.

8.26 A top is projected with its axis initially inclined to the vertical at a stationary angle $\pi/3$, with the axis rotating about the vertical with angular speed given by $5\sqrt{(mgh/I_1)}/3$, and with spin about its axis given by $s = \sqrt{(mghI_1)}/2I_3$. Show that in the subsequent motion the inclination of the axis to the vertical varies between $\pi/3$ and $2\pi/3$.

8.27 A top is initially projected with its axis horizontal and rotating about the vertical with angular speed $\Omega = \sqrt{(2mgh/I_1)}$. If the spin about the axis is $s = I_1\Omega/I_3$, show that after a time t the inclination of the axis to the vertical θ is given by

$$\tan\frac{\theta}{2} = \operatorname{sech}(\Omega t)$$

8.28 Show that in the steady precessional motion of a top, in which the axis remains at a constant angle θ to the vertical and rotates about the vertical with constant angular speed Ω,

$$I_1\Omega^2\cos\theta - I_3s\Omega + mgh = 0$$

Conclude that steady precessional motion is only possible if

$$I_3^2s^2 \geqslant 4mghI_1\cos\theta$$

and that two precession rates are then possible. Conclude also that, for hanging motion in which $\theta \geqslant \pi/2$, steady precessional motion is always possible.

8.29 A top is given a spin s about its axis and placed with its axis vertical. Show that this position is stable, provided

$$I_3^2s^2 > 4mghI_1$$

8.13. Impulsive motion

The character of impulsive motion has already been discussed in section 6.7 where it was considered with reference to the motion of a particle. It was shown there how the effect of a large force which acts for a very short time period could be described by the concept of its impulse, which may be considered to act at an instant. The main results obtained were given in propositions 6.5 and 6.6 which state that the changes in the linear momentum of a particle and its angular momentum about a point caused by an impulsive force are equal to the impulse applied and its moment about that point respectively. It is necessary here only to extend this discussion with reference to the motion of a rigid body.

Consider a rigid body simultaneously acted on by n impulsive forces, each of which may be represented by a force $\boldsymbol{F}_i$ acting at a point of the body with position vector $\boldsymbol{r}_i$, or $\boldsymbol{\rho}_i$ relative to the centroid. Following the method of section 6.7, we now proceed to integrate the equations of motion. For a rigid body these equations may be considered in the form

$$m\ddot{\bar{\boldsymbol{r}}} = \sum_{i=1}^{n} \boldsymbol{F}_i$$

$$\frac{\mathrm{d}}{\mathrm{d}t}\boldsymbol{L}_\mathrm{c} = \sum_{i=1}^{n} \boldsymbol{\rho}_i \times \boldsymbol{F}_i$$

There may of couse also be other forces acting, but they would be negligible compared to the impulsive forces during the instant at which the impulses are applied, and have therefore been omitted. The time interval during which the impulses act may now be taken to be an infinitesimal interval from t_1 to t_2. Integrating the above equations over this interval gives the results that,

$$m\dot{\bar{\boldsymbol{r}}}(t_2) - m\dot{\bar{\boldsymbol{r}}}(t_1) = \sum_{i=1}^{n} \int_{t_1}^{t_2} \boldsymbol{F}_i \,\mathrm{d}t$$

$$\boldsymbol{L}_\mathrm{c}(t_2) - \boldsymbol{L}_\mathrm{c}(t_1) = \sum_{i=1}^{n} \boldsymbol{\rho}_i \times \int_{t_1}^{t_2} \boldsymbol{F}_i \,\mathrm{d}t$$

In obtaining the second of these equations it should be noticed that the assumption has been made that the position of the body does not alter significantly during the instant of the impulse. The position vectors $\boldsymbol{\rho}_i$ have therefore been regarded as approximately constant over this time interval. The results contained in these two equations may now be stated formally.

***Proposition* 8.8.** *The change in the linear momentum of a rigid body induced by a set of impulses is equal to the sum of those impulses.*

***Proposition* 8.9.** *The change in the angular momentum of a rigid body about its centroid induced by a set of impulses is equal to the sum of the moments of those impulses about the centroid.*

In contrast to propositions 6.5 and 6.6, which apply to a particle, these two results are independent. They therefore provide two vector equations to describe the change in the motion of a rigid body which has up to six degrees of freedom. They are therefore sufficient to describe completely the effects of a set of impulses on the motion of a rigid body.

Although the above approach is sufficient, it is not always necessary to consider the angular momentum of the body relative to the centroid. Relative to any fixed point in space, the equation for the rate of change of angular momentum about that point is given by proposition 7.6 in the form

$$\frac{\mathrm{d}}{\mathrm{d}t}\boldsymbol{L}=\sum_{i=1}^{n}\boldsymbol{r}_i\times\boldsymbol{F}_i$$

Integrating this over the instant of the impulse gives

$$\boldsymbol{L}(t_2)-\boldsymbol{L}(t_1)=\sum_{i=1}^{n}\boldsymbol{r}_i\times\int_{t_1}^{t_2}\boldsymbol{F}_i\,\mathrm{d}t$$

This result may be stated as follows.

***Proposition* 8.10.** *The change induced by a set of impulses in the angular momentum of a rigid body about any point in space is equal to the sum of the moments of the impulses about that point.*

It should be noticed that this result applies to any point, which need not necessarily be a fixed point of the body. Of course, the basic equation of motion, which is based on proposition 7.6, also applies to an arbitrary point. However, if this is not a fixed point of the body, then it is usually not convenient to consider finding an expression for the rate of change of the angular momentum of the body about that point. For impulsive motion, on the other hand, the position of the body is considered to be constant during the instant of the impulse, and the instantaneous change in the angular momentum of the body relative to any point can easily be obtained.

Although proposition 8.10 applies generally, it is particularly appropriate when a body is constrained to rotate about a fixed point or axis. In such cases, however, it should also be remembered that any external impulses applied to the body will also induce an impulsive reaction which acts on the body at the fixed point, or at points along the axis.

8.14. Impulses: examples

Example 8.19

A uniform rod of mass m_1 and length $2a$ rests on a horizontal table. A particle of mass m_2 slides along the table with speed V and strikes the rod at one end in a direction perpendicular to its length. If the coefficient of restitution between the particle and the rod is e, find the velocity of the particle and the velocity and angular velocity of the rod immediately after the collision.

When the particle strikes the rod there is clearly an impulsive interaction between them. The impulse acting on the rod sets it in motion. This motion will be composed initially of a velocity in the same direction as the initial motion of the particle, and a rotation in the horizontal plane. In its subsequent motion frictional forces have to be considered, but the problem here is only to find the initial motion generated by the collision. The motion of the particle, on the other hand, is either slowed down by the collision or its direction of motion may be reversed.

The parameters defining the initial configuration have already been specified. It is now convenient after the collision to denote the speed of the centroid of the rod by V_1 and its angular speed by ω, and to denote the speed of the particle by V_2, assuming that its direction of motion is unchanged. If in fact the direction of the particles motion is reversed by the collision, then V_2 would be negative. In problems of this type it is often convenient to draw two diagrams illustrating the situations immediately before and after the collision, as in figure 8.26. In the diagrams here the magnitude of the impulsive interaction between the particle and the rod is denoted by P.

For the rod, the linear momentum gained is equal to the impulse acting on it, and the angular momentum gained about the centroid is equal to the moment of the impulse about the centroid. These statements are expressed by the equations

$$m_1 V_1 = P$$

$$\tfrac{1}{3} m_1 a^2 \omega = aP$$

The change in linear momentum of the particle is also equal to the impulse acting on it, and therefore

$$m_2 V_2 - m_2 V = -P$$

These equations are sufficient to express V_1, V_2 and ω in terms of V and P. However, as in example 6.5, the need to know the impulse P can be

Fig. 8.26. Collision of a particle and a rod, (a) before the impulse, (b) after the impulse.

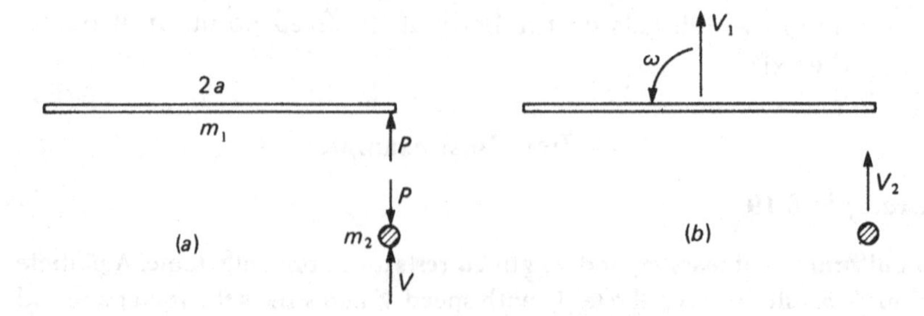

avoided if the coefficient of restitution between the particle and the rod is known. This is the ratio of the speed of separation to the speed of approach at the point of contact and is assumed to be a constant characteristic of the bodies concerned. Here it leads to the equation

$$(V_1 + a\omega) - V_2 = eV$$

These four equations are sufficient to determine the required parameters

$$V_1 = \frac{(1+e)m_2}{(m_1+4m_2)}V \qquad \omega = \frac{3(1+e)m_2 V}{(m_1+4m_2)a} \qquad V_2 = \frac{(4m_2 - em_1)}{(m_1+4m_2)}V$$

It may be observed that V_2 is positive or negative according as e is less than or greater than $4m_2/m_1$. Also, if $e=0$, then $V_2 = V_1 + a\omega$, as required.

It is also possible to show that the kinetic energy lost in the impact is

$$\tfrac{1}{2}m_2V^2 - \tfrac{1}{2}m_2V_2^2 - \tfrac{1}{2}m_1V_1^2 - \tfrac{1}{6}m_1a^2\omega^2 = \frac{(1-e^2)m_1m_2V^2}{2(m_1+4m_2)}$$

from which it may be confirmed that if $e=1$ the total kinetic energy is unaltered.

Example 8.20

Two uniform rods AB and BC, each of mass m and length $2a$, are smoothly hinged together at B and rest on a smooth horizontal table in a straight line. Determine the initial motion generated by an impulse P which acts at A in a direction perpendicular to the rods.

In this situation two distinct rigid bodies are joined together and the connection between them is here assumed to be frictionless, for simplicity. In the traditional approach described in the present chapter, each rod has to be treated separately. Since the impulse is perpendicular to the rods, it is clear that there is no initial motion in a direction aligned with the rods. It is then convenient to denote the initial speed of the centroid of AB by v_1 and its initial angular speed by ω_1, where the appropriate directions are indicated in figure 8.27. Similarly, the intial speed of the centroid of BC and its angular speed may be denoted by v_2 and ω_2.

Fig. 8.27. The motion of two hinged rods generated by an impulse P.

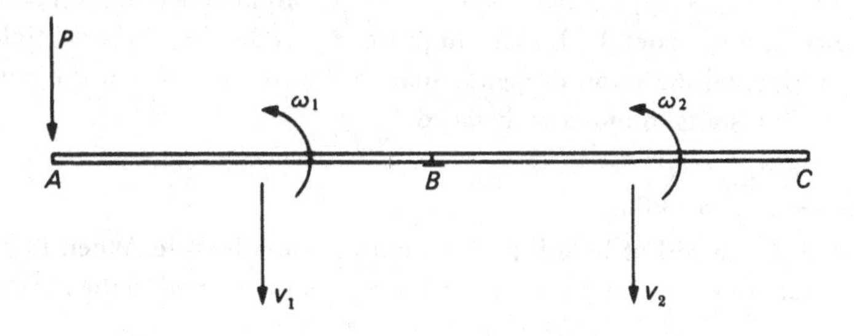

Fig. 8.28. The impulsive interaction between two hinged rods.

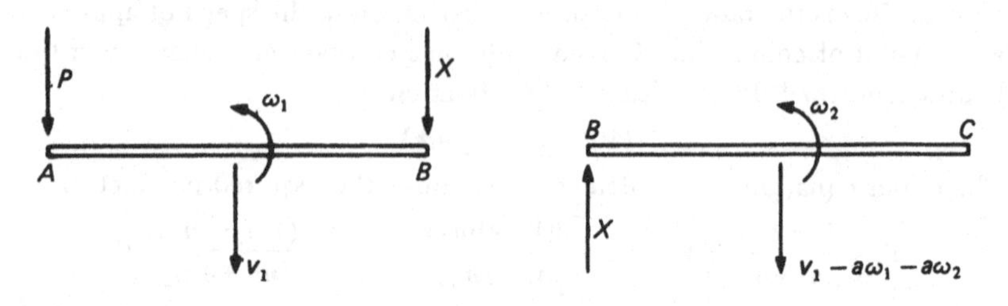

It is, however, necessary to impose the constraint that the velocity of the end B of each rod is the same. This can be expressed by the equation

$$v_2 = v_1 - a\omega_1 - a\omega_2$$

The two rods must now be treated separately. There must, however, be an interaction between them at B. Here this is an impulsive interaction and, since there is no initial motion along the direction of the rods, this must be perpendicular to them. The magnitude of this impulsive interaction may be denoted by X, as indicated in figure 8.28.

The equations for the impulsive motion of each rod are now the following.

$$mv_1 = P + X$$
$$\tfrac{1}{3}ma^2\omega_1 = aP - aX$$
$$m(v_1 - a\omega_1 - a\omega_2) = -X$$
$$\tfrac{1}{3}ma^2\omega_2 = -aX$$

From these it follows that

$$v_1 = \frac{5}{4}\frac{P}{m} \qquad \omega_1 = \frac{9}{4}\frac{P}{ma}$$
$$v_2 = -\frac{1}{4}\frac{P}{m} \qquad \omega_2 = -\frac{3}{4}\frac{P}{ma}$$

which determine the initial motion of the rods after the impulse P.

This example will again be considered using a different method in example 9.6.

Exercises

8.30 A rod of mass m_1 has two particles of mass m_2 attached to each end and rests on a horizontal table. An impulse P acts on one of the particles in a horizontal direction perpendicular to the rod. Show that the particle acted on starts to move with speed

$$\frac{4(m_1 + 3m_2)P}{(m_1 + 2m_2)(m_1 + 6m_2)}$$

8.31 A uniform rod of length $2a$ falls onto a smooth table. When it strikes the table its speed is V; it is not rotating but is inclined to the horizontal at an angle α. if the coefficient of restitution between the rod and the table is e, show that the rod begins to turn with angular speed ω where

$$\omega = \frac{3(1+e)\cos\alpha}{(1+3\cos^2\alpha)} \frac{V}{a}$$

Show also that the instantaneous loss in kinetic energy is

$$\frac{(1-e^2)}{(1+3\cos^2\alpha)} \frac{1}{2} mV^2$$

8.32 Two identical uniform rods are smoothly hinged together at a pair of ends and rest on a horizontal table in a straight line. An impulse acts at the free end of one of the rods in a direction perpendicular to the rods. Show that the kinetic energy generated is $\frac{7}{4}$ times greater than would have occurred if the rods had been rigidly connected.

8.33 Three identical rods AB, BC and CD are smoothly hinged at B and C and rest on a horizontal table in a straight line. An impulse acts at the joint B in a horizontal direction perpendicular to the rods. Show that the rods start to move with angular velocities which are in the ratio $7:-6:2$.

9

Analytical dynamics

The approach to classical dynamics proposed so far has been a more or less direct application of Newton's laws of motion. In such an approach the motion of a body or particle can be predicted on the assumption of a given set of external forces which act on it, simply by integrating the equations of motion. However, for complex systems of particles or rigid bodies, it is not always easy to determine appropriate equations for each component, let alone perform the required integration. In practice, using this approach, it is found that each individual type of problem requires its own particular insights and techniques.

In this chapter the Lagrangian approach to classical dynamics is developed. This approach is based upon two scalar properties of a system, its kinetic energy and work. It leads to a powerful and general method for the solution of dynamical problems which is found to be particularly useful in the analysis of mechanical systems which contain a number of rigid bodies that are connected in some way, but which may move relative to each other. In the traditional approach each component would have to be treated separately in terms of the forces acting on it. However, the Lagrangian approach enables such a system to be considered as a whole.

The aim here is to develop a general approach which may be applied to any dynamical system. It is found that the equations of motion can be presented in a standard and convenient form. In fact, so general is the approach that it may easily be extended to include the field equations of other physical theories. However, it is developed in this chapter only with reference to the theory of classical dynamics.

9.1. Generalised coordinates

Following the methods of classical dynamics, all mechanical systems are assumed to be composed of particles. Sometimes these may be considered to be combined to form a finite number of rigid bodies, or they may be considered to satisfy some other system of constraints. In order to develop a general approach to the subject, therefore, it is necessary to consider a general system composed of N particles. The position of each particle at any time relative to a given frame of reference, may be repesented by the set of vectors $\boldsymbol{r}_i$ $(i = 1, 2, \ldots, N)$.

Since the position of each particle can be determined by three coordinates, the configuration of the whole system can be determined by $3N$ coordinates. These coordinates, however, may be chosen for convenience. They may be distances, angles, or in fact any function of these. In addition, it may be convenient to consider the position, and hence the coordinates,of some particles relative to some other point of the system whose position has already been determined. In complete generality, a set of coordinates must therefore be regarded as the parameters of a system which uniquely determine its configuration at any time. Such parameters are referred to as *generalised coordinates*, and may be denoted by q_i, $(i = 1, 2, \ldots, n)$, where it has initially been assumed that $n = 3N$. Since these parameters determine the position of the particles of the system at any time, these positions may be expressed in the general form

$$\boldsymbol{r}_i = \boldsymbol{r}_i(q_1, q_2, \ldots, q_n, t) \qquad (i = 1, 2, \ldots, N)$$

It should be emphasised that the generalised coordinates may be chosen with almost complete freedom. They may, for example, be amplitudes of Fourier series expansions or certain functions of physical coordinates. However, they must be finite, single-valued and differentiable with respect to time.

If there were no constraints restricting the motion of the particles, then the system would possess $3N$ degrees of freedom. However, it must be assumed in general that the system is subject to a number of constraints which accordingly reduce the number of degrees of freedom. These need not be specified at this stage, but it can be assumed that they may include a number of holonomic constraints of the form

$$f(q_i, q_2, \ldots, q_n, t) = 0$$

Such constraints may be used to reduce the number of generalised coordinates. When considering the motion of a rigid body, for example, N, the number of particles composing it, may be assumed to be arbitrarily large. However, the constraints of rigidity may be used to reduce the number of generalised coordinates that describe the configuration to six, which is the maximum number of degrees of freedom. In general it may therefore be assumed that some of the constraints of the system have been used to reduce the number of generalised coordinates, and that n is therefore less than $3N$. However, it must also be assumed that a number of constraints still remain to be considered.

The concept of generalised coordinates may now be defined formally.

***Definition* 9.1.** *The generalised coordinates* (q_i) *which may be used to describe a dynamical system are any set of n parameters which completely determine the configuration of the system at any time.*

Of course, to determine the system completely, n cannot be less than the number of degrees of freedom of the system. On the other hand, if a number of redundant parameters are used, n may sometimes be greater than $3N$. However, it may usually be assumed that n is less than $3N$, particularly when N is large.

It should be noticed that the time dependence which appears explicitly in the above expression for the position of the particles of the system may arise either from the definition of the generalised coordinates, or from the introduction of rheonomic constraints. Of course, if a rheonomic or moving constraint has been used to reduce the number of generalised coordinates, then an explicit time dependence has necessarily been introduced. On the other hand, if only scleronomic constraints have been used, then it is always possible to choose a set of generalised coordinates such that the position vectors of the particles of the system depend only on those coordinates and have no explicit time dependence. It is not always convenient, however, to make such a restriction. In general a time dependence must also be permitted in the introduction of the generalised coordinates. This may be interpreted in terms of a time dependence in the set of coordinate axes, and a set of generalised coordinates introduced in this way is thus often referred to as a 'moving coordinate system'.

When the position of each of the particles of the system is determined by the n generalised coordinates, their velocities can be written as

$$\dot{\mathbf{r}}_i = \sum_{j=1}^{n} \frac{\partial \mathbf{r}_i}{\partial q_j} \dot{q}_j + \frac{\partial \mathbf{r}_i}{\partial t} \qquad (i = 1, 2, \ldots, N)$$

The terms $\dot{q}_i$ $(i = 1, 2, \ldots, n)$ are referred to as the *generalised velocities.* Apart from the fact that these are scalar components, they do not even always have the dimensions of velocity, since the generalised coordinates themselves do not necessarily have the dimensions of length. For example, if q_i is an angular coordinate, then $\dot{q}_i$ is an angular velocity. However, it can be seen that the term 'generalised velocity' is an appropriate one.

Before proceeding it is necessary to introduce the following notation.

Notation. *For purposes of partial differentiation the generalised coordinates $q_1, q_2, \ldots, q_n$ the generalised velocities, $\dot{q}_1, \dot{q}_2, \ldots, \dot{q}_n$ and time t are regarded as independent variables.*

This has been introduced purely for mathematical convenience and has no physical significance.

Using this notation, the following two lemmas which will be required in later calculations, can be obtained directly from the above expression for the velocity of the particles.

***Lemma* 9.1**

$$\frac{\partial \dot{\mathbf{r}}_i}{\partial \dot{q}_j} = \frac{\partial \mathbf{r}_i}{\partial q_j} \qquad (i = 1, 2, \ldots, N;\ j = 1, 2, \ldots, n)$$

***Lemma* 9.2**

$$\frac{\partial}{\partial q_j}[\dot{\mathbf{r}}_i] = \frac{\mathrm{d}}{\mathrm{d}t}\left[\frac{\partial \mathbf{r}_i}{\partial q_j}\right] \qquad (i = 1, 2, \ldots, N;\ j = 1, 2, \ldots, n)$$

The general approach is now essentially to rewrite the equations for the motion of the system in terms of the generalised coordinates. However, this is not done directly at this point. Instead, it is convenient to proceed indirectly by first considering the expression for kinetic energy.

9.2. Kinetic energy and the generalised momentum components

The kinetic energy of a system of N particles is given by

$$T = \tfrac{1}{2} \sum_{i=1}^{N} m_i \dot{\mathbf{r}}_i^2$$

Using the above expression this can now be written as

$$T = \tfrac{1}{2} \sum_{i=1}^{N} m_i \left[\sum_{j=1}^{n} \frac{\partial \mathbf{r}_i}{\partial q_j} \dot{q}_j + \frac{\partial \mathbf{r}_i}{\partial t}\right]^2$$

and this can be expanded in terms of the generalised coordinates in the form

$$T = \tfrac{1}{2}\sum_{j=1}^{n}\sum_{k=1}^{n} a_{jk}\dot{q}_j\dot{q}_k + \sum_{j=1}^{n} a_j\dot{q}_j + \tfrac{1}{2}a_0$$

where

$$a_{jk} = \sum_{i=1}^{N} m_i \frac{\partial \mathbf{r}_i}{\partial q_j}\cdot\frac{\partial \mathbf{r}_i}{\partial q_k} \qquad (j, k = 1, 2, \ldots, n)$$

$$a_j = \sum_{i=1}^{N} m_i \frac{\partial \mathbf{r}_i}{\partial q_j}\cdot\frac{\partial \mathbf{r}_i}{\partial t} \qquad (j = 1, 2, \ldots, n)$$

$$a_0 = \sum_{i=1}^{N} m_i \frac{\partial \mathbf{r}_i}{\partial t}\cdot\frac{\partial \mathbf{r}_i}{\partial t}$$

It can be seen from these definitions that the coefficients a_{jk}, a_j and a_0 are in general all functions of the generalised coordinates and time. It should be noticed that the coefficients a_{jk} are symmetric in the indices, and also that the definition of kinetic energy implies that the above expression for it is positive.

This result may now be summarised as follows.

***Proposition* 9.1.** *The kinetic energy of a dynamical system can be written as a quadratic function of the generalised velocities with coefficients which are in general functions of the generalised coordinates and time.*

It should, however, be understood that in practical applications it is usually appropriate first to choose a set of generalised coordinates, and then to evaluate the kinetic energy of the system in terms of them. Then, when an expression for the kinetic energy has been obtained, the coefficients a_{jk}, a_j and a_0 are already determined. Thus it is not necessary to evaluate these coefficients from first principles using the above definitions and summing over all N particles.

At this point it is appropriate to define a suitably generalised concept of momentum.

***Definition* 9.2.** *The generalised momentum component p_i associated with the generalised coordinate q_i is defined by the equation*

$$p_i = \frac{\partial T}{\partial \dot{q}_i} \qquad (i = 1, 2, \ldots, n)$$

The appropriateness of this definition can easily be demonstrated by considering a number of examples. In simple cases it can be shown that if q_i is a linear coordinate, then p_i is a component of linear momentum in the direction of that coordinate. Similarly, it can be shown that if q_i is an angular coordinate, then p_i is the associated angular momentum component. In fact, the above definition includes the concepts of both linear and angular momentum that have been considered in previous chapters.

Since the kinetic energy is a quadratic function of the generalised velocities, it follows from the definition that the generalised momentum components can be written as a linear combination of the generalised velocities.

It should perhaps be pointed out that, as in the previous section, a number of constraints may have been used to reduce the number of generalised coordinates to n. However, it has generally been assumed that there may remain some further constraints which have yet to be considered. It therefore follows that the generalised velocities, and hence the generalised momentum components, are not necessarily independent.

In the particular case of a holonomic system in which all the constraints have been used, the system has n degrees of freedom and the generalised velocities are independent. It can also be seen that the above expression for the kinetic energy is now positive definite, and thus the expressions for the generalised momentum components as linear combinations of the generalised velocities are nonsingular. It follows therefore, in this case, that the generalised momentum components are independent, and it is also always possible to obtain the inverse relations which express the generalised velocities as linear combinations of the generalised momentum components.

It should also be pointed out that all of the above results take a considerably simplified form if the positions of the particles are functions of the generalised coordinates alone and do not depend explicitly on time. This occurs when no moving coordinates have been introduced, and when no rheonomic or moving constraints have been used to reduce the number of generalised coordinates, although there may possibly remain a number of rheonomic constraints which have yet to be considered. In this case

$$\mathbf{r}_i = \mathbf{r}_i(q_1, q_2, \ldots, q_n) \qquad (i = 1, 2, \ldots, N)$$

and the kinetic energy can be expanded as the homogeneous quadratic function

$$T = \tfrac{1}{2} \sum_{i=1}^{n} \sum_{j=1}^{n} a_{ij} \dot{q}_i \dot{q}_j$$

The generalised momentum components are obtained by differentiating this with respect to $\dot{q}_i$ giving

$$p_i = \sum_{j=1}^{n} a_{ij}\dot{q}_j \qquad (i = 1, 2, \ldots, n)$$

which is a linear combination of the generalised velocities, as expected.

In this particular case it may also be observed that the kinetic energy can be expressed in the form

$$T = \tfrac{1}{2} \sum_{i=1}^{n} p_i \dot{q}_i$$

This form is consistent with the result obtained in section 8.11 that the kinetic energy of a rigid body is divisible into component parts, of which the translational kinetic energy is equal to one-half of the scalar product of its velocity with its linear momentum, and its rotational kinetic energy is equal to one-half of the scalar product of its angular velocity vector with its angular momentum.

9.3. Virtual work and the generalised force components

Following the Lagrangian approach, the motion of a system of particles can be analysed in terms of generalised coordinates, generalised velocities and generalised momentum components. But before proceeding to do so, it is convenient to introduce an appropriate set of 'generalised' quantities which represent the forces which act on the system.

Generalised force components may be introduced by first considering the total work done by all the forces when the system is artificially given an arbitrary small displacement at some instant of time. Such displacements are referred to as *virtual displacements.* They are not actual displacements, since a real physical displacement would require a finite time interval, and would also be required to be the effect of the forces which are acting. They must be regarded merely as a theoretical device which can only be considered to apply to a theoretical model of the system.

***Definition* 9.3.** *A virtual displacement is an arbitrary displacement of a theoretical model of a system, made artificially without any consideration of what may cause it, and it may be made without variation of the time parameter.*

A virtual displacement of a system of particles can be represented by the set of vectors $\delta \boldsymbol{r}_i$ $(i = 1, 2, \ldots, N)$. However, the position of each particle at any time is uniquely determined by the n generalised coordinates which at this stage are not necessarily independent. In addition, the virtual displacements must be consistent with the constraints of the system. It follows that the displacement of each particle can be expressed in terms of small virtual variations of the n generalised coordinates δq_i. Bearing in mind that the time parameter is held constant, the displacement of each particle can be written in the form

$$\delta \boldsymbol{r}_i = \sum_{j=1}^{n} \frac{\partial \boldsymbol{r}_i}{\partial q_j} \delta q_j \qquad (i = 1, 2, \ldots, N)$$

The work done by the forces which act on the particles in an arbitrary virtual displacement may now be considered. This is referred to as the virtual work to distinguish it from the case when actual displacements are being made.

***Definition* 9.4.** *Virtual work is the work done by the forces acting on the particles of a system when they are subjected to a virtual displacement.*

This may be given by the expression

$$\delta W = \sum_{i=1}^{N} \boldsymbol{F}_i \cdot \delta \boldsymbol{r}_i$$

where $\boldsymbol{F}_i$ is the resultant of all the forces which act on each particle of the system. In terms of the virtual variations of the generalised coordinates, this may be written as

$$\delta W = \sum_{i=1}^{n} Q_i \, \delta q_i$$

where the coefficients Q_i are given by

$$Q_i = \sum_{j=1}^{N} \boldsymbol{F}_j \cdot \frac{\partial \boldsymbol{r}_j}{\partial q_i} \qquad (i = 1, 2, \ldots, n)$$

These coefficients may be regarded as the generalised force components associated with each of the generalised coordinates, since $\sum_{i=1}^{n} Q_i \, \delta q_i$ is the work done in the virtual displacement. Clearly, if any q_i has the dimensions of length, then the associated Q_i has the dimensions of force. Similarly, if q_i is an angle, then Q_i is the moment of a force. It can therefore be seen that the n components are appropriate generalisations of the force components. These may now be defined formally.

Definition 9.5. *If $\boldsymbol{F}_i$ $(i = 1, 2, \ldots, N)$ are the resultant forces acting on the N particles of a system situated at the points $\boldsymbol{r}_i$ $(i = 1, 2, \ldots, N)$, and if the system is described uniquely at any time by n generalised coordinates q_i $(i = 1, 2, \ldots, n)$, then the generalised force components associated with each of the generalised coordinates are defined by the equations*

$$Q_j = \sum_{i=1}^{N} \boldsymbol{F}_i \cdot \frac{\partial \boldsymbol{r}_i}{\partial q_j} \qquad (j = 1, 2, \ldots, n)$$

It is immediately clear from this definition that the components of the resultant forces $\boldsymbol{F}_i$ which do no work in an arbitrary virtual displacement of the system do not contribute to the generalised force components Q_i. For example, in the motion of a rigid body, forces of constraint are assumed to act on each component particle in such a way that the distance between any pair of particles remains constant. Such forces must be aligned with the displacement vector between the two particles, and therefore in any arbitrary virtual displacement of the body these forces of constraint do no work. In fact, there is a whole class of constraints whose associated forces do not contribute to the virtual work associated with an arbitrary virtual displacement. Constraints of this type may thus be classified as *workless constraints.* It should therefore be emphasised that the generalised force components only represent those forces which do work in a virtual displacement. This point will be discussed again in greater detail in sections 9.7 and, particularly, 9.8.

Although the above equation provides an adequate definition of the generalised force components, it does not usually provide a convenient method for evaluating them in practice. It is usually more convenient to obtain the generalised force components by calculating directly the work done in an arbitrary virtual displacement of the system, and then reading off the appropriate coefficients. However, a distinction has to be made between holonomic and nonholonomic systems, or rather between systems in which all the constraints have been used to reduce the number of coordinates, and systems in which a number of constraints still have to be considered.

For a holonomic system with n degrees of freedom, n independent generalised coordinates may be obtained, and there are then no additional constraints to be considered. In this case an arbitrary virtual displacement of the system can be expressed uniquely in terms of independent variations of the generalised coordinates. It should be remembered that forces which do no work in such a displacement need not be considered. Thus it is only necessary to take the scalar products of the external forces, and possibly any internal forces which do work, with an arbitrary virtual displacement of their points of application. When this is expressed in terms of the n generalised coordinates, the generalised force components are uniquely determined as the corresponding set of coefficients.

The situation is considerably more complicated for nonholonomic systems and other systems in which it is convenient to retain a number of constraints. It may be seen that the difficulty arises because the variations of the generalised coordinates are not independent in these cases. The method that is used in practice is to obtain the generalised force components with the above method by regarding the n generalised coordinates as though they were independent and ignoring the remaining constraints. The constraints are then introduced at a later stage. Such cases will be considered in greater detail in a later section.

Finally, it is appropriate to consider the generalised force components which act in the case of a conservative system. If all the forces which do work in a virtual displacement are conservative, then they may each be derivable from a potential function, and an expression for the potential energy of the whole system may be obtained. In fact, the following results apply not only to conservative systems, but to any system in which the forces which do work are derivable from a potential function. In order to achieve maximum generality, it is therefore appropriate to include a possible time dependence in the potential function

$$V = V(\boldsymbol{r}_1, \boldsymbol{r}_2, \ldots, \boldsymbol{r}_N, t)$$

In this case the forces acting on each particle which do work in a virtual displacement are given by

$$\boldsymbol{F}_i = -\frac{\partial V}{\partial \boldsymbol{r}_i} \qquad (i = 1, 2, \ldots, N)$$

and the generalised force components are given by

$$Q_i = -\sum_{j=1}^{N} \frac{\partial V}{\partial \boldsymbol{r}_j} \cdot \frac{\partial \boldsymbol{r}_j}{\partial q_i} \qquad (i = 1, 2, \ldots, n)$$

Clearly, it is appropriate now to express the potential function in terms of the generalised coordinates

$$V = V(q_1, q_2, \ldots, q_n, t)$$

and with this notation the generalised force components are given by

$$Q_i = -\frac{\partial V}{\partial q_i} \qquad (i = 1, 2, \ldots, n)$$

It should, however, be observed that in this case the explicit time dependence appearing in the potential function may arise from either (a) a time dependence in the potential field, or (b) the use of a moving coordinate system, or (c) the application of rheonomic constraints.

9.4. Lagrange's equations for a holonomic system

It is convenient to introduce Lagrange's equations initially for the restricted case of a holonomic system. In such a system all the constraints are holonomic and, if the system has n degrees of freedom, these constraints may be used to reduce to n the number of generalised coordinates which completely determine the system. It is assumed here that this has been done, so that the configuration of the system at any time is uniquely determined in terms of n independent generalised coordinates.

In such a system the generalised momentum components are also independent and are given by

$$p_j = \frac{\partial T}{\partial \dot{q}_j} = \sum_{i=1}^{N} m_i \dot{\mathbf{r}}_i \cdot \frac{\partial \dot{\mathbf{r}}_i}{\partial \dot{q}_j} \qquad (j = 1, 2, \ldots, n)$$

Using lemma 9.1 this can be rewritten as

$$\frac{\partial T}{\partial \dot{q}_j} = \sum_{i=1}^{N} m_i \dot{\mathbf{r}}_i \cdot \frac{\partial \mathbf{r}_i}{\partial q_j} \qquad (j = 1, 2, \ldots, n)$$

Then, making use of lemma 9.2, the rate of change of the generalised momentum components is therefore given by the equation

$$\frac{\mathrm{d}}{\mathrm{d}t}\left[\frac{\partial T}{\partial \dot{q}_j}\right] = \sum_{i=1}^{N} m_i \ddot{\mathbf{r}}_i \cdot \frac{\partial \mathbf{r}_i}{\partial q_j} + \sum_{i=1}^{N} m_i \dot{\mathbf{r}}_i \cdot \frac{\partial \dot{\mathbf{r}}_i}{\partial q_j} \qquad (j = 1, 2, \ldots, n)$$

This may immediately be simplified by noticing that

$$\sum_{i=1}^{N} m_i \dot{\mathbf{r}}_i \cdot \frac{\partial \dot{\mathbf{r}}_i}{\partial q_j} = \frac{\partial T}{\partial q_j} \qquad (j = 1, 2, \ldots, n)$$

So far the equations contain only kinematical quantities. However, the equation of motion of each particle of the system may be written in the form

$$m_i\ddot{\boldsymbol{r}}_i = \boldsymbol{F}_i \qquad (i = 1, 2, \ldots, N)$$

where $\boldsymbol{F}_i$ are the resultant forces which act on each particle. These forces are now substituted into the above equations, which then contain information on the dynamical behaviour of the system. It can immediately be seen that the term containing these forces is, in fact, the corresponding generalised force component.

$$\sum_{i=1}^{N} m_i\ddot{\boldsymbol{r}}_i \cdot \frac{\partial \boldsymbol{r}_i}{\partial q_j} = \sum_{i=1}^{N} \boldsymbol{F}_i \cdot \frac{\partial \boldsymbol{r}_i}{\partial q_j} = Q_j \qquad (j = 1, 2, \ldots, n)$$

The above equations may thus be written in the form

$$\frac{\mathrm{d}}{\mathrm{d}t}\left(\frac{\partial T}{\partial \dot{q}_i}\right) - \frac{\partial T}{\partial q_i} = Q_i \qquad (i = 1, 2, \ldots, n)$$

In a holonomic system with n degrees of freedom, the n generalised momentum components are independent, and the generalised force components may be uniquely determined. These equations therefore provide n independent equations which govern the dynamical behaviour of the system. These are, in fact, Lagrange's equations of motion written in one of their principal forms. Since the generalised momentum components are a linear combination of the generalised velocities, these equations are a set of n second-order ordinary differential equations for the generalised coordinates. They are therefore sufficient to describe completely the possible motions of the system.

***Proposition* 9.2.** *The dynamical behaviour of a holonomic system with n degrees of freedom may be determined completely by Lagrange's equations of motion in the form*

$$\frac{\mathrm{d}}{\mathrm{d}t}\left[\frac{\partial T}{\partial \dot{q}_i}\right] - \frac{\partial T}{\partial q_i} = Q_i \qquad (i = 1, 2, \ldots, n)$$

where T is the kinetic energy of the system and Q_i $(i = 1, 2, \ldots, n)$ are the generalised force components corresponding to the n independent generalised coordinates q_i $(i = 1, 2, \ldots, n)$.

One obvious feature of Lagrange's equations is that they can be applied using any arbitrary coordinate system. Once a set of coordinates has been chosen, the kinetic energy must be expressed in terms of that set and the corresponding generalised force components must be determined. These are then substituted into the above standard equations which apply equally well to any holonomic system. By contrast, in the Newtonian approach, if the coordinates are not cartesian, difficulties often arise in the evaluation of the components of acceleration. However, even using the Lagrangian approach, it is found that in any particular problem some coordinate systems give rise to equations which are easier to integrate than others. Thus part of the skill in using Lagrange's equations is in the choice of an appropriate system of coordinates.

It can easily be shown that for a simple system, such as a single particle or a single rigid body, Lagrange's equations reduce to the same equations as could have been obtained from a more direct application of the equations of motion. In such cases the particular scalar equations for the motion may usually be obtained just as easily by either method.

The practical advantage of Lagrange's equations, however, can be seen in more complicated systems which contain several particles or rigid bodies, particularly when these are connected in some way. In a direct application of the more traditional equations of motion, each component would have to be treated separately in terms of the forces acting on it, and these must include the forces of constraint relating to the other components. However, in the Lagrangian approach the system may be treated as a whole and only the forces which do work in a virtual displacement need to be considered.

Lagrange's equations, as stated in proposition 9.2, may easily be interpreted. The leading term in each equation is simply the rate of change of each generalised momentum component. The right-hand sides are the generalised force components which allow for the forces which do work in a virtual displacement of the system. Finally, the terms $\partial T/\partial q_i$ $(i = 1, 2, \ldots, n)$ may be interpreted as the fictitious forces which arise through the choice of the coordinate system. In a cartesian coordinate system these would be zero. However, in a curvilinear coordinate system they may be regarded as the generalised fictitious forces which arise from the possible motion in the other generalised coordinate directions. It is these terms which allow for the complicated expressions for acceleration which would appear in the Newtonian approach. Lagrange's equations may thus be interpreted as stating that the rate of change of the generalised momentum components are each equal to the sum of the corresponding generalised force component and the corresponding generalised fictitious force component.

$$\dot{p}_i = Q_i + \frac{\partial T}{\partial q_i} \qquad (i = 1, 2, \ldots, n)$$

An alternative form of Lagrange's equations can be developed in the case of a system which is both conservative and holonomic. This case, however, may immediately be generalised to include all holonomic systems in which the generalised force components are derivable from a potential function $V(q_1, q_2, \ldots, q_n, t)$ by the conditions

$$Q_i = -\frac{\partial V}{\partial q_i} \qquad (i = 1, 2, \ldots, n)$$

In this case Lagrange's equations take the form

$$\frac{\mathrm{d}}{\mathrm{d}t}\left(\frac{\partial T}{\partial \dot{q}_i}\right) - \frac{\partial T}{\partial q_i} = -\frac{\partial V}{\partial q_i} \qquad (i = 1, 2, \ldots, n)$$

At this point it is convenient to define a new function L, known as the *Lagrangian*, in the following way.

***Definition* 9.6.** *The Lagrangian function for a dynamical system is its kinetic energy minus the potential function from which the generalised force components are determined.*

$$L = T - V$$

Now the kinetic energy of the system is a quadratic function of the generalised velocities with coefficients which may be functions of the generalised coordinates and possibly time. Also, the potential function is a function of the generalised coordinates and also, possibly, time. Thus

$$L = L(q_1, q_2, \ldots, q_n, \dot{q}_1, \dot{q}_2, \ldots, \dot{q}_n, t)$$

It may now be observed that, since the potential function is independent of the generalised velocities, the generalised momentum components may be written in either of the alternative forms

$$p_i = \frac{\partial T}{\partial \dot{q}_i} = \frac{\partial L}{\partial \dot{q}_i} \qquad (i = 1, 2, \ldots, n)$$

and Lagrange's equations may be written in this case in their more familiar form.

***Proposition* 9.3.** *For a holonomic system having n degrees of freedom, in which the generalised force components are derivable from a* *potential function, Lagrange's equations may be written in terms of the Lagrangian function and n generalised coordinates in the form*

$$\frac{\mathrm{d}}{\mathrm{d}t}\left(\frac{\partial L}{\partial \dot{q}_i}\right)-\frac{\partial L}{\partial q_i}=0 \qquad (i=1,2,\ldots,n)$$

In this form the generalised force components and the generalised fictitious force components which arise from the choice of coordinates have been combined into the single set of components $\partial L/\partial q_i$. These are equal to the rate of change of the generalised momentum components, and Lagrange's equations may be written as

$$\dot{p}_i=\frac{\partial L}{\partial q_i} \qquad (i=1,2,\ldots,n)$$

No particular physical significance is attached to the Lagrangian function at this stage. Its importance lies solely in its mathematical convenience. However, such convenience is considerable. It can be seen that in any holonomic mechanical system for which a Lagrangian function can be obtained, the particular equations for the motion of the system can be obtained simply by applying the standard form of Lagrange's equations to the particular Lagrangian. Thus in order to determine the possible motions of the system it is necessary only to specify this single scalar function.

When the forces which do work in a virtual displacement of a holonomic system are derivable from a potential function, then it is always convenient to obtain the Lagrangian function and to use Lagrange's equations in the form in which they are stated in proposition 9.3. However, for systems in which only some of the generalised force components are derivable from a potential function, it is sometimes convenient to consider a combination of the results of propositions 9.2 and 9.3. In this case it is possible to obtain a kind of partial Lagrangian from the kinetic energy of the system and the potential function associated with some of the generalised force components then have to be included explicitly in Lagrange's equations of motion in a similar way to that stated in proposition 9.2. There is, however, no necessity to adopt this approach and it is here assumed that such cases are already covered by a direct application of proposition 9.2.

9.5. First integrals of Lagrange's equations

It has already been pointed out in previous chapters that a particular set of equations of motion sometimes have a number of simple first integrals. These can usually be interpreted physically. So far the integrals considered have expressed the possible constancy of linear momentum components, angular momentum components and energy. It is the purpose of this section to obtain the equivalent set of first integrals which can sometimes be obtained in the Lagrangian formulation.

It is convenient to start with the integrals of momentum. Of course, in the Lagrangian approach the generalised momentum components may include components both of linear and angular momentum according to the choice of the generalised coordinates. Now, Lagrange's equations state that the rate of change of the generalised momentum components are equal to the corresponding generalised force components which must include the fictitious forces which arise from the choice of coordinates. For a holonomic system with n degrees of freedom, Lagrange's equations may be written in either of the forms

$$\text{(a)} \qquad \dot{p}_i = Q_i + \frac{\partial T}{\partial q_i} \qquad (i = 1, 2, \ldots, n)$$

$$\text{(b)} \qquad \dot{p}_i = \frac{\partial L}{\partial q_i} \qquad (i = 1, 2, \ldots, n)$$

It can immediately be seen that if the right-hand side of any of these equations is zero, then the corresponding generalised momentum component is a constant of the motion. This can be stated as follows.

***Proposition* 9.4.** *If, in a holonomic system with n degrees of freedom, either (a) one of the n independent generalised coordinates does not appear explicitly in the expression for kinetic energy, and the corresponding generalised force component is zero, or (b) the Lagrangian function which characterises the system does not contain explicitly one of the n generalised coordinates, then the corresponding generalised momentum component is a constant of the motion.*

It can be seen that this result is easy to apply and can immediately indicate the constants of linear and angular momentum which may occur in the system. However, it can also be seen that this result is, to some extent, dependent upon a suitable choice of generalised coordinates. This again illustrates the point that part of the skill in applying the Lagrangian approach is in the choice of appropriate coordinates.

The generalised coordinate corresponding to a constant generalised momentum component is known as an *ignorable coordinate.*

Such a coordinate does not appear explicitly in the equations of motion. If q_i is such a coordinate, then the equations are unaffected if q_i is replaced by $q_i + c$, where c is a constant. It can therefore be seen that the origin of an ignorable coordinate can be chosen arbitrarily.

If a system contains a large number of ignorable coordinates, and hence of constants of momentum, it is sometimes convenient to modify Lagrange's equations using a method developed by E. J. Routh† to analyse the stability of states of steady motion. In this approach, a modified Lagrangian function is introduced which is known as the *Routhian.* However, only in a few special situations does this approach have advantages over the standard Lagrangian approach. It has been developed purely for its mathematical convenience and does not involve any new physical concepts. It therefore does not need further consideration here.

It has now been shown that first integrals of linear and angular momentum can easily be obtained in the Lagrangian formulation, and it only remains to consider the energy integral. Such an integral will, of course, only occur if all the forces which do work are conservative. In such cases, the generalised force components are derivable from a potential function. It is therefore sufficient only to consider systems which can be characterised by a Lagrangian function.

For a holonomic system characterised by a Lagrangian function, Lagrange's equations of motion may each be multiplied by the corresponding generalised velocity. Summing all such products gives the equation

$$\sum_{i=1}^{n} \dot{q}_i \frac{\mathrm{d}}{\mathrm{d}t}\left(\frac{\partial L}{\partial \dot{q}_i}\right) - \sum_{i=1}^{n} \dot{q}_i \frac{\partial L}{\partial q_i} = 0$$

The second of these sums may be eliminated using the result that

$$\frac{\mathrm{d}L}{\mathrm{d}t} = \sum_{i=1}^{n} \frac{\partial L}{\partial q_i}\dot{q}_i + \sum_{i=1}^{n} \frac{\partial L}{\partial \dot{q}_i}\ddot{q}_i + \frac{\partial L}{\partial t}$$

and the resulting equation may be rearranged in the form

$$\frac{\mathrm{d}}{\mathrm{d}t}\left(\sum_{i=1}^{n} \dot{q}_i \frac{\partial L}{\partial \dot{q}_i} - L\right) = -\frac{\partial L}{\partial t}$$

From this it is immediately clear that if the Lagrangian function does not contain time explicitly, then the term in brackets is a constant of the motion. This result is known as *Jacobi's integral.*

† See L. A. Pars, *A Treatise on Analytical Dynamics* (Heinemann, 1965).

Proposition 9.5. *If a holonomic system with n degrees of freedom can be characterised by a Lagrangian L which is a function of, at most, n independent generalised coordinates and their corresponding generalised velocities so that it does not depend on time explicitly, then* $\sum_{i=1}^{n} \dot{q}_i \partial L/\partial \dot{q}_i - L$ *is a constant of the motion.*

In order to interpret this result it is convenient to consider a holonomic system which contains no rheonomic constraints. Then, assuming that a moving coordinate system has not been introduced, its configuration can be determined at any time by n independent generalised coordinates

$$r_i = r_i(q_1, q_2, \ldots, q_n) \qquad (i = 1, 2, \ldots, N)$$

In this case it has already been shown that the kinetic energy of the system is a homogeneous quadratic function of the generalised velocities which does not depend explicitly on time and can be written as

$$T = \tfrac{1}{2} \sum_{i=1}^{n} \dot{q}_i p_i$$

If the potential function does not explicitly depend on time either, the Lagrangian function is therefore also independent of time. In this case Jacobi's integral is a constant of the motion and can be written in the form

$$\begin{aligned} \sum_{i=1}^{n} \dot{q}_i \frac{\partial L}{\partial \dot{q}_i} - L &= 2T - (T - V) \\ &= T + V \end{aligned}$$

It can therefore be seen that Jacobi's integral in this case is identical to the familiar energy integral for such a system.

In order to clarify the relation between Jacobi's integral and the energy integral it is necessary to relax the conditions which led to the above result. This may now be done by considering the three ways in which time may be introduced explicitly into the Lagrangian function.

The first case to consider is that in which there is a time dependence in the forces which do work in a virtual displacement of the system. Even when such forces are derivable from a potential function, this potential must be time dependent. In this way time enters explicitly into the Lagrangian function and so the system does not possess a Jacobi integral. On the other hand, it has already been shown in sections 6.2 and 6.3 that systems in which the potential function is time dependent are not conservative and therefore do not possess an energy integral either.

The second case to consider is that of a system which contains at least one rheonomic constraint. It may be assumed that the associated forces of constraint are nonzero. Although such forces do no work in a virtual displacement of the system, it will be shown in section 9.8 that they necessarily do work in an actual displacement. Thus in the Newtonian approach it can be seen that the system is not conservative and an energy integral does not exist. In the Lagrangian approach, on the other hand, the rheonomic constraints must be used to reduce the number of generalised coordinates and so the configuration of the system must necessarily depend explicitly on time as well as the n generalised coordinates. In this case a time dependence thus enters explicitly into the expression for the kinetic energy and hence into the Lagrangian. It may therefore also be concluded that systems which contain a rheonomic constraint possess neither an energy integral nor a Jacobi integral.

The final case to consider is that in which a moving coordinate system has been adopted. It may be assumed that the system itself is conservative so that an energy integral exists. However, in this case, a time dependence has been introduced in the definition of the generalised coordinates. This will usually introduce a time dependence into the Lagrangian function and so a Jacobi integral can not be obtained in the form defined above. However, since the time dependence only appears in the mathematical description of the system and not in the physical system itself, it must be possible to find some combination of Lagrange's equations of motion which would give the energy integral.

There are, however, a number of exceptions in this final case, since it is possible for a moving coordinate system to be introduced in such a way that the kinetic energy, and hence the Lagrangian, have no explicit time dependence. In this case, the Lagrangian takes the general form

$$L = \tfrac{1}{2}\sum_{i=1}^{n}\sum_{j=1}^{n} a_{ij}\dot{q}_i\dot{q}_j + \sum_{i=1}^{n} a_i\dot{q}_i + \tfrac{1}{2}a_0 - V$$

where each term is independent of time. It can then be shown that the expression for Jacobi's integral is

$$\sum_{i=1}^{n} \dot{q}_i\frac{\partial L}{\partial \dot{q}_i} - L = \tfrac{1}{2}\sum_{i=1}^{n}\sum_{j=1}^{n} a_{ij}\dot{q}_i\dot{q}_j - \tfrac{1}{2}a_0 + V$$

A simple example of such a system is that of a particle which moves in a conservative force field relative to a set of cartesian coordinates which are rotating with constant angular velocity. In this case the position of the particle relative to an inertial frame is a function of the coordinates of the rotating frame and time. It follows, therefore, that its kinetic energy includes the linear and constant terms in addition to the terms which are quadratic in the generalised velocities. However, the coefficients in this case do not depend on time. The Lagrangian therefore has no explicit time dependence and so Jacobi's integral, which now takes the above form, is a constant of the motion. However, although Jacobi's integral is not the same as $T+V$, it may still be shown to be the same as the energy integral. Since a rotating frame of reference is used, it is necessary in the Newtonian approach to include the associated fictitious forces. As was shown at the end of section 6.4, if the frame is rotating with constant angular velocity, then an energy integral may still be obtained, provided the term $-\frac{1}{2}m\,(\boldsymbol{\omega}\times\boldsymbol{r})^2$ is included in the potential energy. This term can be regarded as the potential energy associated with the centrifugal force. In the Lagrangian approach, on the other hand, it is precisely this expression which appears as the term $-\frac{1}{2}a_0$. The fact that this term appears here in the kinetic energy also illustrates the point that, in the Lagrangian approach, the terms $\partial T/\partial q_i$ can be interpreted as the fictitious forces associated with the choice of coordinates.

It can now be concluded that, except when a set of moving coordinates are used, Jacobi's integral is identical to the familiar energy integral. Even in the exceptional case with moving coordinates, the two integrals are identical if they exist, although every such case has to be considered separately.

Finally, to summarise this section, it may be pointed out that for a holonomic system the Lagrangian function necessarily contains all the n independent generalised velocities. If, however, it is independent of one of the generalised coordinates, $\partial L/\partial q_i=0$, where q_i is an ignorable coordinate, then the corresponding generalised momentum component $\partial L/\partial \dot{q}_i$ is a constant of the motion. In addition, if the Lagrangian does not depend on time explicitly, $\partial L/\partial t=0$, and Jacobi's integral $\sum_{i=1}^{n}\dot{q}_i\,\partial L/\partial\dot{q}_i-L$ is a constant of the motion.

9.6. Holonomic systems: examples

Example 9.1: *a double compound pendulum*

Consider here only the particular case of a uniform rod which is free to move about an end, and which has another identical rod smoothly connected to the other end. Consider further the particular case in which motion occurs in a fixed vertical plane.

This restricted system has two degrees of freedom, and it is convenient to describe the motion in terms of two parameters, $q_1 = \theta$ and $q_2 = \phi$, which are the angles at which the two rods are inclined to the vertical (see figure 9.1).

The only external forces which act on the system are the reactional forces at the fixed point and the gravitational forces. Denoting the mass of the rods by m and their lengths by $2a$, the kinetic energy of the system in motion is given by

$$\begin{aligned} T &= \tfrac{2}{3}ma^2\dot{\theta} + \tfrac{1}{6}ma^2\dot{\phi}^2 + \tfrac{1}{2}ma^2(\dot{\phi}^2 + 4\dot{\theta}^2 + 4\dot{\theta}\dot{\phi}\cos(\phi-\theta)) \\ &= \tfrac{2}{3}ma^2\{4\dot{\theta}^2 + 3\cos(\phi-\theta)\dot{\theta}\dot{\phi} + \dot{\phi}^2\} \end{aligned}$$

This expression is a clear illustration of a system where the kinetic energy is a homogeneous quadratic in the generalised velocities with coefficients which are functions of the generalised coordinates. This system is clearly conservative and its potential energy may be taken as

$$V = -mga\,(3\cos\theta + \cos\phi)$$

The Lagrangian function for this system is

$$L = \tfrac{2}{3}ma^2\{4\dot{\theta}^2 + 3\cos(\phi-\theta)\dot{\theta}\dot{\phi} + \dot{\phi}^2\} + mga(3\cos\theta + \cos\phi)$$

Substituting this directly into Lagrange's equations, as stated in proposition 9.3, give the two equations for the motion

$$\tfrac{16}{3}\ddot{\theta} + 2\cos(\phi-\theta)\ddot{\phi} - 2\sin(\phi-\theta)\dot{\phi}^2 + 3\frac{g}{a}\sin\theta = 0$$

$$2\cos(\phi-\theta)\ddot{\theta} + \tfrac{4}{3}\ddot{\phi} + 2\sin(\phi-\theta)\dot{\theta}^2 + \frac{g}{a}\sin\phi = 0$$

These equations are fairly complicated and no first integral is immediately

Fig. 9.1. The double compound pendulum.

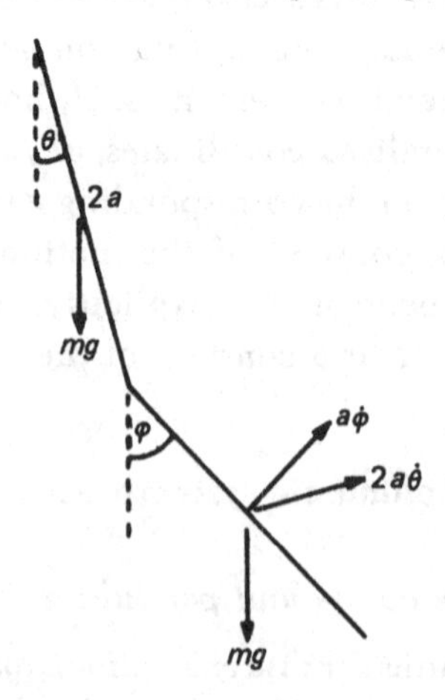

obvious. However, it may be observed that the Lagrangian function does not contain time explicitly, which implies that Jacobi's integral can be obtained here, and since the kinetic energy is a homogeneous quadratic this can be expressed in the form $T+V=\text{const.}$

Although the above equations look complicated, they have in fact been obtained very easily. In contrast to this, it may be seen that the traditional approach of applying the equations of motion described here in chapter 8 would have been difficult. It would have been necessary to introduce the interactive forces between the rods and, although the energy integral could immediately have been found, the derivation of the full equations of motion would have been laborious. This example thus illustrates the practical advantages of the Lagrangian method when considering systems of rigid bodies which are hinged or connected in some way.

The case of small oscillations of this system about its position of stable equilibrium will be considered later in example 9.4.

Example 9.2: *The spinning top*

The motion of a top has already been considered in example 8.18. It is the purpose of this example simply to show that the equations for its motion can be obtained more easily using the Lagrangian approach.

The notation of example 8.18 may immediately be adopted, in which the orientation of the top at any time is described in terms of Euler's angular coordinates θ, ϕ and ψ. The kinetic energy may be quoted as

$$T=\tfrac{1}{2}I_1\dot{\theta}^2+\tfrac{1}{2}I_1\dot{\phi}^2\sin^2\theta+\tfrac{1}{2}I_3(\dot{\psi}+\dot{\phi}\cos\theta)^2$$

and the potential energy is

$$V=mgh\cos\theta$$

It may immediately be observed that the Lagrangian function $L=T-V$ does not contain time explicitly. And since the kinetic energy is a homogeneous quadratic in the velocities, proposition 9.5 implies that

$$T+V=\text{const}$$

It is also clear that the Lagrangian function does not depend explicitly on either of the coordinates ψ or ϕ. These are therefore ignorable coordinates and, according to proposition 9.4, there exist corresponding integrals of momentum $\partial L/\partial\dot{\psi}$ and $\partial L/\partial\dot{\phi}$ which can be written as

$$I_3(\dot{\psi}+\dot{\phi}\cos\theta)=I_3s$$

$$I_1\dot{\phi}\sin^2\theta+I_3s\cos\theta=L_z$$

These three integrals are sufficient to describe the system. However it is sometimes convenient to use the equation of motion obtained from Lagrange's equations with $q_1=\theta$. This can be written as

$$I_1\ddot{\theta}=(I_1\dot{\phi}^2\sin\theta-I_3s\dot{\phi}+mgh)\sin\theta$$

It may be observed that the above equations have been obtained slightly more conveniently using the Lagrangian method than by taking components of the equation $\dot{\boldsymbol{L}}+\boldsymbol{\Omega}\times\boldsymbol{L}=\boldsymbol{M}$. However, having obtained these equations, the Lagrangian approach adds nothing to the subsequent analysis.

Example 9.3: *normal modes of oscillation*

The purpose here is to develop a general method to describe the small oscillations of a dynamical system about a configuration of stable equilibrium.

Consider a general conservative holonomic system with n degrees of freedom and a position of stable equilibrium. It is always possible to choose a system of n generalised coordinates which are zero when the system is in the equilibrium configuration. The parameters thus measure the displacement of the system from equilibrium. It is appropriate here to consider only small oscillations, so it may generally be assumed that the generalised coordinates are small quantities.

If the equilibrium configuration is static, choosing coordinates as described above does not introduce any explicit time dependence. The kinetic energy must therefore be a homogeneous quadratic in the generalised velocities which can be written as

$$T=\tfrac{1}{2}\sum_{i=1}^{n}\sum_{j=a}^{n}a_{ij}\dot{q}_i\dot{q}_i$$

In general the coefficients a_{ij} are functions of the generalised coordinates. However, since we are only considering small oscillations, it is appropriate to make the approximation that these coefficients are constants. The expression for kinetic energy is therefore of second order in small quantities.

Since the system is conservative a potential function exists which is stationary in the equilibrium position. Moreover, since the equilibrium position has been assumed to be stable, the potential energy in that configuration is a minimum. Ignoring an arbitrary constant, it follows that, for small displacements from equilibrium, the potential function can be approximated by a quadratic function of the generalised coordinates which can be written as

$$V=\tfrac{1}{2}\sum_{i=1}^{n}\sum_{j=1}^{n}b_{ij}q_iq_j$$

where the coefficients b_{ij} are constants.

Thus, for small oscillations about the equilibrium position, the Lagrangian function may be approximated by

$$L=\tfrac{1}{2}\sum_{i=1}^{n}\sum_{j=1}^{n}(a_{ij}\dot{q}_i\dot{q}_j-b_{ij}q_iq_j)$$

and it follows that the equations of motion are

$$\sum_{j=1}^{n}(a_{ij}\ddot{q}_j+b_{ij}q_j)=0\qquad(i=1,2,\ldots,n)$$

It may be pointed out here that these are just the linearised equations for motion near the equilibrium position. In some situations it is convenient to obtain these by other methods. The Lagrangian approach used above is not necessary.

If the motion is indeed an oscillatory motion about the equilibrium position, it is appropriate to consider a trial solution of the form

$$q_i = c_i \sin(\omega t + \varepsilon) \qquad (i = 1, 2, \dots, n)$$

where $c_1, c_2, \dots, c_n$, ω and ε are constants. Substituting these expressions into the equations of motion reduces them to the form

$$\sum_{j=1}^{n} (-\omega^2 a_{ij} + b_{ij}) c_j = 0 \qquad (i = 1, 2, \dots, n)$$

These can be regarded as n linear equations for the coefficients c_i. A necessary and sufficient condition for nontrivial solutions of these equations to exist, is that the determinant of the coefficients $-\omega^2 a_{ij} + b_{ij}$ must be zero. For any given sets of constants a_{ij} and b_{ij}, this condition is an nth order equation in ω^2. If the equilibrium configuration is in fact stable, this equation must have n real positive roots for ω^2. In this case there are n possible values for the frequency of oscillation ω, with each root corresponding to a distinct possible mode.

Each distinct mode of oscillation is characterised by $n+1$ constants ω, $c_1, c_2, \dots, c_n$, although the amplitudes $c_1, c_2, \dots, c_n$ may be multiplied by an arbitrary constant, and the phase ε is also arbitrary. A general solution of the linearised equations of motion, however, must be an arbitrary linear superposition of all possible modes.

Example 9.4

Consider again the double compound pendulum described in example 9.1 in the case of small oscillations about the equilibrium position.

The double pendulum clearly has a configuration of stable equilibrium in which both rods hang vertically and, using the same parameters, θ and ϕ are both zero. It is therefore appropriate to consider motion in which θ and ϕ are small. Ignoring an arbitrary constant, the Lagrangian function in this case may be approximated by

$$L = \tfrac{2}{3}ma^2(4\dot{\theta}^2 + 3\dot{\theta}\dot{\phi} + \dot{\phi}^2) - \tfrac{1}{2}mga(3\theta^2 + \phi^2)$$

which leads to the equations of motion

$$\tfrac{16}{3}\ddot{\theta} + 2\ddot{\phi} + 3\frac{g}{a}\theta = 0$$

$$2\ddot{\theta} + \tfrac{4}{3}\ddot{\phi} + \frac{g}{a}\phi = 0$$

Substituting the trial solutions

$$\theta = c_1 \sin(\omega t + \varepsilon)$$
$$\phi = c_2 \sin(\omega t + \varepsilon)$$

these equations can conveniently be expressed in the matrix form

$$\begin{pmatrix} -\frac{16}{3}\omega^2+3\frac{g}{a} & -2\omega^2 \\ -2\omega^2 & -\frac{4}{3}\omega^2+\frac{g}{a} \end{pmatrix}\begin{pmatrix} c_1 \\ c_2 \end{pmatrix}=0$$

Nonzero solutions of these equations exist only if the determinant of the matrix is zero. This condition leads to the equation for the frequencies

$$28\omega^4-84\omega^2\frac{g}{a}+27\frac{g^2}{a^2}=0$$

Thus the two possible frequencies are given by

$$\omega_1^2=3\left(\frac{1}{2}-\frac{1}{\sqrt{7}}\right)\frac{g}{a} \qquad \omega_2^2=3\left(\frac{1}{2}+\frac{1}{\sqrt{7}}\right)\frac{g}{a}$$

Substituting ω_1^2 back into the above equations gives

$$c_2=\frac{(2\sqrt{7}-1)}{3}c_1$$

which indicates that the two rods swing back and forth in the same directions, with ϕ being always slightly larger than θ.

Similarly substituting back for the other normal frequency ω_2^2 implies that in this mode

$$c_2=-\frac{(1+2\sqrt{7})}{3}c_1$$

Thus, in this second mode, the two rods swing in opposite directions, with the magnitude of ϕ being more than twice the magnitude of θ.

Exercises

9.1 One end A of a uniform rod AB of mass m and length $2a$ can slide along a smooth vertical wire through a fixed point O. The end B is attached to O by a light string of length $2a$ so that, when the string is taut, OB makes an angle θ with the downward vertical. The vertical plane OBA rotates about OB with angular speed $\dot{\phi}$. Show that the kinetic energy of the rod is

$\frac{2}{3}ma^2\{(1+6\sin^2\theta)\dot{\theta}^2+\sin^2\theta\dot{\phi}^2\}$

Obtain the equations of motion and expressions for the energy integral and the constant angular momentum about OA.

If initially A is below 0 with $\theta=\pi/3$, $\dot{\theta}=0$ and $\dot{\phi}=\sqrt{(12g/a)}$ show that the end A begins to rise and that it then oscillates between its initial configuration and the configuration in which AB is horizontal.

9.2 Consider a thin uniform disc of mass m and radius a which is spinning on a smooth horizontal surface. Describe the orientation of the disc in terms of Euler's angular coordinates θ, ϕ and ψ, and describe the position of its centroid in terms of the cartesian coordinates x, y and $z = a \sin\theta$. Show that the Lagrangian function for the system is

$$L = \frac{m}{2}\{\dot{x}^2 + \dot{y}^2 + (\tfrac{1}{4} + \cos^2\theta)a^2\dot{\theta}^2 + \tfrac{1}{4}a^2\sin^2\theta\dot{\phi}^2 + \tfrac{1}{2}a^2(\dot{\psi} + \dot{\phi}\cos\theta)^2 - 2ga\sin\theta\}$$

Obtain the integrals of momentum associated with the ignorable coordinates x, y, ϕ and ψ, and show that the remaining equation of motion can be written in the form

$$(1 + 4\cos^2\theta)\ddot{\theta} = \cos\theta\sin\theta(4\dot{\theta}^2 + \dot{\phi}^2) - 2s\dot{\phi}\sin\theta - 4\frac{g}{a}\cos\theta$$

where $s = \dot{\psi} + \dot{\phi}\cos\theta$.

Hence show that the motion in which the disc rotates about a vertical diameter with angular speed Ω is stable if $\Omega^2 > 4g/a$.

9.3 A uniform rod of length $2a$ is suspended from a fixed point by a light string of length $5a/12$ which is attached to one end of the rod. Show that, for small oscillations in a vertical plane about the equilibrium position, the frequency of oscillation in the two normal modes are $\sqrt{(3g/5a)}$ and $\sqrt{(12g/a)}$.

9.4 A uniform solid sphere of radius a is suspended from a fixed point by a light string of length $6a/5$ which is attached to a point on the surface of the sphere. Show that, for small oscillations in a vertical plane the periods of the normal modes are $2\pi\sqrt{(a/5g)}$ and $4\pi\sqrt{(3a/5g)}$. Describe the modes.

9.5 Show that the kinetic energy of a particle relative to a set of cartesian axes which are rotating with angular speed ω about the z axis is

$$T = \tfrac{1}{2}m\{\dot{x}^2 + \dot{y}^2 + \dot{z}^2 + 2\omega(x\dot{y} - y\dot{x}) + \omega^2(x^2 + y^2)\}$$

Conclude that, although moving coordinates are used, the kinetic energy does not necessarily depend on time explicitly.

9.7. The fundamental equation

In the previous two sections Lagrange's equations were considered for a holonomic system which is described by a set of independent generalised coordinates. It is now appropriate to return to the more general situation and to consider systems which contain a number of constraints.

Even for systems which are subject to constraints, it is usually possible to obtain an expression for the kinetic energy and hence the generalised momentum components. However, these components are not independent in this case. In addition, although it may be theoretically possible to obtain the corresponding generalised force components using definition 9.5, it is not possible in practice to evaluate these uniquely by considering the work done in an arbitrary virtual displacement of the system, since the displacement must satisfy the constraints. Thus it is not possible, in general, to obtain a set of equations of motion in Lagrangian form in the simple way described in section 9.4. An alternative approach therefore needs to be considered.

To obtain a more general approach it is convenient to return to the equations of motion for every particle of the system. These may be expressed in the form

$$m_i\ddot{\boldsymbol{r}}_i - \boldsymbol{F}_i = 0 \qquad (i = 1, 2, \ldots, N)$$

where $\boldsymbol{F}_i$ is the sum of all the forces which act on each particle. It thus follows trivially that for any arbitrary virtual displacement of the system $\delta\boldsymbol{r}_i$ $(i = 1, 2, \ldots, N)$, which is consistent with the constraints

$$(m_i\ddot{\boldsymbol{r}}_i - \boldsymbol{F}_i) \cdot \delta\boldsymbol{r}_i = 0 \qquad (i = 1, 2, \ldots, N)$$

and hence that

$$\sum_{i=1}^{N} (m_i\ddot{\boldsymbol{r}}_i - \boldsymbol{F}_i) \cdot \delta\boldsymbol{r}_i = 0$$

It is this equation which may be regarded as the *fundamental equation* of analytical dynamics, although it is sometimes stated in a slightly different form.

***Proposition* 9.6.** *For an arbitrary virtual displacement of a system, the accelerations of each of the component particles at positions* $\boldsymbol{r}_i$ $(i = 1, 2, \ldots, N)$ *and the forces* $\boldsymbol{F}_i$ *which act upon them satisfy the equation*

$$\sum_{i=1}^{N} (m_i\ddot{\boldsymbol{r}}_i - \boldsymbol{F}_i) \cdot \delta\boldsymbol{r}_i = 0$$

It is immediately clear that the components of the resultant forces $\boldsymbol{F}_i$ which do no work in a virtual displacement of the system do not contribute to the above expression. In the previous discussions of systems of several particles and also of rigid body dynamics, it was found to be convenient to distinguish between the internal and external forces of the system. In analytical dynamics, however, it is convenient to make the alternative distinction between those forces which do work in a virtual displacement and those which do not. Those forces which do no work need not be included in the fundamental equation.

It is usually found in practice that the forces which do no work in a virtual displacement are those which are associated with the constraints of the system. Similarly, the remaining forces, which may be described as the given forces, usually do work in a virtual displacement of the system. The fundamental equation is therefore often stated as containing only the given forces, as distinct from the forces of constraint. In this way it may be based upon a principle that the forces of constraint do no work. There must, however, be some restrictions on such an approach, since exceptions sometimes occur in which problems are stated in terms of constraints whose associated forces do contribute to the virtual work. In fact, the term 'constraint' can be used very loosely. Thus, before basing the fundamental equation on a principle that the forces of constraint do no work, it is necessary to specify exactly the classes of constraints that are acceptable. This, in fact, is considered in the next section. However, it is sufficient here to continue with the fundamental equation in the above form, in which the resultant forces $\boldsymbol{F}_i$ include all the forces in the system.

Since it is required that the virtual displacements be consistent with the constraints of the system, it is possible to express them in terms of the variations of the n generalised coordinates. The fundamental equation can therefore be written in the form

$$\sum_{j=1}^{N} \sum_{i=1}^{N} (m_i\ddot{\boldsymbol{r}}_i - \boldsymbol{F}_i) \cdot \frac{\partial \boldsymbol{r}_i}{\partial q_j} \delta q_j = 0$$

This equation can now be simplified by using a result obtained in section 9.4, and by introducing the generalised force components as given by definition 9.5. The resulting equation can be written in the form

$$\sum_{j=1}^{n} \left\{ \frac{\mathrm{d}}{\mathrm{d}t}\left(\frac{\partial T}{\partial \dot{q}_j}\right) - \frac{\partial T}{\partial q_j} - Q_j \right\} \delta q_j = 0$$

This is an alternative form of the fundamental equation and is of great importance.

***Proposition* 9.7.** *The fundamental equation can be written in the form*

$$\sum_{i=1}^{n} \left\{ \frac{\mathrm{d}}{\mathrm{d}t}\left(\frac{\partial T}{\partial \dot{q}_i}\right) - \frac{\partial T}{\partial q_i} - Q_i \right\} \delta q_i = 0$$

where T is the kinetic energy of the system expressed in terms of n generalised coordinates q_i $(i = 1, 2, \ldots, n)$, Q_i are the corresponding generalised force components, and δq_i represent an

arbitrary virtual displacement of the system which is consistent with any remaining constraints.

It can easily be seen that Lagrange's equations for a holonomic system can immediately be obtained from this result. If the system is holonomic and all the constraints have been used to reduce the number of generalised coordinates, then the n remaining generalised coordinates are independent. In this case the virtual variations δq_i ($i = 1, 2, \ldots, n$) are unconstrained and may be varied arbitrarily. It can therefore be seen that their coefficients in the above expansion must necessarily be zero and these give Lagrange's equations in the form

$$\frac{\mathrm{d}}{\mathrm{d}t}\left(\frac{\partial T}{\partial \dot{q}_i}\right) - \frac{\partial T}{\partial q_i} = Q_i \qquad (i = 1, 2, \ldots, n)$$

However for a system which is still subject to a number of constraints, the variations δq_i ($i = 1, 2, \ldots, n$) are not independent. It is not possible, therefore, to deduce that their coefficients in the above expansion must be zero. Thus Lagrange's equations can not be obtained in the simple way described above for any nonholonomic system. In such cases the constraints have to be considered in greater detail. This is the subject of the next section.

9.8. Systems subject to constraints

The main purpose of this section is to obtain the equations of motion in Lagrangian form for nonholonomic systems. The approach may, however be generalised to include any system in which the generalised coordinates are subject to a number of constraints. It is therefore assumed here that the system being considered is described in terms of n generalised coordinates, and is subject to k constraints where $k < n$, so that it possesses $n - k$ degrees of freedom.

It may initially be assumed that some of the constraints are holonomic. These may be written in the form

$$f_i(q_1, q_2, \ldots, q_n, t) = 0 \qquad (i = 1, 2, \ldots, p)$$

where $p \leq k$. Such constraints also place restrictions on the generalised velocity components which it can be seen must satisfy the conditions

$$\sum_{j=1}^{n} \frac{\partial f_i}{\partial q_j}\dot{q}_j + \frac{\partial f_i}{\partial t} = 0 \qquad (i = 1, 2, \ldots, p)$$

This in fact is simply an alternative way of representing the same constraints. Yet another way of writing the constraints is in the *Pfaffian form* involving the differentials

$$\sum_{j=1}^{n} \frac{\partial f_i}{\partial q_j} \mathrm{d}q_j + \frac{\partial f_i}{\partial t} \mathrm{d}t = 0 \qquad (i = 1, 2, \ldots, p)$$

The differentials $\mathrm{d}q_j$ $(j = 1, 2, \ldots, n)$ appearing in the Pfaffian form of the constraints may be regarded as the set of *possible infinitesimal displacements* which are permitted in the system. These are of course not the same as the actual displacements since these are uniquely determined by the forces which are acting. Neither are they the same as the set of virtual displacements since they are assumed to occur in time $\mathrm{d}t$. The virtual displacements δq_i $(i = 1, 2, \ldots, n)$ necessarily occur without variation of time, and since these must also be consistent with the above constraints they must clearly satisfy the conditions

$$\sum_{j=1}^{n} \frac{\partial f_i}{\partial q_j} \delta q_j = 0 \qquad (i = 1, 2, \ldots, p)$$

It can be seen that these conditions on the virtual displacements are identical to the conditions on the possible displacements whenever all the constraints are scleronomic and therefore satisfy $\partial f_i/\partial t = 0$ $(i = 1, 2, \ldots, p)$. However, in any system which contains a rheonomic constraint, the virtual and possible displacements necessarily differ. This is a special case of the result to be obtained as proposition 9.8.

Having briefly considered holonomic constraints, it is necessary now to consider nonholonomic constraints. Since these have been defined in a negative way, there can be no exhaustive representation of them. In practice, however, by far the most common nonholonomic constraints that are considered are those which can be expressed as linear relations between the generalised velocities. They may therefore be written in the form

$$\sum_{j=1}^{n} A_{ij}\dot{q}_j + A_i = 0 \qquad (i = 1, 2, \ldots, k)$$

Clearly these include the p holonomic constraints for which $A_{ij} = \partial f_i/\partial q_j$, $A_i = \partial f_i/\partial t$ $(i = 1, 2, \ldots, p)$. However, the remaining $k - p$ nonholonomic constraints are not integrable and therefore cannot be used to reduce the number of independent variables.

It has already been seen that it is convenient to distinguish clearly between holonomic constraints which are, or are not, time dependent. For nonholonomic systems of the above type it is similarly convenient to distinguish between constraints in which the term A_i is zero or nonzero. However, it is not appropriate to adapt the terms scleronomic and rheonomic to distinguish these cases, and a new terminology is introduced.

***Definition* 9.7.** *A constraint which may be written in the form*

$$\sum_{j=1}^{n} A_{ij}\dot{q}_j + A_i = 0 \quad \textit{for some value of } i$$

is called catastatic if $A_i = 0$, or acatastatic if $A_i \neq 0$

Using this terminology, a system may be described as catastatic when all the constraints can be written in the above form with every $A_i = 0$ $(i = 1, 2, \ldots, k)$. Otherwise, it may be called acatastatic.

The above constraints may also be written in the Pfaffian form

$$\sum_{j=1}^{n} A_{ij}\,\mathrm{d}q_j + A_i\,\mathrm{d}t = 0 \qquad (i = 1, 2, \ldots, k)$$

which expresses the constraints on the possible displacements of the system. It can also be seen, by permitting no variation in time, that the virtual displacements of the system must satisfy the constraints

$$\sum_{j=1}^{n} A_{ij}\delta q_j = 0 \qquad (i = 1, 2, \ldots, k)$$

The following result may now immediately be deduced from these equations.

***Proposition* 9.8.** *In a catastatic system in which all the constraints can be written in the form $\sum_{j=1}^{n} A_{ij}\,\mathrm{d}q_j = 0$ $(i = 1, 2, \ldots, k)$, the class of virtual displacements is identical to the class of possible displacements.*

It is convenient now to return to the fundamental equation in the form stated in proposition 9.7, namely

$$\sum_{j=1}^{n} \left\{ \frac{\mathrm{d}}{\mathrm{d}t}\left(\frac{\partial T}{\partial \dot{q}_j}\right) - \frac{\partial T}{\partial q_j} - Q_j \right\} \delta q_j = 0$$

In the present case, however, the virtual variations δq_j $(j = 1, 2, \ldots, n)$ are not independent but, if all the constraints are of the above form, they must satisfy the k conditions

$$\sum_{j=1}^{n} A_{ij}\delta q_j = 0 \qquad (i = 1, 2, \ldots, k)$$

It is possible in this case to combine these equations and to derive equations of motion from them by using Lagrange multiplier techniques. This method involves introducing k new parameters $\lambda_1, \lambda_2, \ldots, \lambda_k$ with which to multiply, respectively, all the terms of each constraint. Since each constraint remains zero, they may all be included in the fundamental equation in the form

$$\sum_{j=1}^{n}\left\{\frac{\mathrm{d}}{\mathrm{d}t}\left(\frac{\partial T}{\partial \dot{q}_j}\right)-\frac{\partial T}{\partial q_j}-Q_j-\sum_{i=1}^{k}\lambda_i A_{ij}\right\}\delta q_j=0$$

Now the Lagrange multipliers can be chosen arbitrarily. In particular it is possible to choose them so that k of the terms in the above sum are zero. The remaining equation then contains multiples of only $n-k$ of the variations δq_j. However, the system has $n-k$ degrees of freedom, and therefore the remaining variations can be chosen arbitrarily, indicating that their coefficients must be zero. In this way it is possible to obtain the n equations which are the equations of motion in Lagrangian form.

$$\frac{\mathrm{d}}{\mathrm{d}t}\left(\frac{\partial T}{\partial \dot{q}_j}\right)-\frac{\partial T}{\partial q_j}=Q_j+\sum_{i=1}^{k}\lambda_i A_{ij} \qquad (j=1,2,\ldots,n)$$

It should be noticed that there are now effectively $n+k$ variables, the generalised coordinates $q_1, q_2, \ldots, q_n$ and the Lagrange multipliers $\lambda_1, \lambda_2, \ldots, \lambda_k$. These are required to satisfy $n+k$ equations, the n equations of motion above and the k equations of constraint. This result may now be summarised as follows.

***Proposition* 9.9.** *For a system described by n generalised coordinates q_j $(j=1,2,\ldots,n)$ which are subject to k constraints of the form*

$$\sum_{j=1}^{n} A_{ij}\dot{q}_j+A_i=0 \qquad (i=1,2,\ldots,k)$$

it is possible to choose k multipliers λ_i $(i=1,2,\ldots,k)$ so that, in addition to the above k equations of constraint, the motion of the system is determined by the equations

$$\frac{\mathrm{d}}{\mathrm{d}t}\left(\frac{\partial T}{\partial \dot{q}_j}\right)-\frac{\partial T}{\partial q_j}=Q_j+\sum_{i=1}^{k}\lambda_i A_{ij} \qquad (j=1,2,\ldots,n)$$

where T is the kinetic energy of the system and Q_j are the generalised force components corresponding to the generalised coordinates.

It should be noticed that, if the constraints were removed, this result would be identical to proposition 9.2 which states Lagrange's equations for a holonomic system with n degrees of freedom. It is in fact possible to use the identical expressions for the kinetic energy and the generalised force components, even in this case. This essentially, means that the kinetic energy and the generalised force components can be evaluated as though the constraints were not applied. They would then be uniquely determined. Following this method it can then be seen that the additional terms $\lambda_i A_{ij}$ appear when the constraints are included, and since these are added to the generalised force components they may be interpreted as the *generalised forces of constraint.* Furthermore, the constraints immediately imply that

$$\sum_{j=1}^{n} \lambda_i A_{ij} \delta q_j = 0 \qquad (i = 1, 2, \ldots, k)$$

which indicates that, if the constraints are of this type, the forces of constraint do no work in a virtual displacement of the system.

If the generalised force components can be derived from a potential function, it is possible to introduce the Lagrangian function $L = T - V$. Then, following the analogy with proposition 9.3, the n equations of motion in proposition 9.9 can be replaced by the equations

$$\frac{\mathrm{d}}{\mathrm{d}t}\left(\frac{\partial L}{\partial \dot{q}_j}\right) - \frac{\partial L}{\partial q_j} = \sum_{i=1}^{k} \lambda_i A_{ij} \qquad (j = 1, 2, \ldots, n)$$

Clearly it is also possible to generalise proposition 9.4. If for any of the n equations the generalised force component Q_j, the corresponding generalised fictitious force component $\partial T/\partial q_j$, and all the corresponding forces of constraint $\lambda_i A_{ij}$ $(i = 1, 2, \ldots, k)$ are zero, then a first integral of momentum can be obtained.

An energy integral can also be obtained in some situations. It is easy to show in this case that

$$\frac{\mathrm{d}}{\mathrm{d}t}\left(\sum_{j=1}^{n} \dot{q}_j \frac{\partial L}{\partial \dot{q}_j} - L\right) = -\frac{\partial L}{\partial t} + \sum_{i=1}^{k} \lambda_i A_i$$

It can therefore be deduced that, if the Lagrangian function does not contain time explicitly and the system is catastatic, then Jacobi's integral is a constant of the motion. This result is a generalisation of proposition 9.5.

This also clarifies the point that a holonomic system with a rheonomic constraint cannot have an energy integral. Although the associated forces of constraint do no work in a virtual displacement,

$$\sum_{j=1}^{n} \lambda_i A_{ij} \delta q_j = 0$$

they necessarily do work in an actual displacement as with any other acatastatic constraint

$$\sum_{j=1}^{n} \lambda_i A_{ij}\, \mathrm{d}q_j = -\lambda_i A_i\, \mathrm{d}t$$

Since this term is nonzero in any rheonomic or acatastatic system, Jacobi's integral cannot be obtained in these cases.

Finally it must be pointed out that, although the system has only $n-k$ degrees of freedom, the approach described above involves $n+k$ equations for $n+k$ variables. It is found in practice that this approach is lengthy and sometimes difficult to use. However, there is an alternative set of analytic techniques which have been developed by Gibbs and Appell† which are often much easier to use in the analysis of nonholonomic systems.

9.9. Lagrange's equations for impulsive motion

Having obtained the various forms of Lagrange's equations, it is now appropriate to apply this approach to the subject of impulsive motion. The character of impulsive motion has already been discussed in section 6.7 and therefore need not be repeated here. There, and in section 8.13, the equations for such motion were obtained simply by integrating the equations of motion over the instant of the impulse. The same method may be followed here.

It is sufficient here to consider only holonomic systems which are described in terms of n independent generalised coordinates. Since the forces associated with impulsive motion are nonconservative, Lagrange's equations of motion should be considered in the form stated in proposition 9.2.

$$\frac{\mathrm{d}}{\mathrm{d}t}\left(\frac{\partial T}{\partial \dot{q}_i}\right)-\frac{\partial T}{\partial q_i}=Q_i \qquad (i=1,2,\ldots,n)$$

Integrating this over the short time interval from t_1 to t_2 in which the impulse acts gives

$$\left[\frac{\partial T}{\partial \dot{q}_i}\right]_{t_1}^{t_2}-\int_{t_1}^{t_2}\frac{\partial T}{\partial q_i}\,\mathrm{d}t=\int_{t_1}^{t_2}Q_i\,\mathrm{d}t \qquad (i=1,2,\ldots,n)$$

Now the terms $\partial T/\partial q_i$ have been interpreted as the generalised fictitious forces which arise from the choice of coordinates. Clearly, these forces are bounded and are small compared to the impulsive forces over the instant in which the impulses act. Their integral over this instant can therefore be neglected. Thus

$$p_i(t_2)-p_i(t_1)=\int_{t_1}^{t_2}Q_i\,\mathrm{d}t \qquad (i=1,2,\ldots,n)$$

† See L. A. Pars, *A Treatise on Analytical Dynamics* (Heinemann, 1965).

It is now possible to regard the integrals of the generalised force components as the *generalised impulse components.* This can be demonstrated as follows. The work done by the forces in a virtual displacement of the system is given by

$$\delta W = \sum_{i=1}^{N} \boldsymbol{F}_i \cdot \delta \boldsymbol{r}_i = \sum_{i=1}^{n} Q_i \delta q_i$$

According to the theoretical idealisation, the time interval in which the impulses act may be considered to be infinitesimal. The virtual displacement may also be considered to be independent of time and so this equation may be integrated to give

$$\int_{t_1}^{t_2} \delta W \, \mathrm{d}t = \sum_{i=1}^{N} \left(\int_{t_1}^{t_2} \boldsymbol{F}_i \, \mathrm{d}t \right) \cdot \delta \boldsymbol{r}_i = \sum_{i=1}^{n} \left(\int_{t_1}^{t_2} Q_i \, \mathrm{d}t \right) \delta q_i$$

Thus it is possible to define the generalised impulses as follows, using the analogy with definition 9.5.

***Definition* 9.8.** *If $\int_{t_1}^{t_2} \boldsymbol{F}_i \, \mathrm{d}t$ $(i = 1, 2, \ldots, N)$ are the impulses which act on the N particles of a system situated at the points $\boldsymbol{r}_i$ $(i = 1, 2, \ldots, N)$ in the infinitesimal time interval $t_1 \leqslant t \leqslant t_2$, and if the system is described uniquely by n generalised coordinates q_i $(i = 1, 2, \ldots, n)$, then the generalised impulses which correspond to each of the generalised coordinates are defined by the equations*

$$Q_j^* = \int_{t_1}^{t_2} Q_j \, \mathrm{d}t = \sum_{i=1}^{N} \left(\int_{t_1}^{t_2} \boldsymbol{F}_i \, \mathrm{d}t \right) \cdot \frac{\partial \boldsymbol{r}_i}{\partial q_j} \qquad (j = 1, 2, \ldots, n)$$

These components may be obtained simply by evaluating the work done by the impulses in a virtual displacement. When this is expressed in terms of variations of the generalised coordinates, the generalised impulses are then the appropriate coefficients.

Lagrange's equations for impulsive motion may now be stated as follows.

***Proposition* 9.10.** *For a holonomic system described by n independent generalised coordinates, the changes in the generalised momentum components caused by a set of impulses are each equal to the corresponding generalised impulse components.*

$$p_i(t_2) - p_i(t_1) = Q_i^* \qquad (i = 1, 2, \ldots, n)$$

This clearly is a generalisation of previous results for impulsive motion.

In practice, for the impulsive motion of a single particle or a single rigid body, the Lagrangian approach is just as easy to apply as that considered previously. However, the Lagrangian approach, as usual, offers considerable simplification for systems containing several rigid bodies which are constrained in some way. Using this approach the impulsive interactions between the bodies do not need to be considered and the system may be treated as a whole.

The approach introduced briefly here may also be extended to include nonholonomic systems and systems subject to impulsive constraints. The equations for these cases can be derived by first obtaining another alternative form of the fundamental equation.

9.10. Nonholonomic and impulsive motion: examples

Example* 9.5: *the rolling penny

Consider the rolling motion of a thin uniform disc on a rough horizontal plane. It may be assumed that, apart from the gravitational force, the only external forces acting on the disc are the reactional forces from the plane which act at the point of contact. In rolling motion the point of contact is instantaneously stationary, and the forces acting at that point therefore do no work. It follows that the rolling motion considered here is conservative.

It is appropriate here to use Euler's angular coordinates θ, ϕ, ψ to describe the orientation of the disc, and cartesian coordinates x, y, z to define the position of its centroid. θ is the inclination of the disc to the horizontal plane, ϕ defines the angular rotation about the vertical and ψ the angular rotation about the axis of symmetry which is perpendicular to the plane of the disc. These angles are illustrated in figure 9.2.

The six coordinates defined above are clearly not independent. To start with, the fact that the disc rolls on a horizontal plane implies that

$$z = a \sin \theta$$

Fig. 9.2. A disc rolling on a horizontal plane.

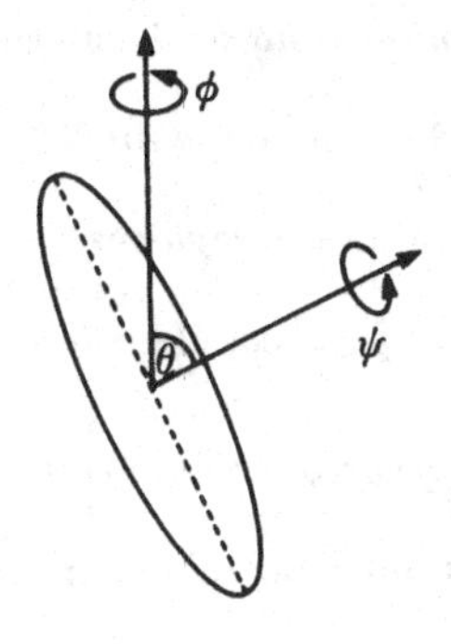

This is a holonomic constraint that can be used to eliminate the parameter z. In terms of the remaining five coordinates, the Lagrangian function for the motion of the disc can be expressed as

$$L=\tfrac{1}{2}m(\dot{x}^2+\dot{y}^2)+\tfrac{1}{8}ma^2(1+4\cos^2\theta)\dot{\theta}^2+\tfrac{1}{8}ma^2\sin^2\theta\dot{\phi}^2$$
$$+\tfrac{1}{4}ma^2(\dot{\psi}+\dot{\phi}\cos\theta)^2-mga\sin\theta$$

It is now necessary to introduce the condition that the disc is rolling rather than sliding. It is convenient to define ϕ as the angle between the vertical plane containing the axis of symmetry of the disc and the point of contact, and the vertical plane $y=0$. By considering the motion of the point of contact it can then be seen that the rolling condition leads to the two equations of constraint

$$\cos\phi\dot{x}+\sin\phi\dot{y}-a\sin\theta\dot{\theta}=0$$
$$\sin\phi\dot{x}-\cos\phi\dot{y}-a\dot{\psi}-a\cos\theta\dot{\phi}=0$$

These equations are clearly nonintegrable and are therefore nonholonomic. The rolling motion of a disc is therefore nonholonomic. It can be described in terms of five coordinates with two nonintegrable constraints, and therefore has three degrees of freedom.

It may immediately be seen that the nonholonomic constraints are linear equations in the generalised velocities, and are therefore of the type described in section 9.8. It may also be seen that the system is catastatic. Labelling the coordinates as $q_1=x$, $q_2=y$, $q_3=\theta$, $q_4=\phi$, $q_5=\psi$, the coefficients in the constraint equations can be taken to be

$$A_{11}=\cos\phi \quad A_{12}=\sin\phi \quad A_{13}=-a\sin\theta \quad A_{14}=0, \quad A_{15}=0$$
$$A_{21}=\sin\phi \quad A_{22}=-\cos\phi \quad A_{23}=0 \quad A_{24}=-a\cos\theta \quad A_{25}=-a$$

It is now necessary to introduce two Lagrange multipliers λ_1 and λ_2. Then, using a slight modification of proposition 9.9 the equations of motion may be considered in the form

$$\frac{\mathrm{d}}{\mathrm{d}t}\left(\frac{\partial L}{\partial \dot{q}_j}\right)-\frac{\partial L}{\partial q_j}=\sum_{i=1}^{2}\lambda_i A_{ij} \qquad (j=1,2,\ldots,5)$$

These can be written out explicitly as

$$m\ddot{x}=\lambda_1\cos\phi+\lambda_2\sin\phi$$
$$m\ddot{y}=\lambda_1\sin\phi-\lambda_2\cos\phi$$

$$\frac{\mathrm{d}}{\mathrm{d}t}\left\{\tfrac{1}{4}ma^2(1+4\cos^2\theta)\dot{\theta}\right\}+ma^2\cos\theta\sin\theta\dot{\theta}^2-\tfrac{1}{4}ma^2\sin\theta\cos\theta\dot{\phi}^2$$
$$+\tfrac{1}{2}ma^2\sin\theta\dot{\theta}(\dot{\psi}+\dot{\phi}\cos\theta)+mga\cos\theta=-\lambda_1 a\sin\theta$$

$$\frac{\mathrm{d}}{\mathrm{d}t}\left\{\tfrac{1}{4}ma^2\sin^2\theta\dot{\phi}+\tfrac{1}{2}ma^2\cos\theta(\dot{\psi}+\dot{\phi}\cos\theta)\right\}=-\lambda_2 a\cos\theta$$

$$\frac{\mathrm{d}}{\mathrm{d}t}\left\{\tfrac{1}{2}ma^2(\dot{\psi}+\dot{\phi}\cos\theta)\right\}=-\lambda_2 a$$

These five equations must now be solved in conjunction with the two equations of constraint. It can thus be seen that, although there are only three degrees of freedom, we have here seven equations for the seven variables x, y, θ, ϕ, ψ, λ_1 and λ_2.

From the first two equations of motion, it can immediately be seen that

$$\lambda_1 = m(\ddot{x}\cos\phi + \ddot{y}\sin\phi)$$
$$\lambda_2 = m(\ddot{x}\sin\phi - \ddot{y}\cos\phi)$$

It may then be observed that these are in fact the components of the reactional force on the disc at the point of contact, in the horizontal directions perpendicular to and along the tangent to the disc at that point.

It is now possible to use the derivatives of the equations of constraint to eliminate the variables x and y. Writing

$$\omega_3 = \dot{\psi} + \dot{\phi}\cos\theta$$

the resulting equations are

$$\lambda_1 = ma(\sin\theta\ddot{\theta} + \cos\theta\dot{\theta}^2 + \omega_3\dot{\phi})$$
$$\lambda_2 = ma(\dot{\omega}_3 - \sin\theta\dot{\theta}\dot{\phi})$$

These expressions may now be substituted into the remaining equations of motion to give

$$5\ddot{\theta} = \sin\theta\cos\theta\dot{\theta}^2 - 6\sin\theta\omega_3\dot{\phi} - 4\frac{g}{a}\cos\theta$$
$$\sin\theta\ddot{\phi} = 2(\omega_3 - \dot{\phi}\cos\theta)\dot{\theta}$$
$$3\dot{\omega}_3 = 2\sin\theta\dot{\theta}\dot{\phi}$$

These equations now describe the motion in terms of the three variables θ, ϕ and ω_3. It is clear, however, that general analytic solutions of these equations do not exist. Numerical methods and approximation techniques are now required in any further analysis of the problem. However the general features of the motion of nonholonomic systems have been demonstrated.

Example 9.6

Consider again example 8.20, in which two rods AB and BC that are smoothly hinged at B are acted on by an impulse P at the end A. This time use the Lagrangian method.

Fig. 9.3. The motion of two hinged rods generated by an impulse P.

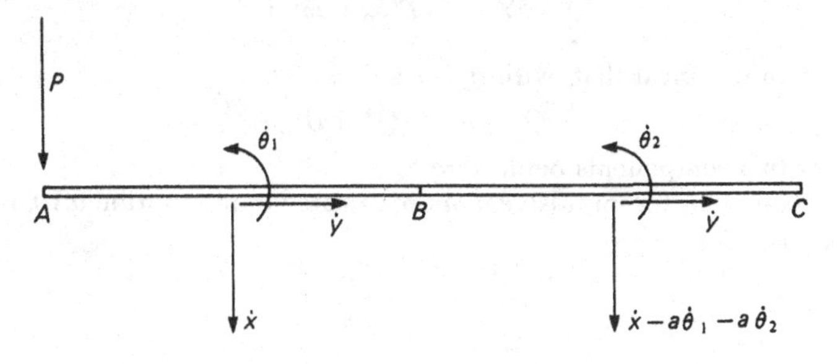

Using the same notation as example 8.20, the initial speed and angular speed gained by two rods are given by v_1 and ω_1 and $v_2 = v_1 - a\omega_1 - a\omega_2$ and ω_2. It is, however, convenient here to replace v_1 by $\dot{x}$, and ω_1 and ω_2 by $\dot{\theta}$ and $\dot{\theta}_2$, as illustrated in figure 9.3.

In their general motion on a horizontal plane, these two rods have four degrees of freedom. These can be described by the parameters x, y, θ_1 and θ_2, where y is the displacement of the centroid of AB in the direction of its initial length. Previously it had been assumed that the impulse was such that $\dot{y}$ remained zero immediately after the action of the impulse. This assumption may be relaxed here to illustrate the method used.

It is, however, only necessary to consider the velocity components of the rods in their configuration at the instant of the impulse. Thus the velocity components of the centroid of BC may be taken to be $\dot{x} - a\dot{\theta}_1 - a\dot{\theta}_2$ and $\dot{y}$. The kinetic energy of the rods in this configuration is given by

$$\begin{aligned} T &= \tfrac{1}{2}m\dot{x}^2 + \tfrac{1}{2}m\dot{y}^2 + \tfrac{1}{6}ma^2\dot{\theta}_1^2 \\ &\quad + \tfrac{1}{2}m(\dot{x} - a\dot{\theta}_1 - a\dot{\theta}_2)^2 + \tfrac{1}{2}m\dot{y}^2 + \tfrac{1}{6}ma^2\dot{\theta}_2^2 \\ &= m(\dot{x}^2 + \dot{y}^2 + \tfrac{2}{3}a^2\dot{\theta}_1^2 + \tfrac{2}{3}a^2\dot{\theta}_2^2 - \dot{x}a\dot{\theta}_1 - \dot{x}a\dot{\theta}_2 + a^2\dot{\theta}_1\dot{\theta}_2) \end{aligned}$$

The momentum components are, therefore,

$$\frac{\partial T}{\partial \dot{x}} = m(2\dot{x} - a\dot{\theta}_1 - a\dot{\theta}_2)$$

$$\frac{\partial T}{\partial \dot{y}} = 2m\dot{y}$$

$$\frac{\partial T}{\partial \dot{\theta}_1} = ma(-\dot{x} + \tfrac{4}{3}a\dot{\theta}_1 + a\dot{\theta}_2)$$

$$\frac{\partial T}{\partial \dot{\theta}_2} = ma(-\dot{x} + a\dot{\theta}_1 + \tfrac{4}{3}a\dot{\theta}_2)$$

These are the expressions for the momentum gained by the action of the impulses, the initial momentum components here being all zero.

To find expressions for the generalised impulse components it is necessary to consider a virtual displacement of the system. In such a displacement the end A is displaced in the direction of the applied impulse by an amount $\delta x + a\delta\theta_1$. The work done by the impulse is therefore

$$\int_{t_1}^{t_2} \delta W \, \mathrm{d}t = P(\delta x + a\delta\theta_1)$$

from which it is clear that, with $q_1 = x$ and $q_3 = \theta_1$

$$Q_1^* = P \qquad Q_3^* = aP$$

the other two components being zero.

The equations for impulsive motion, as stated in proposition 9.10, in this case are, therefore,

$$m(2\dot{x} - a\dot{\theta}_1 - a\dot{\theta}_2) = P$$

$$2m\dot{y} = 0$$

$$ma(-\dot{x}+\tfrac{4}{3}a\dot{\theta}_1+a\dot{\theta}_2)=aP$$
$$ma(-\dot{x}+a\dot{\theta}_1+\tfrac{4}{3}a\dot{\theta}_2)=0$$

From which it is clear that $\dot{y}=0$, as had been anticipated, and

$$\dot{x}=\frac{5}{4}\frac{P}{m}\qquad \dot{\theta}_1=\frac{9}{4}\frac{P}{ma}\qquad \dot{\theta}_2=-\frac{3}{4}\frac{P}{ma}$$

as before.

After a little practice it is found that solutions to problems of this type can be obtained more quickly using the Lagrangian method, since the impulsive interactions between the rods do not have to be taken into account explicitly.

Exercises

9.6 Show that the motion of a thin uniform disc of radius a, which is rotating with angular speed Ω about a vertical diameter on a rough horizontal plane, is stable if $\Omega^2>4g/5a$.

9.7 Show that the motion of a thin uniform disc of radius a, which is rolling with speed V along a rough horizontal plane in a straight line, is stable if $V>\sqrt{(ag/3)}$.

9.8 Repeat exercise 8.33 using the Lagrangian method.

10

Variational principles

The approaches to the subject of classical mechanics considered so far have relied heavily on the mathematical techniques associated with the study of differential equations. Both in the vectorial approach to Newtonian mechanics, and in the analytic approach to Lagrangian dynamics, the motion of a system is ultimately described in a mathematical model in terms of a set of differential equations. Historically, however, the study of differential equations has proceeded in parallel with the study of the calculus of variations. It was thus natural in the development of the subject that the techniques associated with the calculus of variations should also be applied to the problems of classical dynamics. The variational principles of dynamics obtained in this way have in fact always been considered to be of great importance, and they certainly include a number of very interesting results.

The advantage of the variational approach is basically that it considers some property of a system over its entire motion. The aim is to find some integral, taken over the whole motion, which has a stationary value with respect to a certain class of permissible variations. Such a principle enables the motion of a system to be stated in a most economical way without reference to any particular coordinate system. It also enables motion to be considered in a more metaphysical way, and thus facilitates the development of alternative physical theories.

Many variational principles for classical dynamics have been obtained over the centuries. However, only two of the most important are considered in this chapter.

10.1. Hamilton's principle

In Hamilton's principle for a mechanical system, the variations which are permitted are the set of virtual displacements that have been considered in the previous chapter. The positions of the particles of the system at any time may be considered to be displaced by the arbitrary small amounts $\delta \boldsymbol{r}_i$ $(i=1,2,\ldots,N)$ from the positions $\boldsymbol{r}_i$ $(i=1,2,\ldots,N)$ which they would actually occupy in the motion. These virtual displacements are considered to be made without variation of time, but they may themselves vary with time.

It is appropriate to require that the virtual displacements $\delta \boldsymbol{r}_i$ should be continuous and differentiable functions of time. It is then possible to regard the sequence of positions $\boldsymbol{r}_i+\delta \boldsymbol{r}_i$ as the *varied motion* of the system, in the same way that the position vectors $\boldsymbol{r}_i$ as functions of time determine the *actual motion*. Using this approach, the velocity of the particles in the varied motion at any time may be given by the expressions $\dot{\boldsymbol{r}}_i+\delta \dot{\boldsymbol{r}}_i$ $(i=1,2,\ldots,N)$, where

$$\delta \dot{\boldsymbol{r}}_i = \frac{\mathrm{d}}{\mathrm{d}t}\,\delta \boldsymbol{r}_i \qquad (i=1,2,\ldots,N)$$

It should, however, be pointed out that, using this approach, the virtual displacements do not necessarily satisfy the constraints of the system. If the system is nonholonomic, then it is possible to show that the varied motion must be such that the nonholonomic constraints are violated. In this case the varied motion cannot be considered to be a geometrically possible motion. This difficulty does not arise, however, for holonomic systems, and in this case it may generally be assumed that the virtual displacements satisfy the equations of constraint and the varied motion is thus a geometrically possible motion.

It is now possible to consider the kinetic energy of the system in the varied motion. The difference between this and the kinetic energy in the actual motion at the same time may be denoted by δT, and is given by

$$\begin{aligned}\delta T &= \tfrac{1}{2}\sum_{i=1}^{N} m_i\{(\dot{\boldsymbol{r}}_i+\delta\dot{\boldsymbol{r}}_i)^2-\dot{\boldsymbol{r}}_i^2\}\\ &= \sum_{i=1}^{N} m_i\dot{\boldsymbol{r}}_i\cdot\delta\dot{\boldsymbol{r}}_i\end{aligned}$$

to first order. The integral of this over the entire motion from an initial time $t=t_0$ to a final time $t=t_1$ is thus given by

$$\int_{t_0}^{t_1}\delta T\,\mathrm{d}t = \left[\sum_{i=1}^{N} m_i\dot{\boldsymbol{r}}_i\cdot\delta\boldsymbol{r}_i\right]_{t_0}^{t_1} - \int_{t_0}^{t_1}\sum_{i=1}^{N} m_i\ddot{\boldsymbol{r}}_i\cdot\delta\boldsymbol{r}_i\,\mathrm{d}t$$

where the right-hand side has been obtained from the above expression using the technique of integration by parts.

It is now convenient to introduce the assumption that the virtual displacement of the system is zero at both the initial and final times. Thus

$$\delta \boldsymbol{r}_i = 0 \quad \text{at } t = t_0 \quad \text{and} \quad t = t_1 \qquad (i = 1, 2, \ldots, N)$$

This implies that the varied motion of the system is between the same initial and final configurations at the same times as in the actual motion. In this case, the first term in the above expression is zero.

The remaining term may now also be simplified, either by inserting the equation of motion to be satisfied by each particle

$$m_i \ddot{\boldsymbol{r}}_i = \boldsymbol{F}_i \qquad (i = 1, 2, \ldots, N)$$

where $\boldsymbol{F}_i$ $(i = 1, 2, \ldots, N)$ are the resultant forces, or by using the fundamental equation in the form stated in proposition 9.6. Using either approach, the above equation may now be expressed in the form

$$\int_{t_0}^{t_1} \left\{ \delta T + \sum_{i=1}^{N} \boldsymbol{F}_i \cdot \delta \boldsymbol{r}_i \right\} \mathrm{d}t = 0$$

This is Hamilton's principle for a general dynamical system.

It can clearly be seen that it is only the forces which do work in a virtual displacement which need to be included in the above equation. If in addition those forces are conservative, or at least are derivable from a potential function, then it is possible to introduce a potential function V such that

$$\delta V = -\sum_{i=1}^{N} \boldsymbol{F}_i \cdot \delta \boldsymbol{r}_i$$

In this case it is convenient to introduce the Lagrangian function

$$L = T - V$$

and it can now be seen that Hamilton's principle for such a system can be expressed in the form

$$\int_{t_0}^{t_1} \delta L \, \mathrm{d}t = 0$$

Further, if the system is also holonomic so that only geometrically possible motions need to be considered, then it is possible to rewrite the above equation as

$$\delta \int_{t_0}^{t_1} L \, \mathrm{d}t = 0$$

This is the usual form for Hamilton's principle, which may now be stated in the following way.

***Proposition* 10.1.** *For a holonomic system whose motion can be described by a Lagrangian function L, the integral* $\int_{t_0}^{t_1} L\,dt$ *has a stationary value when compared with neighbouring geometrically possible motions having the same configurations at the times* t_0 *and* t_1.

In the derivation of the above result it has only been shown that the integral of the Lagrangian has a stationary value when compared with neighbouring motions. In most applications, however, it is usually found that this stationary value is a minimum, although exceptions do occur. It is therefore important to emphasise that Hamilton's principle is not always a strict minimum principle.

It can also be seen from the derivation above that Hamilton's principle is essentially an integrated form of the fundamental equation and, like the fundamental equation, it can be expressed in various forms. It is clearly a consequence of the equations of motion. However, it will be shown in the next section that the equation of motion can also be derived from it. It is therefore also a sufficient condition from which all possible motions of the system can be determined.

An important feature of Hamilton's principle, as stated in proposition 10.1, is that it is stated without reference to any particular coordinate system. This is of particular importance when one comes to generalise the theory of classical dynamics to include the theories of relativity. In these theories it is necessary to state all physical laws in a coordinate independent way and, as has been pointed out, Hamilton's principle, which describes the dynamical motion of a classical system, is already in this form.

Finally, it must be pointed out that, although Hamilton's principle is of considerable theoretical interest, it is of little value in practical applications.

10.2. Deductions from Hamilton's principle

The main purpose of this section is to show how the equations of motion can be deduced from Hamilton's principle, using the techniques of the calculus of variations. Although these can be obtained in total generality, it is sufficient here to consider only the case in which the forces which do work in a virtual displacement are derivable from a potential function. In this case, a Lagrangian function for the system can be obtained, and Hamilton's principle can be expressed in the form

$$\int_{t_0}^{t_1} \delta L \, \mathrm{d}t = 0$$

which is applicable to both holonomic and nonholonomic systems.

Since a Lagrangian function has been introduced it is appropriate to assume that the system is described in terms of n generalised coordinates. The virtual displacements of the system that are considered in Hamilton's principle can then be expressed in terms of the virtual variations δq_i $(i=1, 2, \ldots, n)$ which are now assumed to be continuous and differentiable functions of time. This reduction is possible because the virtual displacements may be considered to satisfy the holonomic constraints, and it is these constraints that it may be assumed have been used to reduce the number of generalised coordinates from $3N$ to n. However, it may still be assumed at this stage that the system is also required to satisfy a number of nonholonomic constraints. In this case, there are less than n degrees of freedom, and the varied motion does not satisfy the nonholonomic constraints.

Now, since the variation considered compares points on the varied and actual motions at the same time, Hamilton's principle can be stated as

$$\int_{t_0}^{t_1} \sum_{i=1}^{n} \left\{ \frac{\partial L}{\partial q_i} \delta q_i + \frac{\partial L}{\partial \dot{q}_i} \delta \dot{q}_i \right\} \mathrm{d}t = 0$$

It also follows from the fact that the variation is contemporaneous that

$$\delta \dot{q}_i = \frac{\mathrm{d}}{\mathrm{d}t} \delta q_i \qquad (i=1, 2, \ldots, n)$$

and, therefore, the second set of terms in the above equation can be integrated by parts and the equation takes the form

$$\left[\sum_{i=1}^{n} \frac{\partial L}{\partial \dot{q}_i} \delta q_i \right]_{t_0}^{t_1} - \int_{t_0}^{t_1} \sum_{i=1}^{n} \left\{ \frac{\mathrm{d}}{\mathrm{d}t} \left(\frac{\partial L}{\partial \dot{q}_i} \right) - \frac{\partial L}{\partial q_i} \right\} \delta q_i \, \mathrm{d}t = 0$$

One of the conditions of Hamilton's principle, however, is that there is no displacement of the system at the times t_0 and t_1. Thus Hamilton's principle for a system with a Lagrangian function is equivalent to the equation

$$\int_{t_0}^{t_1} \sum_{i=1}^{n} \left\{ \frac{\mathrm{d}}{\mathrm{d}t} \left(\frac{\partial L}{\partial \dot{q}_i} \right) - \frac{\partial L}{\partial q_i} \right\} \delta q_i \, \mathrm{d}t = 0$$

In the case of a holonomic system having n degrees of freedom the virtual variations are independent. In this case all the terms in the integrand must be zero and this condition simply gives Lagrange's equations of motion in the form stated in proposition 9.3.

$$\frac{\mathrm{d}}{\mathrm{d}t}\left(\frac{\partial L}{\partial \dot{q}_i}\right)-\frac{\partial L}{\partial q_i}=0 \qquad (i=1,2,\ldots,n)$$

Lagrange's equations for nonconservative and nonholonomic systems can also be obtained in a similar way.

These results show that Hamilton's principle stated in an appropriate form is both a necessary and a sufficient condition for the possible motions of a dynamical system. It thus provides an alternative approach to the subject, which contrasts with the more familiar approaches that are stated in terms of differential equations. However it must be admitted that in practical problems a direct application of Hamilton's principle is not as convenient as the approaches which use equations of motion.

The alternative approach using Hamilton's principle is, however, of considerable theoretical interest. In fact, it enables the whole subject to be considered in a new way.

It is possible to regard the Lagrangian function for a system as being of fundamental importance. The possible motions of the system can be obtained from it either by using Lagrange's equations, or by using Hamilton's principle. It is thus possible to consider alternative or more general theories simply by changing the form of the Lagrangian function. In fact, it is known that the theories of electrodynamics, quantum mechanics and general relativity can all be expressed in terms of appropriate Lagrangian type functions whose integrals have stationary values with respect to virtual variations which are zero at the end points. Thus, by changing the form of the Lagrangian function, new theories can be put forward. And since one of the basic requirements of relativistic theories is that they should be stated in a coordinate independent way, generalised forms of Hamilton's principle are particularly important. This technique has featured significantly in theoretical research in recent years.

It is interesting to compare such an approach with that which regards the concept of energy as being of fundamental importance. Both are based upon results which initially apply strictly to the subject of classical dynamics. On the one hand, it is the fact that the sum of the kinetic energy of a system and its potential energy is sometimes a constant, which is considered to be of fundamental significance. And on the other hand, it is the fact that the integral of the Lagrangian has a stationary value when compared with neighbouring possible motions, which is considered to have physical significance. Both approaches can be generalised to include concepts in other theories, and both have been used in the development of new theories. However, both approaches are essentially metaphysical and are certainly not implied by the results of classical dynamics.

10.3. The principle of least action

A different type of variational principle may now be considered which is based upon a different class of variations. In Hamilton's principle the variations considered were the class of virtual displacements which compare points on the actual and varied motions at the same time. In addition, it was required that the initial and final configurations should be the same at the same times. The innovation introduced here enables points on the varied motion to be compared with corresponding points on the actual motion that do not necessarily occur at the same time.

With the introduction of variations in time as well as position a new notation is called for. It may be assumed that a system is described in terms of n generalised coordinates q_i $(i=1,2,\ldots,n)$ which are not necessarily independent. Then, when a point on the varied motion is compared with the corresponding point on the actual motion, the difference in the coordinate values may be denoted by Δq_i $(i=1,2,\ldots,n)$ and the difference in time by Δt. These variations can be related to the coordinate differences between the motions at the same time according to the equations

$$\Delta q_i = \delta q_i + \dot{q}_i \Delta t \qquad (i=1,2,\ldots n)$$

It is convenient to refer to Δ variations in which corresponding points on neighbouring motions are compared, where the correspondence permits differences in time as well as position. These may be contrasted with δ variations in which points on neighbouring motions are compared at the same time.

Attention may now be restricted to conservative dynamical systems. In this case a potential function exists and hence a Lagrangian function can be obtained.

The integral of the Lagrangian function for the system may now be considered between an initial time t_0 and a final time t_1. This can be compared with its integral along a neighbouring motion in which variations of the configuration and time are permitted even at the end points. If the variations are small it is possible to show that

$$\Delta\int_{t_0}^{t_1} L\,\mathrm{d}t=\int_{t_0}^{t_1}\delta L\,\mathrm{d}t+\Big[L\Delta t\Big]_{t_0}^{t_1}$$

It should be noticed that Hamilton's principle does not imply that the integral of δL vanishes, since in this case the variation of the end points in nonzero. However, it has already been shown in the previous section that

$$\int_{t_0}^{t_1}\delta L\,\mathrm{d}t=\left[\sum_{i=1}^{n}\frac{\partial L}{\partial\dot q_i}\delta q_i\right]_{t_0}^{t_1}-\int_{t_0}^{t_1}\sum_{i=1}^{n}\left\{\frac{\mathrm{d}}{\mathrm{d}t}\left(\frac{\partial L}{\partial\dot q_i}\right)-\frac{\partial L}{\partial q_i}\right\}\delta q_i\,\mathrm{d}t$$

It can immediately be seen that the integral on the right-hand side must be zero since, if the system is holonomic, Lagrange's equations of motion imply that all components of the integrand are zero. On the other hand, even for a nonholonomic system, the fundamental equation, as stated in proposition 9.7, immediately implies that the integrand is zero. Thus, as a consequence of the equations of motion, the above identity can be written as

$$\Delta\int_{t_0}^{t_1} L\,\mathrm{d}t=\left[\sum_{i=1}^{n}\frac{\partial L}{\partial\dot q_i}\delta q_i+L\Delta t\right]_{t_0}^{t_1}$$

Alternatively, this can be expressed in terms of Δ variations in the form

$$\Delta\int_{t_0}^{t_1} L\,\mathrm{d}t=\left[\sum_{i=1}^{n}\frac{\partial L}{\partial\dot q_i}\Delta q_i-\left(\sum_{i=1}^{n}\dot q_i\frac{\partial L}{\partial\dot q_i}-L\right)\Delta t\right]_{t_0}^{t_1}$$

To obtain the principle of least action, three further conditions must now be applied. It is first assumed that there are no variations in the positions of the system at the initial and final times. Thus

$$\Delta q_i=0\quad\text{at } t=t_0\quad\text{and}\quad t=t_1\qquad(i=1,2,\ldots,n)$$

This implies that the actual and varied motions are between the same initial and final configurations, although the times at which they are in these states may be varied slightly. With this assumption the above expression is considerably simplified.

The second set of assumptions to be made are that the Lagrangian function should not depend on time explicitly, and that all the remaining constraints should be catastatic. Under these conditions Jacobi's integral, which is the coefficient of Δt in the above expression, is a constant of the motion which may be interpreted as the constant total energy of the system.

Finally, it is assumed that the varied motion is such that the total energy of the system remains unaltered. In this case, Jacobi's integral has the same constant value in both the varied and the actual motions.

Under this assumption

$$\left[\left(\sum_{i=1}^{n} \dot{q}_i \frac{\partial L}{\partial \dot{q}_i} - L\right)\Delta t\right]_{t_0}^{t_1} = \Delta \int_{t_0}^{t_1} \left(\sum_{i=1}^{n} \dot{q}_i \frac{\partial L}{\partial \dot{q}_i} - L\right) \mathrm{d}t$$

Under this set of conditions, the above equation can be reduced to the form

$$\Delta \int_{t_0}^{t_1} \sum_{i=1}^{n} \dot{q}_i \frac{\partial L}{\partial \dot{q}_i} \mathrm{d}t = 0$$

This result may now be stated formally after first defining the term 'action'.

***Definition* 10.1.** *For a system which can be described by a Lagrangian L which is a function of n generalised coordinates $q_1, q_2, \ldots, q_n$ and their corresponding generalised velocities, but which does not depend on time explicitly, the action over the period $t_0 \leq t \leq t_1$ is defined to be the integral* $\int_{t_0}^{t_1} \sum_{i=1}^{n} \dot{q}_i(\partial L/\partial \dot{q}_i)\,\mathrm{d}t$

It may immediately be observed that, if the kinetic energy of the system is expressed as a homogeneous quadratic in the generalised velocities, then the action is simply twice the integral of the kinetic energy.

The above result can now be stated as follows.

***Proposition* 10.2.** *For a system which can be described by a Lagrangian function which does not depend on time explicitly, the action has a stationary value when compared with neighbouring motions between the same initial and final configurations in which Jacobi's integral has the same constant value.*

This result is known as the *principle of least action* and is frequently associated with the names of Maupertuis and Euler. However, as with Hamilton's principle, it has only been demonstrated to be a stationary principle, but in applications to situations in which the above conditions are satisfied it is in fact usually found that the action has a minimum value. However, exceptions do occur and even a local minimum is not necessarily the least possible value.

Although it will not be demonstrated here, it should still be pointed out that, like Hamilton's principle, the principle of least action is also a sufficient condition for the motion of the system. It is therefore possible to derive equations of motion from it, although the technique involved is more complicated in this case.

Finally, it is necessary to mention that a number of generalisations of the principle of least action have also been obtained. The most significant of these was obtained by Jacobi and essentially enables the last of the conditions imposed above to be relaxed. By expressing the action in a slightly different form, it permits the actual motion to be compared with a varied motion in which Jacobi's integral has a different value. It is, however, still required that Jacobi's integral should be a constant in each motion. In addition to being of greater generality, Jacobi's form of the principle of least action enables the motion to be represented in a neat mathematical way in terms of geodesics in the configuration space.

11

Hamilton–Jacobi theory

Up until this point the theory of classical dynamics has been described more or less in the form in which it had been developed by the end of the eighteenth century. The analytic approach of Lagrangian dynamics had been developed, as well as the vectorial approach of Newtonian mechanics. In practice, either of these approaches may be applied to a large class of practical problems, but ultimately they both yield sets of second-order differential equations which it is necessary to solve. Sometimes it is possible to obtain some simple first integrals of these equations, but in many situations complete integrals in analytic form are difficult or impossible to obtain. Thus, although these approaches enable the motion of many systems to be described in terms of equations of motion, certain mathematical problems prevent their complete analysis.

Because of this situation the development of the theory of classical dynamics has always been associated with the development of appropriate mathematical techniques. In particular, some of the techniques developed have enabled the equations of motion to be formulated in new ways, and thus they have contributed to a deeper understanding of the theory itself. Such results must therefore be included in a discussion of the foundations of the subject.

It is the purpose of this chapter to describe some of the alternative approaches to the subject which were developed in the nineteenth century. Of particular interest are the Hamiltonian approach and the development of the Hamilton–Jacobi theory. Many other important advances of this period, however, such as the development of phase space analysis and the theory of canonical transformations, are of a mainly mathematical character and are therefore not discussed here apart from an occasional passing comment.

The new approaches to be described here can be applied to general dynamical systems, but for the sake of simplicity attention is restricted in this chapter to holonomic systems which can be described in terms of n independent generalised coordinates, and in which all the generalised force components are derivable from a potential function. Such a system can be described by a Lagrangian function

$$L = L(q_1, \ldots, q_n, \dot{q}_1, \ldots, \dot{q}_n, t)$$

and must satisfy the equations of motion which can be written in the Lagrangian form

$$\frac{\mathrm{d}}{\mathrm{d}t}\left(\frac{\partial L}{\partial \dot{q}_i}\right) - \frac{\partial L}{\partial q_i} = 0 \qquad (i = 1, 2, \ldots, n)$$

11.1. Hamilton's equations of motion

The Hamiltonian approach can be considered as a further development of analytical dynamics. It is therefore appropriate to initially consider a system which is described in terms of its Lagrangian function.

Of fundamental importance in the Hamiltonian approach are the generalised momentum components which have been defined by the equations

$$p_1 = \frac{\partial L}{\partial \dot{q}_i} \qquad (i = 1, 2, \ldots, n)$$

It has previously been emphasised that the Lagrangian is a quadratic function of the generalised velocities, and so the generalised momentum components are linear in the velocities. However, since the system is here assumed to be holonomic, the generalised velocities can also be expressed as linear combinations of the generalised momentum components. Thus, in any expression involving the dynamical variables, it is possible to eliminate either the generalised velocities or the generalised momentum components, and thus to work with only one of these sets of quantities. In the Lagrangian approach the generalised velocities are always used. However, in the Hamiltonian approach these are always eliminated in favour of the generalised momentum components.

The basic approach to these variables may now be stated formally.

Notation. *In the Hamiltonian formalism the generalised coordinates* $q_1, q_2, \ldots, q_n$, *the generalised momentum components* $p_1, p_2, \ldots, p_n$ *and time* t *are considered to be independent variables to be connected only through the equations of motion.*

It is important to recognise that this approach differs significantly from that of Lagrange. In the Lagrangian formalism the generalised coordinates and velocities are regarded as independent variables only for purposes of partial differentiation, and it is always recognised that the generalised velocities are the derivatives of the coordinates. Thus, since their derivatives appear in the equations of motion, these become second-order differential equations in the coordinates. In direct contrast to this the generalised coordinates and momentum components are regarded as totally independent variables in the Hamiltonian approach. Thus, the Hamiltonian formalism involves twice as many basic parameters as the Lagrangian and the derivatives of these, which appear in the equations of motion, yield only first-order differential equations.

It is now convenient to introduce a new function which is known as the *Hamiltonian.*

***Definition* 11.1.** *The Hamiltonian function H is defined such that*

$$H = \sum_{i=1}^{n} \dot{q}_i p_i - L$$

where L is the Lagrangian, but in which the generalised velocity components are eliminated in favour of the generalised momentum components so that

$$H = H(q_1, \ldots, q_n, p_1, \ldots, p_n, t)$$

It can be seen that, although the initial definition of the Hamiltonian contains the generalised velocities explicitly, since the system is holonomic it is always possible to substitute for these to obtain the Hamiltonian in terms of the correct variables.

It has been shown previously that all the dynamical properties of a system can be summarised in the single statement of its Lagrangian. The Hamiltonian is a similar descriptive function. Once it is given for any system, the equations for the motion can immediately be obtained from a standard form which will be obtained below. However, the initial definition of the Hamiltonian function should be recognised as being identical to the form of Jacobi's integral, as stated in proposition 9.5. It can therefore sometimes be interpreted as the expression for the energy of the system. In particular, if the kinetic energy of the system is a homogeneous quadratic in the generalised velocities which does not depend on time, then

$$H = T + V$$

If, in addition, the potential function does not depend explicitly on time either, then the Hamiltonian is a constant of the motion. This result will be obtained again later using the Hamiltonian formalism consistently.

The equations of motion in Hamiltonian form can now be obtained by considering the differential derived from the above definition

$$\mathrm{d}H=\sum_{i=1}^{n}\dot{q}_i\mathrm{d}p_i+\sum_{i=1}^{n}p_i\,\mathrm{d}\dot{q}_i-\sum_{i=1}^{n}\frac{\partial L}{\partial q_i}\mathrm{d}q_i-\sum_{i=1}^{n}\frac{\partial L}{\partial \dot{q}_i}\mathrm{d}\dot{q}_i-\frac{\partial L}{\partial t}\mathrm{d}t$$

The components involving $\mathrm{d}\dot{q}_i$ clearly cancel and, using the equations of motion in Lagrangian form, this can be written as

$$\mathrm{d}H=\sum_{i=1}^{n}\dot{q}_i\,\mathrm{d}p_i-\sum_{i=1}^{n}\dot{p}_i\,\mathrm{d}q_i-\frac{\partial L}{\partial t}\mathrm{d}t$$

This differential now involves the correct variables and it therefore follows that

$$\dot{q}_i=\frac{\partial H}{\partial p_i}\qquad \dot{p}_i=-\frac{\partial H}{\partial q_i}\qquad \frac{\partial H}{\partial t}=-\frac{\partial L}{\partial t}\qquad (i=1,2,\ldots,n)$$

The last of these identities is of little practical value since the functions H and L depend on different sets of variables. However it can be seen that if the Lagrangian does not depend on time explicitly, then neither does the Hamiltonian, and in this case the Hamiltonian is a constant of the motion. The remaining identities are in fact *Hamilton's equations of motion.* These equations are clearly independent and have been obtained using all n of Lagrange's equations. They are therefore sufficient to describe the motion of the system.

***Proposition* 11.1.** *The dynamical behaviour of a holonomic system with n degrees of freedom may be determined completely by Hamilton's equations of motion in the form*

$$\dot{q}_i=\frac{\partial H}{\partial p_i}\qquad \dot{p}_i=-\frac{\partial H}{\partial q_i}\qquad (i=1,2,\ldots,n)$$

where H is the Hamiltonian function for the system.

Hamilton's equations can immediately be interpreted in the following way. The terms $-\partial H/\partial q_i$ may be considered to be the generalised force components which include the fictitious forces associated with the choice of coordinates. These are exactly equivalent to the terms $\partial L/\partial q_i$ in the Lagrangian formalism. The equations $\dot{p}_i=-\partial H/\partial q_i$ therefore simply state that the rate of change of the generalised momentum components are equal to the corresponding generalised force components. The remaining equations $\dot{q}_i = \partial H/\partial p_i$ simply express the generalised velocities as linear combinations of the generalised momentum components. However, since the Hamiltonian approach does not consider the generalised velocities as such, these are regarded as the necessary equations for the rates of change of the generalised coordinates.

It can thus be seen that Hamilton's equations are a set of $2n$ first-order differential equations. This is in contrast to Lagrange's equations which are a set of n second-order differential equations. In fact such a reduction of the equations of motion to first order does give considerable mathematical advantage, even though it requires twice as many equations.

In a strictly Lagrangian approach the solutions of the equations of motion can be analysed as trajectories in the n-dimensional configuration space spanned by the generalised coordinates. In contrast, the Hamiltonian approach enables the solutions to be analysed as paths in the $2n$-dimensional phase space spanned by both the generalised coordinates and the momentum components. Such an approach is extremely useful in that it facilitates the qualitative description of a system, even in cases where exact analytic solutions cannot be obtained. The mathematical techniques associated with this approach are now well known and phase space analysis has been developed into a general qualitative theory of ordinary differential equations. Such an approach is particularly well suited to an analysis of the stability of particular solutions of the equations of motion. However, the mathematical techniques of this theory can now also be applied to Lagrange's equations of motion for a particular system by first formally rewriting them as a set of $2n$ first-order equations. Thus the distinctive approach of Hamiltonian theory is not necessary for a generalised phase space analysis, and in practice it is found that Hamilton's equations have little advantage over those of Lagrange when applied to particular problems of classical mechanics.

It may be recalled that Lagrange's equations were formulated in a coordinate independent way. Thus if one set of generalised coordinates is transformed into another set, the form of Lagrange's equations remains the same. A similar result is also true of Hamilton's equations. However, in this case twice as many basic parameters are used and so a larger class of transformations may be considered. It is found that, if the transformations between one set of parameters and another satisfies a certain condition, then the form of Hamilton's equations remains the same. Such transformations are referred to as *contact* or *canonical* transformations. They may thus alternatively be defined as that group of transformations under which the form of Hamilton's equations remains invariant. It may be noticed that under such transformations the coordinates and momentum components do not necessarily remain distinct. For example, a generalised coordinate can assume the role of a momentum component or vice versa. It is not necessary to describe the mathematical techniques in detail here, but it must be pointed out that the application of transformation theory is of considerable theoretical interest in the development of classical dynamics, and that it also has a number of important practical applications. However, it is again found that the practical applications are not dependent upon the Hamiltonian approach.

It must thus be realised that Hamilton's equations are little used in treating the practical problems of classical mechanics. The importance of the Hamiltonian approach, however, lies rather in the fact that it introduces a new conceptual framework. The basic approach of regarding the coordinates and momentum components as independent variables, as well as the introduction of the Hamiltonian function, are important in the development of classical dynamics and are foundational in the modern theory of quantum mechanics. Thus the Hamiltonian approach to classical mechanics also forms a convenient base for the study of quantum mechanics and statistical mechanics. In fact, a general Hamiltonian theory may be developed which includes a large number of traditionally distinct theories.

11.2. Integrals of Hamilton's equations

It has frequently been emphasised that, whenever possible, it is important to obtain first integrals of the equations of motion. These can in fact be obtained at least as easily in the Hamiltonian formalism as in any other. For example, it can immediately be seen from the equation $\dot{p}_i = \partial H/\partial q_i$ that if the Hamiltonian function does not depend explicitly on a particular coordinate, then that coordinate is ignorable and the corresponding momentum component is a constant of the motion.

In the Hamiltonian approach it is also possible, at least in principle, to use a contact transformation in the variables in such a way that the equations of motion are transformed into forms which can immediately be integrated. It is in fact even possible to make a transformation such that the Hamiltonian function takes a form which does not depend explicitly, either on any of the coordinates, or on any of the momentum components, and therefore these must all be constants. However, it is not obvious how such a transformation may be chosen without first obtaining explicit solutions of the equations of motion. An attempt to solve this problem leads directly to the Hamilton–Jacobi equation which is introduced later using a different approach.

An energy integral, when one exists, can also easily be obtained in this case, particularly since the Hamiltonian function is formally identical to Jacobi's integral. However, it is convenient here to proceed immediately to obtain a more general result.

Consider any function of the Hamiltonian variables

$$F = F(q_1, \ldots, q_n, p_1, \ldots, p_n, t)$$

Its total derivative is

$$\frac{\mathrm{d}F}{\mathrm{d}t} = \sum_{i=1}^{n} \left(\frac{\partial F}{\partial q_i} \dot{q}_i + \frac{\partial F}{\partial p_i} \dot{p}_i \right) + \frac{\partial F}{\partial t}$$

and using Hamilton's equations, this can be written in the form

$$\frac{\mathrm{d}F}{\mathrm{d}t} = [F, H] + \frac{\partial F}{\partial t}$$

where the Poisson bracket of the function F with the Hamiltonian has been inserted according to the definition.

***Definition* 11.2.** *The Poisson bracket of two functions* A *and* B *of the generalised coordinates, the generalised momentum components and time is given by the equation*

$$[A, B] = \sum_{i=1}^{n} \left(\frac{\partial A}{\partial q_i} \frac{\partial B}{\partial p_i} - \frac{\partial A}{\partial p_i} \frac{\partial B}{\partial q_i} \right)$$

In the case when F does not depend on time explicitly the following result can immediately be deduced.

***Proposition* 11.2.** *Any function of the generalised coordinates and generalised momentum components whose Poisson bracket with the Hamiltonian function for a system is zero is a constant of the motion.*

Since the Poisson bracket of any function with itself is clearly zero, it follows immediately that, if the Hamiltonian function does not depend on time explicitly, then it is a constant of the motion.

This result is essentially the same as that stated in proposition 9.5 and therefore can similarly be interpreted in terms of energy.

It can be shown using standard mathematical results that Poisson brackets can also be used to generate further constants of motion, although this approach should not be pursued indefinitely as the constants generated are not necessarily independent.

It can also be shown that Poisson brackets are invariant with respect to contact transformations. In fact, contact transformations can alternatively be defined in terms of the set of fundamental Poisson bracket relations

$$[q_i, q_j]=0 \qquad [p_i, p_j]=0 \qquad [q_i, p_j]=\delta_{ij} \qquad (i, j=1, 2, \ldots, n)$$

Finally it may also be pointed out that the equations of motion can conveniently be written in terms of Poisson brackets in the form

$$\dot{q}_i=[q_i, H] \qquad \dot{p}_i=[p_i, H] \qquad (i=1, 2, \ldots, n)$$

11.3. The Hamiltonian approach: examples

Example **11.1:** *the spinning top*

Consider again the motion of a top. This has already been discussed using a vectorial approach in example 8.20, and again using the Lagrangian approach in example 9.2. The distinctive features of the Hamiltonian approach can therefore be illustrated by considering this situation yet again.

Using the same notation, the kinetic energy of the system is

$$T=\tfrac{1}{2}I_1\dot{\theta}^2+\tfrac{1}{2}I_1\sin^2\theta\dot{\phi}^2+\tfrac{1}{2}I_3(\dot{\psi}+\dot{\phi}\cos\theta)^2$$

The momentum components corresponding to each of the generalised coordinates are therefore

$$p_\theta=I_1\dot{\theta}$$
$$p_\phi=I_1\sin^2\theta\dot{\phi}+I_3\cos\theta(\dot{\psi}+\dot{\phi}\cos\theta)$$
$$p_\psi=I_3(\dot{\psi}+\dot{\phi}\cos\theta)$$

The potential energy is

$$V=mgh\cos\theta$$

Now, since the kinetic energy is a homogeneous quadratic in the velocities, it is clear that in this case the Jacobi integral, and hence the Hamiltonian function, is formally identical to $T+V$. To obtain the Hamiltonian, however, it is necessary to express this in terms of the generalised momentum components. The result is

$$H=\frac{1}{2I_1}p_\theta^2+\frac{1}{2I_1\sin^2\theta}(p_\phi-\cos\theta p_\psi)^2+\frac{1}{2I_3}p_\psi^2+mgh\cos\theta$$

Substituting this directly into Hamilton's equations, as stated in proposition 11.1, gives the equations of motion in the form

$$\dot{\theta} = \frac{1}{I_1} p_\theta$$

$$\dot{\phi} = \frac{1}{I_1 \sin^2 \theta} (p_\phi - \cos \theta p_\psi)$$

$$\dot{\psi} = \frac{1}{I_3} p_\psi - \frac{\cos \theta}{I_1 \sin^2 \theta} (p_\phi - \cos \theta p_\psi)$$

$$\dot{p}_\theta = \frac{1}{I_1 \sin^3 \theta} (p_\phi - \cos \theta p_\psi)(\cos \theta p_\phi - p_\psi) + mgh \sin \theta$$

$$\dot{p}_\phi = 0$$

$$\dot{p}_\psi = 0$$

From these it is immediately clear that p_ϕ and p_ψ are constant integrals of momentum.

These equations, however, are no easier to solve than those obtained by other methods, and after a little manipulation they may be shown to be equivalent. It is only the Hamiltonian approach which is illustrated in this example.

Example 11.2: *angular momentum Poisson bracket relations*

Consider motion of a particle which is free to move in space, relative to a set of fixed cartesian axes. Clearly, the linear momentum components are

$$p_x = m\dot{x} \qquad p_y = m\dot{y} \qquad p_z = m\dot{z}$$

Also the angular momentum of the particle about the origin is given by

$$\boldsymbol{L} = \boldsymbol{r} \times m\boldsymbol{v}$$

and this has components

$$L_x = yp_z - zp_y, \qquad L_y = zp_x - xp_z, \qquad L_z = xp_y - yp_x$$

It can now be shown that these components satisfy the Poisson bracket relation

$$[L_x, L_y] = L_z$$

This may be demonstrated either by substituting the components into the expansion used to define the Poisson bracket, or by using the algebraic properties of Poisson brackets and the fundamental Poisson bracket identities. It may also be observed that it is possible to cyclicly permute the coordinates in the above identity.

Further identities of this type are indicated in exercise 11.4.

Exercises

11.1 Consider the motion of a particle relative to a set of cartesian axes which are rotating with constant angular speed ω about the z axis. If the particle is subject only to a force which is derivable from a potential function $V(x, y, z)$ show that the Hamiltonian function is

$$H=\frac{1}{m}(p_x^2+p_y^2+p_z^2)+\omega(yp_x-xp_y)+V$$

Show also that $\frac{1}{2}mv^2-\frac{1}{2}m\omega^2(x^2+y^2)+V$ is a constant of the motion, v being the speed of the particle relative to the rotating coordinates.

11.2 Show that, for any function of Hamiltonian variables $F = F(q_1, \ldots, q_n, p_1, \ldots, p_n, t)$

$$\frac{\partial F}{\partial q_i}=-[p_i, F] \qquad \frac{\partial F}{\partial p_i}=[q_i, F] \qquad (i=1, 2, \ldots, n)$$

11.3 For any functions A, B and C of Hamiltonian variables, derive the following algebraic properties of Poisson brackets

$$[A, B]=-[B, A]$$
$$[A+B, C]=[A, C]+[B, C]$$
$$[AB, C]=[A, C]B+A[B, C]$$

11.4 For the motion of a particle relative to a fixed set of cartesian axes, show that the linear momentum components and the components of angular momentum about the origin satisfy the Poisson bracket relations

$$[L_x, L_y]=L_z \qquad [L_x, L_z]=-L_y$$
$$[x, L_y]=z \qquad [x, L_z]=-y \qquad [x, L_x]=0$$
$$[p_x, L_y]=p_z \qquad [p_x, L_z]=-p_y \qquad [p_x, L_x]=0$$
$$[L_x, L^2]=0$$

11.4. Hamilton's principal function

It is convenient initially to introduce Hamilton's principal function S as the integral of the Lagrangian function along the motion: $S=\int_{t_0}^{t_1} L\,dt$. This may immediately be recognised as the function which has featured prominently in the variational principles of classical mechanics, such as proposition 10.1. However, in this case its functional dependence needs to be specified more precisely.

The Lagrangian is in general a function of the generalised coordinates, the velocities and also possibly time, and the coordinates and velocities themselves for any particular solution are functions of time and a set of $2n+1$ parameters which specify that particular solution. These parameters may, for example, be the set of initial positions and velocities and the initial time. Using an obvious notation, the generalised coordinates can be expressed in this case in the form

$$q_i=q_i(q_{10}, \ldots, q_{n0}, \dot{q}_{10}, \ldots, \dot{q}_{n0}, t_0, t) \qquad (i=1, 2, \ldots, n)$$

The generalised velocity components, and hence also the Lagrangian function, may be expressed as functions of the same variables. Thus, once the solutions of the equations of motion are known, the integral of the Lagrangian along the motion may be evaluated in a form which is dependent upon the same $2n+1$ parameters and the final time.

It may be noticed, however, that there is considerable arbitrariness in the choice of the $2n+1$ parameters which specify a particular solution of the equations of motion. Any set of $2n+1$ independent parameters are suitable candidates. For example, the initial generalised velocities used above may immediately be replaced by the initial momentum components. However, at this particular point it is convenient to specify particular solutions by the initial and final generalised coordinates and the initial time. These are the parameters that are required in Hamilton's principal function which may now be formally defined.

***Definition* 11.3.** *Hamilton's principal function is the integral of the Lagrangian function $S=\int_{t_0}^{t_1} L\,dt$ evaluated along a particular motion and expressed in terms of the initial and final configurations and times*

$$S=S(q_{10},\ldots,q_{n0},q_{11},\ldots,q_{n1},t_0,t_1)$$

It is now necessary to consider how the principal function varies with respect to its $2n+2$ arguments. In fact, this can immediately be deduced from a result obtained in section 10.3 that, using a Δ variation to compare neighbouring motions,

$$\Delta\int_{t_0}^{t_1} L\,dt=\left[\sum_{i=1}^{n}\frac{\partial L}{\partial \dot{q}_i}\Delta q_i-\left(\sum_{i=1}^{n}\dot{q}_i\frac{\partial L}{\partial \dot{q}_i}-L\right)\Delta t\right]_{t_0}^{t_1}$$

The variations permitted here are precisely those of the arguments of the principal function, and this equation can therefore be written in the form of standard differentials

$$dS=\sum_{i=1}^{n}p_{i1}dq_{i1}-\sum_{i=1}^{n}p_{i0}\,dq_{i0}-H_1\,dt_1+H_0\,dt_0$$

where Hamiltonian notation has also been inserted. From this expression it can immediately be deduced that

$$\frac{\partial S}{\partial q_{i1}}=p_{i1}\qquad \frac{\partial S}{\partial q_{i0}}=-p_{i0}\qquad (i=1,2,\ldots,n)$$

$$\frac{\partial S}{\partial t_1}=-H_1\qquad \frac{\partial S}{\partial t_0}=H_0$$

These identities are discussed in the following section.

11.5. The Hamilton–Jacobi equation

It is convenient at this point to change the notation for the arguments of Hamilton's principal function. First, it may be assumed that the initial time is not varied, so that subsequent times are effectively measured from the initial time. Then, secondly, it is convenient to relabel the initial coordinates and momentum components by $\alpha_1, \ldots \alpha_n$ and $\beta_1, \ldots, \beta_n$ respectively, where

$$\alpha_i = q_{i0} \qquad \beta_i = p_{i0} \qquad (i = 1, 2, \ldots, n)$$

These may in fact be thought of as the $2n$ parameters which, with t_0, uniquely specify any particular solution of the equations of motion. It is now also possible to drop the suffix 1 on the variables at the final point, which is thus described in terms of the coordinates $q_1, \ldots, q_n$, the momentum components $p_1, \ldots, p_n$, and the final time t. However, since particular solutions are now regarded as being determined by the $2n+1$ parameters at the initial point, the final point may therefore be regarded as the state at which the system necessarily arrives at the arbitrary time t. Thus the final point may be considered to describe the motion of the system.

In terms of this notation, Hamilton's principal function

$$S = S(\alpha_1, \ldots, \alpha_n, q_1, \ldots, q_n, t)$$

satisfies

$$\mathrm{d}S = \sum_{i=1}^{n} p_i \,\mathrm{d}q_i - \sum_{i=1}^{n} \beta_i \,\mathrm{d}\alpha_i - H \,\mathrm{d}(t - t_0)$$

and thus

$$\frac{\partial S}{\partial \alpha_i} = -\beta_i \qquad \frac{\partial S}{\partial q_i} = p_i \qquad (i = 1, 2, \ldots, n), \qquad \frac{\partial S}{\partial t} = -H$$

These equations may now be considered in detail.

The first set of equations

$$\frac{\partial S}{\partial \alpha_i} = -\beta_i \qquad (i = 1, 2, \ldots, n)$$

are a set of n identities involving functions of the n generalised coordinates, and $2n$ arbitrary parameters $\alpha_1, \ldots, \alpha_n, \beta_1, \ldots, \beta_n$, and time. They are in fact n independent equations from which it is possible to deduce expressions for the n generalised coordinates in terms of time and the $2n$ parameters which determine any particular solution. It is therefore clear that if the function S is given explicitly, then these identities are the complete integrals of the equations of motion in that they give the configuration of the system at any time.

The second set of equations

$$\frac{\partial S}{\partial q_i} = p_i \qquad (i = 1, 2, \ldots, n)$$

simply define the generalised momentum components in terms of the principal function S. However, if S is known explicitly, the n momentum components can immediately be obtained from these equations. Thus, in this case, both sets of equations actually express the solutions of the equations of motion in Hamiltonian form.

It can therefore be seen that if Hamilton's principal function is known explicitly, then the motion of the system can be determined immediately from its derivatives. This would enable the configuration of a system at any time to be deduced without having to solve any differential equations. However, in order to determine the principal function in the first place, it has already been assumed that the solutions of the equations of motion are known, and so little has actually been achieved. On the other hand, if it were possible to find an alternative way of obtaining the principal function, the solutions of the equations of motion could be deduced directly using the above approach. It is therefore of considerable importance to look for an independent way of obtaining the principal function, and to find this, attention may be directed to the last of the above identities.

The last of the above equations

$$\frac{\partial S}{\partial t} = -H$$

at first sight simply appears to define the Hamiltonian function in terms of the principal function. However, if the Hamiltonian function is known, it can also be considered as a partial differential equation for the principal function, and so may provide an alternative independent way of obtaining it.

It must immediately be observed that the Hamiltonian function is dependent upon the generalised coordinates, the generalised momentum components, and also possibly time. However, the momentum components can be directly expressed in terms of the principal function using the previous set of identities. Once the momentum components have been thus explicitly eliminated, the last identity takes the form

$$\frac{\partial S}{\partial t} + H\left(q_1, \ldots, q_n, \frac{\partial S}{\partial q_i}, \ldots, \frac{\partial S}{\partial q_n}, t\right) = 0$$

This is known as the *Hamilton–Jacobi partial differential equation.* Since the Hamiltonian function is quadratic in the generalised momentum components, this can be seen to be a nonlinear, first-order partial differential equation involving $n+1$ variables and their derivatives of the principal function S.

Clearly Hamilton's principal function $S=S(\alpha_1,\ldots,\alpha_n, q_1,\ldots,q_n, t)$ is a solution of the Hamilton–Jacobi equation. In this case the n constants $\alpha_1,\ldots,\alpha_n$ are the initial values of the generalised coordinates. It may, however, be noticed that these constants can be replaced by any other set of n independent constants which uniquely determine the configuration of the system at the initial time t_0. Strictly speaking, the function S expressed in terms of this more general set of constants should no longer be called the principal function, but it will still satisfy the Hamilton–Jacobi equation. In addition to this, the equations $\partial S/\partial\alpha'_i=-\beta'_i$ still give the integrals of the equations of motion, where $\beta'_1,\ldots,\beta'_n$ are a new set of arbitrary constants corresponding to the new set $\alpha'_1,\ldots,\alpha'_n$.

It can thus be seen that a wide class of complete integrals of the Hamilton–Jacobi equation can be used to generate the solutions of the equations of motion, even when these are not, strictly speaking, the principal function. In view of this it is appropriate to enquire whether or not all complete integrals of the Hamilton–Jacobi equation can be similarly used to determine the motion of a system. In fact, it is possible to prove that this is the case, although a complete proof is not given here.

The Hamilton–Jacobi equation is expressed in terms of $n+1$ variables $q_1,\ldots,q_n$ at t and their corresponding $n+1$ derivatives of the function S. It follows that a general solution of this equation must contain $n+1$ independent arbitrary constants. However, one of these must be a constant to be added to S, and this may be neglected since it is only the derivatives of S that are important. Denoting the remaining set of n constants by $c_1,\ldots,c_n$, a general integral of the Hamilton–Jacobi equation can be expressed as $S=S(c_1,\ldots,c_n, q_1,\ldots,q_n, t)$. It is now possible to prove that the constants of integration $c_1,\ldots,c_n$, and a new set of constants $\beta_1,\ldots,\beta_n$ can be related by a contact transformation to the initial coordinates and momentum components $q_{10},\ldots,q_{n0}, p_{10},\ldots,p_{n0}$ in such a way that the equations $\partial S/\partial c_i=-\beta_i$ express the integrals of the equations of motion. This result is known as the *Hamilton–Jacobi theorem*, and may now be stated in the following way.

***Proposition* 11.3.** *If $S = S(\alpha_1, \ldots, \alpha_n, q_1, \ldots, q_n, t)$ is a complete integral of the Hamilton–Jacobi equation*

$$\frac{\partial S}{\partial t} + H\left(q_1, \ldots, q_n, \frac{\partial S}{\partial q_1}, \ldots, \frac{\partial S}{\partial q_n}, t\right) = 0$$

in which the corresponding momentum components appearing in the Hamiltonian function H are replaced by $\partial S/\partial q_1, \ldots, \partial S/\partial q_n$, where $\alpha_1, \ldots, \alpha_n$ are n arbitrary constants, then the complete integrals of the equations of motion are given by

$$\frac{\partial S}{\partial \alpha_i} = -\beta_i \qquad (i = 1, 2, \ldots, n)$$

where $\beta_1, \ldots, \beta_n$ are n further arbitrary constants. In addition the generalised momentum components are given by

$$p_i = \frac{\partial S}{\partial q_i} \qquad (i = 1, 2, \ldots, n)$$

It should be noticed that the constants $\alpha_1, \ldots, \alpha_n$ are any set of independent constants of integration which are obtained in the process of finding a general solution of the Hamilton–Jacobi equation. In practice, some of these may well turn out to be of physical significance. For example, they may include constants of energy or momentum. Such physical interpretations, however, do not affect the mathematical technique, and by introducing a new set of n constants $\beta_1, \ldots, \beta_n$, the equations $\partial S/\partial \alpha_i = -\beta_i$ may be rearranged to obtain the solutions

$$q_i = q_i(\alpha_1, \ldots, \alpha_n, \beta_1, \ldots, \beta_n, t) \qquad (i = 1, 2, \ldots, n)$$

It should be emphasised that the constants $\alpha_1, \ldots, \alpha_n$ must be independent, and that they must not include a purely additive constant.

The above result is of far-reaching importance. In the context of classical mechanics it provides an alternative method for determining the motion of a system. Instead of having to solve n second-order ordinary differential equations or $2n$ first-order equations, it is necessary to solve only a single first-order partial differential equation. From a general solution of this equation the motion may immediately be deduced. Such an approach is not only of theoretical interest. It also provides a powerful technique for solving practical problems. There are many practical dynamical systems in which an application of the Hamilton–Jacobi approach provides the quickest way of determining the motion.

The Hamilton–Jacobi theory may also be applied to a much wider class of problems that can be stated in terms of differential equations or the calculus of variations. In addition it can also be very easily extended to include geometrical optics and the modern theories of relativity and quantum mechanics.

11.6. Hamilton's characteristic function

In the application of the Hamilton–Jacobi approach in classical mechanics, it is necessary to solve a single nonlinear first-order partial differential equation. This can usually be achieved using the technique known as 'separating the variables' when one variable is separated at a time and each separation introduces an arbitrary constant of integration. In situations where the variables cannot be separated in this way it is usually very difficult to proceed using Hamilton–Jacobi theory.

It is now instructive to consider the obvious case in which an initial attempt to simplify the Hamilton–Jacobi equation can immediately be made. This is the case in which the Hamiltonian function does not depend explicitly on time. In this case the Hamiltonian is a constant of the motion, which can be denoted by E, and the time variable in the Hamilton–Jacobi equation can immediately be separated out to give the equations

$$\frac{\partial S}{\partial t}+E=0 \qquad H\left(q_1,\ldots,q_n,\frac{\partial S}{\partial q_1},\ldots,\frac{\partial S}{\partial q_n}\right)=E$$

Although E is clearly the constant energy of the system, it appears here as a constant of integration, and according to the Hamilton–Jacobi approach it should be denoted by α_1. The solution of the Hamilton–Jacobi equation can now be written in the form

$$S(\alpha_1,\ldots,\alpha_n,q_1,\ldots,q_n,t)=W(\alpha_1,\ldots,\alpha_n,q_1,\ldots,q_n)-\alpha_1 t$$

where the function W is required to satisfy the partial differential equation

$$H\left(q_1,\ldots,q_n,\frac{\partial W}{\partial q_1},\ldots,\frac{\partial W}{\partial q_n}\right)=\alpha_1$$

This result is essentially the first step in the integration of the Hamilton–Jacobi equation, and the integration with respect to the remaining variables is now required. However, this initial step may be made in all cases in which the Hamiltonian function does not depend on time explicitly, and since this is a case which frequently occurs in practice, it is appropriate to single it out for special attention.

Applying the Hamilton–Jacobi theorem directly to this case, the following result is obtained.

***Proposition* 11.4.** *For systems in which the Hamiltonian function H does not depend explicitly on time, the complete integrals of the equations of motion are given by*

$$\frac{\partial W}{\partial \alpha_1} = t - \beta_1$$

$$\frac{\partial W}{\partial \alpha_i} = -\beta_i \qquad (i = 2, 3, \ldots, n)$$

where $W = W(\alpha_1, \ldots, \alpha_n, q_1, \ldots, q_n)$ is a complete integral of the equation

$$H\left(q_1, \ldots, q_n, \frac{\partial W}{\partial q_1}, \ldots, \frac{\partial W}{\partial q_n}\right) = \alpha_1$$

$\alpha_1, \ldots, \alpha_n, \beta_1, \ldots, \beta_n$ are $2n$ arbitrary constants, and the generalised momentum components are

$$p_i = \frac{\partial W}{\partial q_i} \qquad (i = 1, 2, \ldots, n)$$

It can immediately be seen that the equations contained in this proposition do not involve time explicitly, except for the first equation $\partial W/\partial \alpha_1 = t - \beta_1$. The solutions of the equations of motion can therefore be immediately deduced in the form of orbit equations in which each coordinate is expressed in terms of others. The coordinates can finally be expressed as functions of time, at least implicitly, using that first equation.

For systems in which the Hamiltonian does not depend on time it is clearly convenient to work with the function W rather than S. It is this function $W(\alpha_1, \ldots, \alpha_n, q_1, \ldots, q_n)$ which is known as *Hamilton's characteristic function.*

In the same way as the function S can be related to Hamilton's principal function, which is the integral of the Lagrangian along the motion, so the characteristic function also has a physical significance. This can be deduced by noticing that

$$\frac{dW}{dt} = \sum_{i=1}^{n} \frac{\partial W}{\partial q_i} \dot{q}_i = \sum_{i=1}^{n} p_i \dot{q}_i$$

and thus

$$W = \int \sum_{i=1}^{n} p_i \dot{q}_i \, dt$$

The characteristic function can thus be seen to be formally identical to the action of definition 10.1. In the particular case when the kinetic energy is a homogeneous quadratic in the generalised velocities, this is simply twice the integral of the kinetic energy.

However, although the characteristic function can be identified in integral form with the action, it is not necessarily expressed in this form in terms of the required variables. It cannot therefore be evaluated from the action until solutions of the equations of motion are explicitly known. In practice it is therefore preferable to regard the characteristic function simply as a complete integral of the time independent form of the Hamilton–Jacobi equation, from which the solutions of the equations of motion can be deduced.

Finally, it may again be pointed out that the Hamilton–Jacobi theory with its principal and characteristic functions have a significance apart from their relevance to the applications of classical dynamics, in that they can immediately be applied to the subject of geometrical optics. From this it is but a small step to wave mechanics, which is one of the basic approaches of modern quantum theory. In fact, Schrödinger's wave equation can be generated directly from the Hamilton–Jacobi equation by a set of simple but *ad hoc* rules. Strictly, however, it is the Hamilton–Jacobi equation which is obtained from Schrödinger's wave equation in the appropriate classical limit. The alternative matrix mechanics approach to quantum theory can also be obtained as a similar generalisation of the Poisson bracket formulation of classical mechanics. However, all this is a different theory.

11.7. The Hamilton–Jacobi approach: examples

***Example* 11.3**: *simple harmonic motion*

Consider here a system whose equation of motion is of the form

$$m\ddot{x} = -mn^2x$$

where n is a constant. Equations of this type occur, for example, as first approximations for the oscillator motion of one-dimensional systems, where n is the frequency of oscillation. Since the above equation is linear, a complete integral can easily be obtained which can be written in the form

$$x = a\cos(nt + \varepsilon)$$

where a is the amplitude of the oscillation and ε is the phase.

Although this system can be easily analysed by any method, the purpose here is to illustrate the Hamilton–Jacobi approach.

First it is necessary to find an expression for the potential energy associated with the force $-mn^2x$. This may be taken to be

$$V=\tfrac{1}{2}mn^2x^2$$

The momentum associated with the coordinate x is simply

$$p=m\dot{x}$$

and the Hamiltonian function is thus

$$H=\frac{1}{2m}p^2+\frac{mn^2}{2}x^2$$

Then, putting $p=\partial S/\partial x$, the Hamilton–Jacobi equation for this system takes the form

$$\frac{\partial S}{\partial t}+\frac{1}{2m}\left(\frac{\partial S}{\partial x}\right)^2+\frac{mn^2}{2}x^2=0$$

It is now necessary to find a first integral of this equation. This can be done here by looking for solutions of the form

$$S=W(x)+T(t)$$

using the method of separation of variables. It may easily be seen that solutions of this type exist where

$$\frac{1}{2m}\left(\frac{\mathrm{d}W}{\mathrm{d}x}\right)^2+\frac{mn^2}{2}x^2=\alpha$$

and

$$\frac{\mathrm{d}T}{\mathrm{d}t}=-\alpha$$

where α is an arbitrary constant of integration, which can be interpreted as the constant energy of the system. A first integral of the Hamilton–Jacobi equation may thus be written as

$$S=m\int\sqrt{\left(\frac{2\alpha}{m}-n^2x^2\right)}\,\mathrm{d}x-\alpha t$$

Although it is not difficult to evaluate the integral contained in this expression, it is not necessary to do so as it is only the derivatives of S that are required.

According to proposition 11.3, the complete integral of the equation of motion may now be obtained from the equation $\partial S/\partial\alpha=-\beta$, where β is another arbitrary constant. This equation now takes the form

$$\int\frac{\mathrm{d}x}{\sqrt{\left(\frac{2\alpha}{m}-n^2x^2\right)}}-t=-\beta$$

which can be integrated to give

$$-\frac{1}{n}\cos^{-1}\left(\frac{x}{n}\sqrt{\left(\frac{m}{2\alpha}\right)}\right)=t-\beta$$

or

$$x=\frac{1}{n}\sqrt{\left(\frac{2\alpha}{m}\right)}\cos n(t-\beta)$$

This can be seen to be exactly of the form expected, with the energy α being related to the amplitude of the oscillation and the constant β to the phase.

It may be observed that the function $W(x)$ introduced above is in fact Hamilton's characteristic function. The above approach is therefore equivalent to using the time independent equation, as stated in proposition 11.4. This approach will be illustrated explicitly in the following example.

Example 11.4: *inverse square law orbits*

Consider the motion of a particle under the action of a force which is always directed towards a fixed point, and whose magnitude varies inversely as the square of the distance of the particle from that point. This situation has already been discussed in example 6.2, where it was first shown that the particle remains in a fixed plane. Using polar coordinates in the plane of motion, the kinetic and potential energies of the particle are given by

$$T=\tfrac{1}{2}m(\dot{r}^2+r^2\dot{\theta}^2) \qquad V=-\frac{m\mu}{r}$$

The momentum components are therefore

$$p_r=m\dot{r} \qquad p_\theta=mr^2\dot{\theta}$$

and the Hamiltonian function becomes

$$H=\frac{1}{2m}\left(p_r^2+\frac{1}{r^2}p_\theta^2\right)-\frac{m\mu}{r}$$

This system clearly does not depend on time explicitly, so it is possible to use the time independent form of the Hamilton–Jacobi equation as stated in proposition 11.4. In this case this takes the form

$$\frac{1}{2m}\left(\frac{\partial W}{\partial r}\right)^2+\frac{1}{2mr^2}\left(\frac{\partial W}{\partial \theta}\right)^2-\frac{m\mu}{r}=\alpha_1$$

A solution to this may now be obtained by separating the variables by writing $W(r,\theta)=R(r)+\Theta(\theta)$. The resulting equation is

$$r^2\left(\frac{\mathrm{d}R}{\mathrm{d}r}\right)^2-2m^2\mu r-2m\alpha_1 r^2=-\left(\frac{\mathrm{d}\Theta}{\mathrm{d}\theta}\right)^2$$

Putting each side of this equation equal to $-\alpha_2^2$, where α_2 is an arbitrary constant of integration, leads to the equations

$$\frac{\mathrm{d}R}{\mathrm{d}r}=\frac{1}{r}(2m\alpha_1 r^2+2m^2\mu r-\alpha_2^2)^{1/2}$$

$$\frac{\mathrm{d}\Theta}{\mathrm{d}\theta}=\alpha_2$$

The characteristic function may therefore be written as

$$W=\int\frac{1}{r}(2m\alpha_1 r^2+2m^2\mu r-\alpha_2^2)^{1/2}\,\mathrm{d}r+\alpha_2\theta$$

It may now be observed that $\partial W/\partial\theta = \alpha_2$, and therefore α_2 is the constant angular momentum of the particle p_θ. It is already understood that α_1 is the constant energy of the particle.

According to proposition 11.4, the complete integrals of the equations of motion are now given by $\partial W/\partial\alpha_1 = t-\beta_1$, and $\partial W/\partial\alpha_2 = -\beta_2$, where β_1 and β_2 are the two new arbitrary constants. In this case these equations become

$$\int mr(2m\alpha_1 r^2 + 2m^2\mu r - \alpha_2^2)^{-1/2}\,dr = t-\beta_1$$

$$\int \frac{\alpha_2}{r}(2m\alpha_1 r^2 + 2m^2\mu r - \alpha_2^2)^{-1/2}\,dr + \theta = -\beta_2$$

The first of these equations gives r implicitly as a function of t, and β_1 determines the origin of the time parameter t. The second of the above equations gives r as a function of θ, with β_2 determining the origin of θ. Introducing the reciprocal coordinate $u = 1/r$, this latter equation becomes

$$\begin{aligned}\theta+\beta_2 &= -\int\left(\frac{2m\alpha_1}{\alpha_2^2}+\frac{2m^2\mu}{\alpha_2^2}u-u^2\right)^{-1/2}du\\ &= -\cos^{-1}\left\{\left(\frac{\alpha_2^2}{m^2\mu}u-1\right)\left(1+\frac{2\alpha_1\alpha_2^2}{m^3\mu^3}\right)^{-1/2}\right\}\end{aligned}$$

It is now appropriate to introduce new constants l and e where

$$l=\frac{\alpha_2^2}{m^2\mu}\qquad e^2=1+\frac{2\alpha_1\alpha_2^2}{m^3\mu^2}$$

The above equation thus becomes

$$\theta+\beta_2=-\cos^{-1}\left(\frac{lu-1}{e}\right)$$

This may now be rewritten in the more familiar form of the polar equation for the orbit, as previously obtained in example 6.2

$$\frac{l}{r}=1+e\cos(\theta+\beta_2)$$

Exercises

11.5 Consider the motion of a particle of unit mass in a plane under the action of a force which is directed always towards a fixed point and whose magnitude is n^2r, where n is a constant and r is the distance of the particle from that point. Show that Hamilton's principal function in this case is

$$s=-\alpha_1(t-t_0)+\alpha_2(\theta-\theta_0)+\int_{r_0}^{r}\left(2\alpha_1-n^2r^2-\frac{\alpha_2^2}{r^2}\right)^{1/2}dr$$

Hence show that the motion is described by the equations

$$t-\beta_1=\int\left(2\alpha_1-n^2r^2-\frac{\alpha_2^2}{r^2}\right)^{1/2}\mathrm{d}r$$

$$\theta+\beta_2=\int\frac{\alpha^2}{r^2}\left(2\alpha_1-n^2r^2-\frac{\alpha_2^2}{r^2}\right)^{-1/2}\mathrm{d}r$$

where α_1 and α_2 are the energy and angular momentum constants respectively.

By integrating the above equations show that

$$r=\frac{1}{n}\{\alpha_1-\sqrt{(\alpha_1^2-n^2\alpha_2^2)}\cos 2n(t-\beta_1)\}^{1/2}$$

or

$$r=\alpha_2\{\alpha_1-\sqrt{(\alpha_1^2-n^2\alpha_2^2)}\cos 2(\theta+\beta_2)\}^{-1/2}$$

Consider also this system using cartesian coordinates.

11.6 Show that the Hamiltonian function for the motion of a particle of unit mass in a plane, under the action of an inverse cube law central force with potential $-\mu/2r^2$, using polar coordinates in the plane, is

$$H=\frac{1}{2}\left(p_r^2+\frac{1}{r^2}p_\theta^2-\frac{\mu}{r^2}\right)$$

Using the Hamilton–Jacobi approach, show that the orbit of the particle is given by

$$\int\alpha_2(2\alpha_1r^2+\mu+\alpha_2^2)^{-1/2}\frac{1}{r}\,\mathrm{d}r=\theta+\beta_2$$

and that the position in the orbit at any time is given by

$$2\alpha_1r^2+\mu-\alpha_2^2=4\alpha_1^2(t-\beta_1)^2$$

where α_1, α_2, β_1, β_2 are arbitrary constants. Discuss also the physical meaning associated with these constants.

Appendix

List of basic results and definitions

Assumption 2.1 – Representation of time
Assumption 2.2 – Representation of space
    Definition 2.1 – Velocity
    Definition 2.2 – Acceleration
    Definition 2.3 – Angular velocity
    Definition 2.4 – Angular acceleration

    Definition 3.1 – Particle
Assumption 3.1 – The particle representation
Assumption 3.2 – Inertial frame of reference
    Definition 3.2 – Force
    Definition 3.3 – Magnitude and direction of force
    Definition 3.4 – Mass
Assumption 3.3 – Action and reaction
  Proposition 3.1 – Mass ratios determined by accelerations
Assumption 3.4 – Mass ratios are constant
Assumption 3.5 – Forces are independent
  Proposition 3.2 – Law of motion, $m\ddot{\boldsymbol{r}} = \boldsymbol{F}$

Assumption 4.1 – Law of motion relative to class of inertial frames
  Proposition 4.1 – Equation of motion relative to noninertial frames

Assumption 5.1 – Newton's theory of gravitation
Assumption 5.2 – The general relativistic correction

    Definition 6.1 – Work
    Definition 6.2 – Power
    Definition 6.3 – Kinetic energy, T
  Proposition 6.1 – Increase in kinetic energy equals work done
    Definition 6.4 – Irrotational fields
    Definition 6.5 – Conservative fields
    Definition 6.6 – Potential energy, V

Proposition 6.2 – For a conservative field, $\boldsymbol{F} = -\text{grad } V$
Proposition 6.3 – For a conservative field, $T + V = \text{constant}$
Definition 6.7 – Linear momentum
Definition 6.8 – Moments
Definition 6.9 – Angular momentum $\boldsymbol{L}$
Proposition 6.4 – $\frac{\mathrm{d}}{\mathrm{d}t}\boldsymbol{L} = \boldsymbol{r} \times \boldsymbol{F}$
Definition 6.10 – Impulse
Proposition 6.5 – Change of momentum equals impulse
Proposition 6.6 – Change in angular momentum equals moment of impulse

Definition 7.1 – Centroid $\bar{\boldsymbol{r}}$
Proposition 7.1 – Total linear momentum related to centroid
Proposition 7.2 – Kinetic energy related to centroid
Proposition 7.3 – Angular momentum related to centroid
Proposition 7.4 – Particles in constant gravitational field
Proposition 7.5 – $m\ddot{\bar{\boldsymbol{r}}} = \boldsymbol{F}$
Assumption 7.1 – Mutually induced forces are aligned
Proposition 7.6 – $\dot{\boldsymbol{L}} = \boldsymbol{M}$ about fixed point
Proposition 7.7 – $\dot{\boldsymbol{L}} = \boldsymbol{M}$ about centroid
Proposition 7.8 – Variable mass equation
Definition 7.2 – Degrees of freedom
Definition 7.3 – Holonomic and nonholonomic constraints
Proposition 7.9 – Possible configurations restricted by holonomic constraints
Definition 7.4 – Scleronomic and rheonomic constraints
Proposition 7.10 – Energy not constant in rheonomic systems

Definition 8.1 – Rigid body
Definition 8.2 – Moment of inertia
Definition 8.3 – Radius of gyration
Proposition 8.1 – Parallel axis theorem
Proposition 8.2 – Lamina theorem
Definition 8.4 – Moments and products of inertia
Proposition 8.3 – General parallel axis theorem
Definition 8.5 – Principal axes
Definition 8.6 – Principal moment of inertia
Proposition 8.4 – Principal axes and moments as eigenvectors and eigenvalues
Proposition 8.5 – Mutually orthogonal principal axes
Proposition 8.6 – Ellipsoid of inertia
Proposition 8.7 – $\dot{\boldsymbol{L}} + \boldsymbol{\Omega} \times \boldsymbol{L} = \boldsymbol{M}$
Proposition 8.8 – Change in linear momentum due to impulses

Proposition 8.9 – Change in angular momentum about centroid due to impulses
Proposition 8.10 – Impulsive change in angular momentum about any point

Definition 9.1 – Generalised coordinates
Proposition 9.1 – Kinetic energy quadratic in generalised velocities
Definition 9.2 – Generalised momentum components
Definition 9.3 – Virtual displacements
Definition 9.4 – Virtual work
Definition 9.5 – Generalised force components
Proposition 9.2 – Lagrange's equations for nonconservative systems
Definition 9.6 – The Lagrangian function
Proposition 9.3 – Lagrange's equations for conservative systems
Proposition 9.4 – Integrals of momentum
Proposition 9.5 – Jacobi's integral
Proposition 9.6 – Fundamental equation in cartesian coordinates
Proposition 9.7 – Fundamental equations in generalised coordinates
Definition 9.7 – Catastatic and acatastatic constraints
Proposition 9.8 – Virtual displacements are possible in catastatic systems
Proposition 9.9 – Lagrange's equations for nonholonomic systems
Definition 9.8 – Generalised impulse components
Proposition 9.10 – Lagrange's equations for impulsive motion

Proposition 10.1 – Hamilton's principle
Definition 10.1 – Action
Proposition 10.2 – The principle of least action

Definition 11.1 – The Hamiltonian function
Proposition 11.1 – Hamilton's equations of motion
Definition 11.2 – Poisson brackets
Proposition 11.2 – On constants of the motion
Definition 11.3 – Hamilton's principal function
Proposition 11.3 – The Hamilton–Jacobi theorem
Proposition 11.4 – The time independent Hamilton–Jacobi equation

SUGGESTIONS FOR FURTHER READING

A. Textbooks on classical dynamics

V. I. Arnold, *Mathematical Methods of Classical Mechanics* (Springer, 1978).
W. Chester, *Mechanics* (George Allen and Unwin, 1979).
F. Chorlton, *Textbook of Dynamics* (Van Nostrand, 1963).
H. C. Corben & P. Stehle, *Classical Mechanics*, 2nd edition (John Wiley, 1960).
C. E. Easthope, *Three Dimensional Dynamics*, 2nd edition (Butterworths, 1964).
G. R. Fowles, *Analytical Mechanics*, 3rd edition (Holt–Saunders, 1977).
H. Goldstein, *Classical Mechanics* (Addison Wesley, 1950).
D. T. Greenwood, *Principles of Dynamics* (Prentice Hall, 1965).
C. Lanczos, *The Variational Principles of Mechanics*, 3rd edition (Toronto, 1966).
L. Meirovitch, *Methods of Analytical Dynamics* (McGraw-Hill, 1970).
L. A. Pars, *A Treatise on Analytical Dynamics* (Heinemann, 1965).
E. T. Whittaker, *A Treatise on the Analytical Dynamics of Particles and Rigid Bodies* (Cambridge, 1961).

B. Books on topics of related interest

R. B. Angel, *Relativity: the Theory and its Philosophy* (Pergamon, 1980).
M. Bunge, *The Foundations of Physics* (Springer, 1967).
F. Cajori, *Newton's Principia, a Revision of Motte's Translation* (Cambridge, 1934).
J. S. R. Chisholm, *Vectors in Three-dimensional Space* (Cambridge, 1978).
I. B. Cohen, *The Newtonian Revolution* (Cambridge, 1980).
R. Dugas, *A History of Mechanics* (Routledge and Kegan Paul, 1957).
M. Jammer, *Concepts of Force* (Harvard, 1957).
E. Mach, *The Science of Mechanics* (Open Court, 1902).
J. Powers, *Philosophy and the New Physics* (Methuen, 1982).
G. J. Whitrow, *The Natural Philosophy of Time* (Nelson, 1961).

INDEX